PESTICIDES IN SURFACE WATERS

Distribution, Trends, and Governing Factors

T0225554

Pesticides in Surface Waters

Distribution, Trends, and Governing Factors

Steven J. Larson, U.S. Geological Survey, Minneapolis, Minnesota
Paul D. Capel, U.S. Geological Survey, Minneapolis, Minnesota
Michael S. Majewski, U.S. Geological Survey, Sacramento, California

Volume Three of the Series
Pesticides in the Hydrologic System

Robert J. Gilliom, Series Editor
U.S. Geological Survey
National Water Quality Assessment Program

CRC Press
Taylor & Francis Group
Boca Raton London New York

CRC Press is an imprint of the
Taylor & Francis Group, an **informa** business

CRC Press
Taylor & Francis Group
6000 Broken Sound Parkway NW, Suite 300
Boca Raton, FL 33487-2742

First issued in paperback 2020

CRC Press is an imprint of Taylor & Francis Group, an Informa business

No claim to original U.S. Government works

ISBN-13: 978-0-367-45582-8 (pbk)
ISBN-13: 978-1-57504-006-6 (hbk)

Visit the Taylor & Francis Web site at
http://www.taylorandfrancis.com

and the CRC Press Web site at
http://www.crcpress.com

Library of Congress Cataloging-in-Publication Data

Larson, Steven J.
    Pesticides in surface waters : distribution, trends, and governing factors
    / Steven J. Larson, Paul D. Capel, Michael S. Majewski.
        p.   cm. — (Volume three of the series Pesticides in the hydrologic system)
    Includes bibliographical references and index.
    1. Pesticides—Environmental aspects—United States.
    2. Surface waters—Pollution—United States. I. Capel, Paul D. Majewski, Michael S.
    II. Title.   III. Series: Pesticides in the hydrologic system : v. 3.

628. 1'6842—dc20   97-

Library of Congress Card Number 97-

INTRODUCTION TO THE SERIES

Pesticides in the Hydrologic System is a series of comprehensive reviews and analyses of our current knowledge and understanding of pesticides in the water resources of the United States and of the principal factors that influence contamination and transport. The series is presented according to major components of the hydrologic system—the atmosphere, surface water, bed sediments and aquatic organisms, and ground water. Each volume:

- summarizes previous review efforts;
- presents a comprehensive tabulation, review, and analysis of studies that have measured pesticides and their transformation products in the environment;
- maps locations of studies reviewed, with cross references to original publications;
- analyzes national and regional patterns of pesticide occurrence in relation to such factors as the use of pesticides and their chemical characteristics;
- summarizes processes that govern the sources, transport, and fate of pesticides in each component of the hydrologic system;
- synthesizes findings from studies reviewed to address key questions about pesticides in the hydrologic system, such as:

    How do agricultural and urban areas compare?

    What are the effects of agricultural management practices?

    What is the influence of climate and other natural factors?

    How do the chemical and physical properties of a pesticide influence its behavior in the hydrologic system?

    How have past study designs and methods affected our present understanding?

    Are water quality criteria for human health or aquatic life being exceeded?

    Are long-term trends evident in pesticide concentrations in the hydrologic system?

This series is unique in its focus on review and interpretation of reported direct measurements of pesticides in the environment. Each volume characterizes hundreds of studies conducted during the past four decades. Detailed summary tables include such features as spatial and temporal domain studied, target analytes, detection limits, and compounds detected for each study reviewed.

Pesticides in the Hydrologic System is designed for use by a wide range of readers in the environmental sciences. The analysis of national and regional patterns of pesticide occurrence, and their relation to use and other factors that influence pesticides in the hydrologic system, provides a synthesis of current knowledge for scientists, engineers, managers, and policy makers at all levels of government, in industry and agriculture, and in other organizations. The interpretive analyses and summaries are designed to facilitate comparisons of past findings to current and future findings. Data of a specific nature can be located for any particular area of the country. For educational needs, teachers and students can readily identify example data sets that meet their requirements. Through its focus on the United States, the series covers a large portion of the global database on pesticides in the hydrologic system, and international readers will find much that applies to other areas of the world. Overall, the goal of the series is to provide readers from a broad range of backgrounds in the environmental sciences with a synthesis of the factual data and interpretive findings on pesticides in the hydrologic system.

The series has been developed as part of the National Water Quality Assessment Program of the U. S. Geological Survey, Department of the Interior. Assessment of pesticides in the nation's water resources is one of the top priorities for the Program, which began in 1991. This comprehensive national review of existing information serves as the basis for design and interpretation of studies of pesticides in major hydrologic systems of the United States now being conducted as part of the National Water Quality Assessment.

Series Editor

Robert J. Gilliom
U. S. Geological Survey

PREFACE

The use of pesticides in the United States has increased dramatically during the last several decades. Hundreds of different chemicals have been developed for use in agricultural and non-agricultural settings. Concerns about the potential adverse effects of pesticides on the environment and human health have spurred an enormous amount of research into their environmental behavior and fate. Much of this concern has focused on the potential for contamination of the hydrologic system, including surface waters. *Pesticides in Surface Waters* is the first comprehensive summary of research on the occurrence, distribution, and significance of pesticides in surface waters of the United States.

The primary goal of this book is to assess the current understanding of the occurrence and behavior of pesticides in surface waters. To accomplish this, we have compiled and evaluated most of the published studies in which pesticide concentrations in surface waters of the United States have been measured. The primary focus of the literature search was on studies published in the peer-reviewed scientific literature and in reports of government agencies. The literature search covered studies published up to 1993, but many articles and reports published after 1993 were included as they became available. A number of studies—including laboratory studies and studies using microcosms and artificial streams and ponds—also were included in which factors affecting the behavior and fate of pesticides in the environment were investigated. Pertinent studies listed in a series of tables provide concise summaries of study sites, targeted pesticides, and results. Information obtained from these studies is used to develop an overview of the existing knowledge of pesticide contamination of surface waters.

Pesticides in Surface Waters is intended to serve as a resource, text, and reference to a wide spectrum of scientists, students, and water managers, ranging from those primarily interested in the extensive compilations of references, to those looking for interpretive analyses and conclusions. For those unfamiliar with the studies of pesticides in surface waters, it can serve as a comprehensive introduction.

The preparation of this book was made possible by the National Water Quality Assessment (NAWQA) Program of the U.S. Geological Survey (USGS). The authors wish to thank Naomi Nakagaki, who produced nearly all of the maps used in this book, and Theresa Gilchrist for her assistance in organizing and summarizing many of the articles obtained as part of the review. Robert Gilliom of the USGS provided excellent technical advice and guidance in the preparation of this book. Tom Sklarsky, Susan Davis, Yvonne Gobert, and Glenn Schwegmann provided excellent and conscientious editing and manuscript preparation. We are greatly indebted to Dr. Michael Meyer of the USGS and to Dr. R. Peter Richards of Heidelberg College (Ohio) for their thorough reviews of the manuscript. Their suggestions greatly improved the quality of the book.

Steven J. Larson
Paul D. Capel
Michael S. Majewski

EDITOR'S NOTE

This work was prepared by the United States Geological Survey. Though it has been edited for commercial publication, some of the style and usage incorporated is based on the United States Geological Survey's publication guidelines (i.e., *Suggestions to Authors*, 7th edition, 1991). For example, references with more than two authors cited in the text are written as "Smith and others (19xx)," rather than "Smith, et al. (19xx)," decades are written with an apostrophe (e.g., 1980's), and common-use compound adjectives are hyphenated when used as a modifier (e.g., quality-control procedures). Hyphenation and capitalization are repeated when used in an original reference (e.g., State-Wide). For units of measure, the metric system is used except for the reporting of pesticide use, which is commonly expressed in English units. The original system of units is used when data are quoted from other sources. The Abbreviations and Acronyms in the front of the book do not include the names of some models mentioned, either because the name was not formed from first parts of a series of words or because only the name was given in the original source.

Every attempt has been made to design figures and tables as "stand-alone," without the need for repeated cross reference to the text for interpretation of graphics or tabular data. Some exceptions have been made, however, because of the complexity or breadth of the figure or table. In some cases, for example, a figure caption makes reference to a table when the same data are used for both. As an aid in comparison, the same shading patterns are shown in the Explanation of all pesticide usage maps, though each pattern may not necessarily apply to every map. Some of the longer tables are located at the end of the chapter to maintain less disruption of text.

As an organizational aid to the author and reader, chapter headings, figures, and tables are identified in chapter-numbered sequence. The Abbreviations and Acronyms in the front of the book do not include chemical names, which are listed in the Appendix.

CONTENTS

LIST OF FIGURES

LIST OF TABLES

Note: Pages out of sequence indicate that some tables have been placed at the end of the chapter.

CONVERSION FACTORS

Multiply	By	To obtain
centimeter (cm)	0.3937	inch (in)
cubic meter (m^3)	35.31	cubic foot (ft^3)
gram (g)	0.03527	ounce, avoirdupois (oz)
hectare (ha)	2.469	acre
kilogram (kg)	2.205	pound, avoirdupois (lb)
kilometer (km)	0.6214	mile (mi)
liter (L)	0.2642	gallon (gal)
meter (m)	3.281	foot (ft)
square kilometer (km^2)	0.3861	square mile (mi^2)
square meter (m^2)	10.76	square foot (ft^2)

Multiply	By	To obtain
acre	0.405	hectare (ha)
cubic foot (ft^3)	0.02832	cubic meter (m^3)
foot (ft)	0.3048	meter (m)
gallon (gal)	3.7854	liter (L)
inch (in)	2.54	centimeter (cm)
mile (mi)	1.6093	kilometer (km)
ounce, avoirdupois (oz)	28.350	gram (g)
pound, avoirdupois (lb)	0.45359	kilogram (kg)
square foot (ft^2)	0.09290	square meter (m^2)
square mile (mi^2)	2.5900	square kilometer (km^2)

Temperature is given in degrees Celsius (°C), which can be converted to degrees Fahrenheit (°F) by the following equation:

$$°F = 1.8(°C) + 32$$

ABBREVIATIONS AND ACRONYMS

Note: Clarification or additional information is provided in parentheses. Abbreviations for chemical compounds are included in the Appendix.

Computer Models

ACTMO, Agricultural Chemical Transport Model
ARM, Agricultural Runoff Model
CPM, Cornell Pesticide Model
CREAMS, Chemicals, Runoff, and Erosion from Agricultural Fields Management Systems
EXAMS II, Exposure Analysis Modeling Systems
GLEAMS, Ground Water Loading Effects of Agricultural Management Systems
HSPF, Hydrologic Simulation Program—FORTRAN
PRT, Pesticide Runoff Transport
PRZM, Pesticide Root Zone Model
SLSA, Simplified Lake and Stream Analyzer
STREAM, Stream Transport and Agricultural Runoff of Pesticides for Exposure Assessment
SWRRB, Simulator for Water Resources in Rural Basins

Government and Private Agencies and Legislation

FWPCA, Federal Water Pollution Control Administration
FWQA, Federal Water Quality Administration
IEPA, Illinois Environmental Protection Agency
IUPAC, International Union of Pure and Applied Chemistry
NAS, National Academy of Sciences
NAS/NAE, National Academy of Sciences and the National Academy of Engineering
NOAA, National Oceanic and Atmospheric Administration
SDWA, Safe Drinking Water Act
USDA, U.S. Department of Agriculture
USDOI, U.S. Department of the Interior
USEPA, U.S. Environmental Protection Agency
USFS, U.S. Forest Service
USGS, U.S. Geological Survey

Monitoring Programs and Surveys

C/CPAS, Certified/Commercial Pesticide Applicator Survey
NAWQA, National Water Quality Assessment (Program)
NUPAS, National Urban Pesticide Applicator Survey
NURP, National Urban Runoff Program
STORET, STOrage and RETrieval (water quality database maintained by the U.S. Environmental Protection Agency)

Miscellaneous Abbreviations and Acronyms

$atm-m^3/mole$, atmospheres-meters cubed per mole
kg/ha, kilogram(s) per hectare
kg/yr, kilogram(s) per year
lb a.i., pounds(s) active ingredient
$\mu g/g$, microgram(s) per gram
$\mu g/L$, microgram(s) per liter
$\mu g/kg$, microgram(s) per kilogram
mg/kg, milligram(s) per kilogram
mg/L, milligram(s) per liter
$mg/m^2/yr$, milligram(s) per square meter per year
ng/g, nanogram(s) per gram
ng/L, nanogram(s) per liter
nm, nanometer(s)
pg/L, picogram(s) per liter

AGRICOLA, a bibliographic database of the National Agricultural Library (part of the
    Agricultural Research Service of the U.S. Department of Agriculture)
DAR, deethylatrazine/atrazine ratio
DOC, dissolved organic carbon
FCV, final chronic value
GAC, granular activated carbon
h, hour(s)
HA, health advisory
HAL, health advisory level
K_d, distribution coefficient
K_{oc}, organic carbon-normalized distribution coefficient
LC_{50}, the concentration lethal to 50 percent of a test population
LD_{50}, the dosage of a chemical needed to produce death in 50 percent of the treated test animals
MCL, maximum contaminant level
MCLG, maximum contaminant level goal
min, minute(s)
nsg, no standards given
OC, organochlorine insecticide
OP, organophosphorus insecticide
PAC, powdered activated carbon
PAH, polycyclic aromatic hydrocarbon
PCB, polychlorinated biphenyl
pK_a, negative logarithm of the acid-base dissociation constant
ppb, parts per billion
ppm, parts per million
SNARL, Suggested No-Adverse-Response Level

PESTICIDES IN SURFACE WATERS

Distribution, Trends, and Governing Factors

Steven J. Larson, Paul D. Capel, and Michael S. Majewski

ABSTRACT

A comprehensive review was undertaken by the National Water Quality Assessment Program of the U.S. Geological Survey to assess current understanding of the occurrence and distribution of pesticides in surface waters of the United States. Small-scale studies of individual rivers and lakes to large-scale regional and national studies of surface waters from the late 1950's to the early 1990's were reviewed. Of the 118 pesticides and pesticide transformation products targeted in the reviewed studies, 76 have been detected in one or more surface water bodies throughout the United States. Pesticide concentrations generally ranged from nanograms to micrograms per liter. Organochlorine insecticides continue to be detected in surface waters 20 years after their use was banned or severely restricted. A number of currently used pesticides, particularly the triazine and acetanilide herbicides, occurred as seasonal pulses of elevated concentrations in rivers that drain agricultural areas in the central United States. For most pesticides, data from the reviewed studies are not sufficient to assess trends in occurrence, because few studies sampled the same sites consistently for more than 1 or 2 years. Furthermore, where long-term data do exist, trends are difficult to detect because of year-to-year fluctuations in concentrations caused by variable weather. Data relating environmental exposures and the toxicological effects of pesticides are lacking. In addition, standards or criteria for concentrations of many pesticides in surface waters have not been established. As a result, the significance of observed pesticide concentrations, with respect to human and ecosystem health, is not known. Annual mean concentrations of pesticides in surface waters used as sources of drinking water rarely exceeded maximum contaminant levels established by the U.S. Environmental Protection Agency. However, peak concentrations of several herbicides commonly exceeded the maximum contaminant levels for periods of days to weeks in streams of the central United States.

Significant gaps exist in our understanding of the extent and significance of pesticide contamination of surface waters. The results of this analysis indicate a need for long-term monitoring studies in which a consistent study design is used and more of the currently used pesticides and their transformation products are targeted.

CHAPTER 1

Introduction

Approximately 1.1 billion pounds of pesticides currently are used each year in the United States to control many different types of weeds, insects, and other pests in a wide variety of agricultural and non-agricultural settings (Aspelin and others, 1992; Aspelin, 1994). Total pesticide use, and the number of different chemicals applied, have increased substantially since the 1960's, when the first reliable records of pesticide use were established. For example, national use of herbicides and insecticides on cropland and pasture grew from 190 million pounds active ingredient (lb a.i.) in 1964 to 560 million lb a.i. in 1982 (Gilliom and others, 1985) and was estimated to be about 630 million lb a.i. in 1988 (Gianessi and Puffer, 1991, 1992a,b). Increased use of pesticides has resulted in increased crop production, lower maintenance costs, and control of public health hazards. In addition, however, concerns about the potential adverse effects of pesticides on the environment and human health also have grown.

In many respects, the greatest potential for unintended adverse effects of pesticides is through contamination of the hydrologic system, which supports aquatic life and related food chains and is used for recreation, drinking water, and many other purposes. Water is one of the primary mechanisms by which pesticides are transported from applications areas to other parts of the environment, resulting in the potential for movement into and through all components of the hydrologic cycle (Figure 1.1).

Surface waters are particularly vulnerable to contamination by pesticides, because most agricultural and urban areas drain into surface water systems. Once pesticides are in the moving surface water system (streams and rivers), they can be transported downstream and widely dispersed into other rivers, lakes, reservoirs, and ultimately, the oceans. The presence of pesticides in surface waters has been recognized since the 1940's (Butler, 1966). With the discovery of the adverse ecological effects of the pesticide DDT, and the growing awareness of environmental issues in the 1960's, the problem of pesticides in surface waters has become the focus of much greater attention during the last few decades.

1.1 PURPOSE

Pesticides in Surface Waters reviews our present understanding of pesticides in the surface waters of the United States, with an emphasis on the integration and analysis of information from studies conducted across a wide range of spatial and temporal scales. The focus is on pesticides in the water column. Existing information on pesticides in bed sediments and aquatic biota will be assessed in a companion text in this series, *Pesticides in Bed Sediments and Aquatic Biota in Streams* (Nowell, 1996). The main objectives of *Pesticides in Surface Waters* are (1) to evaluate and assess the occurrence and distribution of pesticides in the various matrices

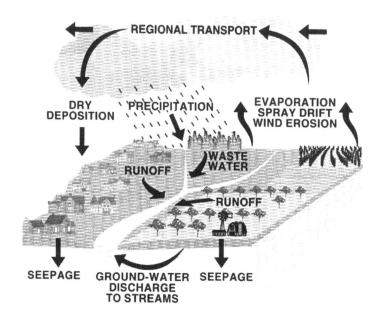

Figure 1.1. Potential routes for pesticide movement into and through components of the hydrologic cycle. Reprinted from Majewski and Capel (1995).

within the water column—water, suspended solids, surface microlayer, and dissolved organic carbon; (2) to evaluate the occurrence and distribution of pesticides in surface waters in relation to pesticide use; (3) to review the factors that affect the behavior and fate of pesticides in surface waters; and (4) to assess the significance of the observed pesticide levels to the health of humans and aquatic biota.

This overview of studies of pesticides in surface waters is one in a series on present knowledge of pesticide contamination of the hydrologic system, which are being conducted as part of the Pesticide National Synthesis project of the U.S. Geological Survey (USGS), National Water Quality Assessment (NAWQA) Program. Other works in the series focus on pesticides in the atmosphere, ground water, and stream bed sediment and aquatic biological tissues. These national topical reviews of published studies on pesticides complement more detailed studies conducted in each NAWQA study area in major hydrologic basins, which are typically 10,000 to 30,000 mi^2, or 25,000 to 75,000 km^2 (Gilliom and others, 1995).

1.2 PREVIOUS REVIEWS

Previous reviews of existing information on various aspects of pesticide contamination of surface waters have been published. A number of these reviews are listed in Table 1.1, along with a brief description of their scope. Most of the reviews focus on a particular pesticide or class of pesticides, a particular body of water, or a particular set of fate or behavior processes. Several of the reviews listed in Table 1.1 are described briefly below, as examples of the types of reviews that have been published previously.

Table 1.1. Selected reviews of pesticide occurrence and behavior in surface waters

Review (see Reference List)	Focus of Review	Topics Discussed		Number of References
		Occurrence and Distribution	Environmental Fate, Transport, or Effects	
Reviews of Environmental Observations				
Wolman, 1971	General discussion of pollution of United States rivers.	√		67
Johnson and Ball, 1972	Historical perspective on pesticide pollution in the Great Lakes.	√		41
Terry and Hughes, 1976	Pollution effects on surface and ground waters.	√		141
Huggett and Bender, 1980	Kepone in the James River.	√		17
Rice and Evans, 1984	Toxaphene in the Great Lakes.	√		90
Strachan and Edwards, 1984	Organochlorine pollutants in Lake Ontario.	√		53
Kutz and Carey, 1986	Pesticides and toxic substances in the environment.	√		11
Logan, 1987	Nonpoint source chemical loadings to Lake Erie.	√		59
Hellawell, 1988	General discussion of toxic substances in rivers and streams.	√		106
Buchman, 1989	Trace contaminants, coastal and estuarine Oregon.	√		102
Ciba-Geigy, 1992a	Atrazine in the Mississippi River, near Baton Rouge–St. Gabriel, Louisiana.	√		3
Ciba-Geigy, 1992b	Atrazine in Chesapeake Bay.	√		32
Ciba-Geigy, 1992c	Atrazine in surface waters of 11 states, 1975–91.	√		9
Ciba-Geigy, 1992d	Atrazine in the Mississippi, Missouri, and Ohio Rivers, 1975–91.	√		7
Ciba-Geigy, 1992e	Influence of agricultural management practices on pesticide runoff.	√ (runoff)		95
Ciba-Geigy, 1992f	Atrazine in surface waters of Illinois, 1975–88.	√		18
Ciba-Geigy, 1994a	Atrazine in surface waters of Iowa, 1975–93.	√		37
Reviews of Environmental Processes and Effects of Pesticides				
Butler, 1966	Pesticides in estuaries and their effects on fisheries.		√	5
Pionke and Chesters, 1973	Pesticide-sediment-water interactions.	√ (runoff)	√	150
Hurlbert, 1975	Secondary effects of pesticides on aquatic ecosystems.		√	197
Faust, 1977	Chemical mechanisms affecting fate of organic pollutants in natural aquatic environments.	√	√	47

Table 1.1. Selected reviews of pesticide occurrence and behavior in surface waters—Continued

Review (see Reference List)	Focus of Review	Topics Discussed		Number of References
		Occurrence and Distribution	Environmental Fate, Transport, or Effects	
Metcalf, 1977	Biological fate and transformation of pollutants in water.		√	50
Wauchope, 1978	Pesticides in agricultural runoff.		√	69
Norris, 1981	Phenoxy herbicides and TCDD in forests.	√	√	212
Lick, 1982	The transport of contaminants in the Great Lakes.		√	116
Willis and McDowell, 1982	Pesticides in agricultural runoff and effects on water quality.	√ (runoff)	√	37
Bedding and others, 1983	Behavior and fate of pesticides in the hydrologic environment. Treatment techniques.		√	210
Biggar and Seiber, 1987	Fate of various pesticides and pesticide classes in the environment.	√	√	more than 500 (book)
Bowmer, 1987	Herbicides in surface water.	√	√	310
Eadie and Robbins, 1987	Role of particulates in movement of contaminants in the Great Lakes.		√	76
Elzerman and Coates, 1987	Equilibria and kinetics of sorption on sediments.		√	137
Leonard, 1988	Herbicides in surface waters.		√	136
Ritter, 1988	Management practices to reduce impacts of nonpoint source pollution from agriculture.		√	34
Berryhill and others, 1989	Impact of conservation tillage and pesticide use on water quality.		√	13
Eidt and others, 1989	Agricultural and forestry use of pesticides—effects on aquatic habitats.		√	69
Bennett, 1990	Fate of pesticides in water and sediment. Assessment techniques.		√	95
Bollag and Liu, 1990	Biological transformation processes of pesticides.		√	217
Green and Karickhoff, 1990	Sorption estimates for modeling.		√	46
Leonard, 1990	Movement of pesticides into surface waters.	√ (runoff)	√	173

Table 1.1. Selected reviews of pesticide occurrence and behavior in surface waters—Continued

Review (see Reference List)	Focus of Review	Topics Discussed		Number of References
		Occurrence and Distribution	Environmental Fate, Transport, or Effects	
Madhun and Freed, 1990	Impact of pesticides on the environment.		✓	246
Miyamoto and others, 1990	The fate of pesticides in aquatic ecosystems. Chemical reactions of pesticide classes.		✓	103
Wolfe and others, 1990	Abiotic transformations in water, sediments, and soil.		✓	210
Day, 1991	Pesticide transformation products in surface waters.		✓	75
Chapra and Boyer, 1992	Fate of various environmental pollutants.		✓	207
Ciba-Geigy, 1992g	Drinking water treatment technology overview.		✓	8
Neary and others, 1993	Fate and effects of pesticides in southern forests.	✓	✓	44
Reviews of Environmental Fate and Behavior of Specific Pesticides				
Weber, 1970	Adsorption of triazines by clay colloids; factors affecting plant availability.		✓	106
Que Hee and Sutherland, 1981	Phenoxy herbicides —characteristics, mode of action, behavior. Summary of environmental occurrence.	✓	✓	more than 500 (book)
Demoute, 1989	Environmental fate and metabolism of pyrethroids.		✓	14
Trotter, 1989	Canadian water quality guidelines for carbofuran; properties, toxicity, and occurrence.	✓	✓	127
Pauli and others, 1990	Canadian water quality guidelines for metribuzin; properties, toxicity, and occurrence.	✓	✓	
Trotter, 1990	Canadian water quality guidelines for atrazine; properties, toxicity, and occurrence.	✓	✓	235
Trotter and others, 1990	Canadian water quality guidelines for glyphosate; properties, toxicity, and occurrence.	✓	✓	120
Howard, 1991	Handbook of environmental fate and exposure data for pesticides.	✓	✓	more than 500 (book)
Kent, 1991	Canadian water quality guidelines for metolachlor; properties, toxicity, and occurrence.	✓	✓	140
Kent and Pauli, 1991	Canadian water quality guidelines for captan; properties, toxicity, and occurrence.	✓	✓	192
Kent and others, 1991	Canadian water quality guidelines for dinoseb; properties, toxicity, and occurrence.	✓	✓	127

Table 1.1. Selected reviews of pesticide occurrence and behavior in surface waters—Continued

Review (see Reference List)	Focus of Review	Topics Discussed		Number of References
		Occurrence and Distribution	Environmental Fate, Transport, or Effects	
Pauli and others, 1991a	Canadian water quality guidelines for cyanazine; properties, toxicity, and occurrence.	√	√	
Pauli and others, 1991b	Canadian water quality guidelines for simazine; properties, toxicity, and occurrence.	√	√	177
Trotter and others, 1991	Aquatic fate and effects of carbofuran.	√	√	110
Fischer and Hall, 1992	Environmental concentrations and toxicity data on diflubenzuron (dimilin).	√	√	92
Kent and others, 1992	Canadian water quality guidelines for triallate; properties, toxicity, and occurrence.	√	√	
Moore, 1992	Canadian water quality guidelines for organotins; properties, toxicity, and occurrence.	√	√	
Huggett and others, 1992	The marine biocide tributyltin.	√	√	43

Howard (1991) has compiled data on the physical and chemical properties and the environmental behavior of 70 pesticide compounds. Included are tabulations of detections of each compound in the different environmental matrices, including surface water. However, a number of the most commonly used agricultural pesticides are not included in Howard's review, including 17 of the 20 highest-use herbicides, 7 of the 20 highest-use insecticides, and 6 of the 10 highest-use fungicides (Gianessi and Puffer, 1991, 1992a,b). Leonard (1990) has thoroughly reviewed the processes involved in the movement of pesticides from agricultural fields to surface waters. Topics considered in Leonard's review include entrainment of pesticides in runoff, the magnitude of runoff losses of various pesticides, the effects of different agricultural practices on runoff losses, and the various computer models used to simulate runoff losses of pesticides. Also included is a tabulation of reported concentrations of pesticides in runoff and seasonal losses of pesticides from agricultural plots. Ciba-Geigy Corporation has reviewed studies in which atrazine concentrations were measured in surface waters, primarily in rivers, streams, and reservoirs in the central United States, in a series of technical reports (Ciba-Geigy, 1992a,b,c,d,f, 1994a). Data from government agencies, utilities, universities, and monitoring programs conducted by Ciba-Geigy and Monsanto Company, are tabulated and cover 1975 to 1993. In these reports, the primary focus is on relating the observed concentrations, and estimated annual mean concentrations at the various sites, to the regulatory criteria for drinking water.

Neary and others (1993) reviewed recent research conducted in the southeastern United States on pesticide use in forests. Results were evaluated from a number of studies that monitored water quality in streams draining forested watersheds where known amounts of pesticides were applied. The authors concluded that current practices result in short-term perturbations in aquatic habitats, and direct effects on aquatic biota are minimal, especially for herbicide use. The indirect and cumulative effects of pesticides used in forests on stream biota are not well known, however, and the authors recommend further study. Day (1991) reviewed studies of the effects of pesticide transformation products on aquatic biota. Data from a number of studies on transformation products indicate that they can be more, less, or similar in toxicity to the parent compounds. Most of the data evaluated were from laboratory studies, and the general lack of data on environmental concentrations of pesticide transformation products was noted. Observed synergistic and interactive effects of pesticides and their transformation products on biota are discussed. Finally, Environment Canada has published reviews on the properties, use, toxicity, and environmental occurrence of a number of pesticides under the general title *Canadian Water Quality Guidelines* (see Table 1.1, 2nd column). Pesticides evaluated in this series include atrazine, captan, carbofuran, cyanazine, dinoseb, glyphosate, metolachlor, metribuzin, organotin compounds, simazine, and triallate.

Together, the reviews listed in Table 1.1 provide a relatively complete overview of the range of factors that affect the sources, transport, and fate of pesticides in surface waters. They do not, however, provide a broad perspective on the occurrence, distribution, and significance of pesticides in surface waters.

1.3 APPROACH

This book focuses primarily on studies of pesticides in the surface waters of the United States. Studies from outside the United States, and laboratory and process studies, were selectively reviewed to help explain particular phenomena or occurrences. The goal was to locate all significant studies within this scope that have been published in an accessible report format, including journal articles, federal and state reports, and university report series. The studies reviewed were located through bibliographic data searches (National Technical Information Service, Chemical Abstracts, AGRICOLA, and Selected Water Resources Abstracts), compilations of state and local agency reports, and bibliographies from reviewed manuscripts. Studies at all spatial scales, from individual sites to multistate regional and national studies, were included. Although all of the reports and papers identified in these databases were evaluated, other studies exist in the literature that were not identified in the bibliographic searches. For example, although many reports from studies conducted by state and local agencies are included, many of the unpublished reports could not be obtained for this book. Many state surface water monitoring programs have expanded their list of analytes to include more pesticides in the 1990's, but much of this data is not yet available. Therefore, the book primarily reflects the information available in the open scientific literature as of the end of 1992.

The studies were evaluated and are presented in four main sections. First, all reviewed studies are characterized and tabulated with selected study features such as location, spatial scale, time frame, number of sites, sampled media, and target analytes. This serves as an overview of the reviewed studies and provides the basis for characterizing the nature, degree, and emphasis of study effort that has accumulated.

Second, a national overview of the occurrence and geographic distribution of pesticides in surface waters is developed from the observations reported in the reviewed studies, with particular emphasis on the large-scale studies. Although limited by the biases inherent in the reviewed studies, this overview provides a perspective on the degree to which contamination of surface waters may be a problem and on past and present assessment and research priorities.

Third, the primary factors that affect pesticide concentrations in surface waters are reviewed. Information on the various sources of pesticides to surface waters and on the behavior and fate of pesticides in surface waters is included in this section. Definitions and terminology used to describe the various processes affecting pesticides in surface waters also are presented. This provides a basis for understanding observed patterns in occurrence and distribution and for addressing specific key topics.

Finally, results from reviewed studies are used to address key topics related to the occurrence of pesticides in surface waters. These topics represent basic points that must be understood to evaluate the causes, degree, and significance of surface water contamination. Some of these topics are addressed more thoroughly than others, reflecting the strengths and weaknesses of existing information. In some cases, gaps in existing knowledge are identified, suggesting future research priorities.

CHAPTER 2

Characteristics of Studies Reviewed

2.1 INTRODUCTION

All studies included in this book investigated pesticide occurrence in one or more water column matrix (water, suspended solids, surface microlayer, and dissolved organic carbon). Studies reviewed are summarized in Tables 2.1, 2.2, and 2.3 (located at end of chapter), according to three main categories: (1) national and multistate monitoring studies, (2) state and local monitoring studies, and (3) process and matrix distribution studies.

National and multistate monitoring studies (Table 2.1) are occurrence surveys for specific compounds or compound classes at several to many locations in multiple states. Relatively few of these large-scale studies have been conducted. The sampling sites included in these studies are shown in Figures 2.1 through 2.4 for the studies conducted in the 1950's–1960's, 1970's, 1980's, and during 1990–1992, respectively. In the early studies (1950's–1970's), the targeted pesticides were primarily the organochlorine insecticides (OCs), and the geographic emphasis was either the entire United States, the western United States, or the Great Lakes. More recent large-scale studies from the 1980's and 1990's have emphasized the current high-use herbicides in the Mississippi River Basin.

State and local monitoring studies (Table 2.2) are occurrence surveys for specific compounds or compound classes, usually at several to many sites within a specific area, and are typically smaller than the state in which they were conducted. This group includes a few studies with one location sampled over several months to years, as well as studies with many locations sampled for several days, weeks, or months. The geographic distribution of reviewed state and local studies is shown in Figure 2.5a.

Process and matrix distribution studies (Table 2.3) generally measured concentrations of one or more pesticides in surface water environments not considered to be ambient or natural. Included are studies of pesticide runoff from field plots, investigations of surface waters to which pesticides have been applied directly for pest control, studies of forest streams immediately after aerial applications of pesticides, and so forth. Field studies that evaluated the water-solid distribution of pesticides also are included in this section. Most of these studies involved relatively specialized sampling at one or several sites for several days, weeks, or months. The geographic distribution of the process and matrix distribution studies reviewed is shown in Figure 2.5b. Laboratory studies, studies using artificial water bodies or ecosystems, and review articles are cited as needed, but are not included in Table 2.3.

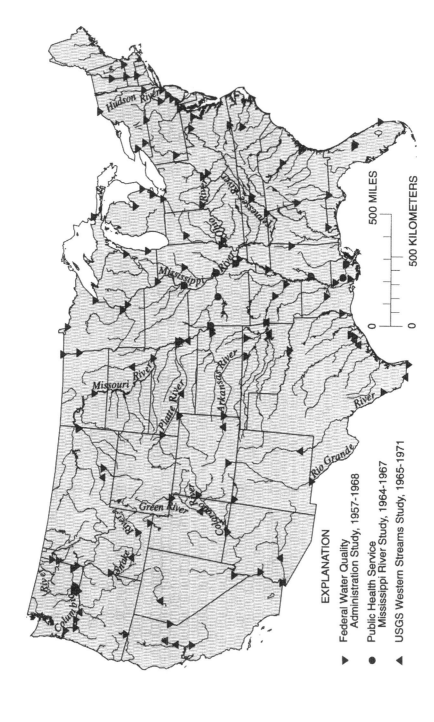

Figure 2.1. Sampling sites of selected national and multistate studies conducted mostly during the 1950's–1960's. References: ▼ – Weaver and others (1965), Breidenbach and others (1967), Green and others (1967), and Lichtenberg and others, 1970; ● – Schafer and others (1969); ▲ – Brown and Nishioka (1967), Manigold and Schulze (1969), and Schulze and others (1973).

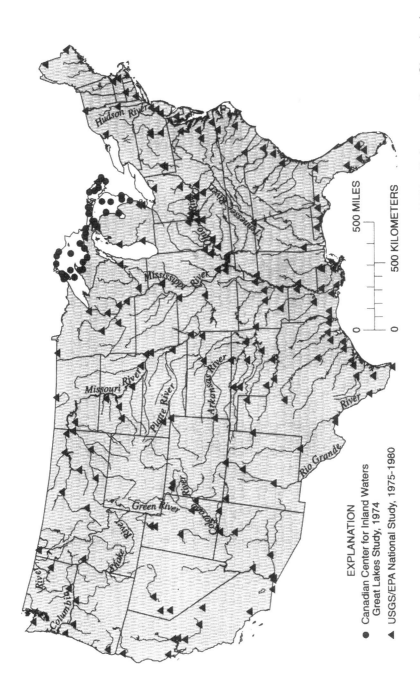

Figure 2.2. Sampling sites of selected national and multistate studies conducted mostly during the 1970's. References: ● — Glooschenko and others (1976); ▲ — Gilliom and others (1985).

EXPLANATION

● Canadian Center for Inland Waters
  Great Lakes Study, 1974

▲ USGS/EPA National Study, 1975-1980

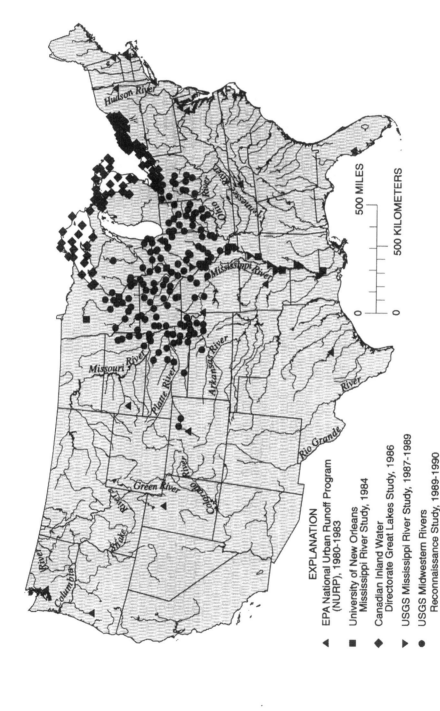

Figure 2.3. Sampling sites of selected national and multistate studies conducted mostly during the 1980's. References: ▲ – Cole and others (1984); ■ – DeLeon and others (1986); ◆ – Stevens and Neilson (1989); ▼ – Pereira and Rostad (1990), Pereira and others (1990, 1992); ● – Goolsby and Battaglin (1993), Thurman and others (1991, 1992), and Goolsby and others (1991a,b).

EXPLANATION

▲ EPA National Urban Runoff Program (NURP), 1980-1983

■ University of New Orleans Mississippi River Study, 1984

◆ Canadian Inland Water Directorate Great Lakes Study, 1986

▼ USGS Mississippi River Study, 1987-1989

● USGS Midwestern Rivers Reconnaissance Study, 1989-1990

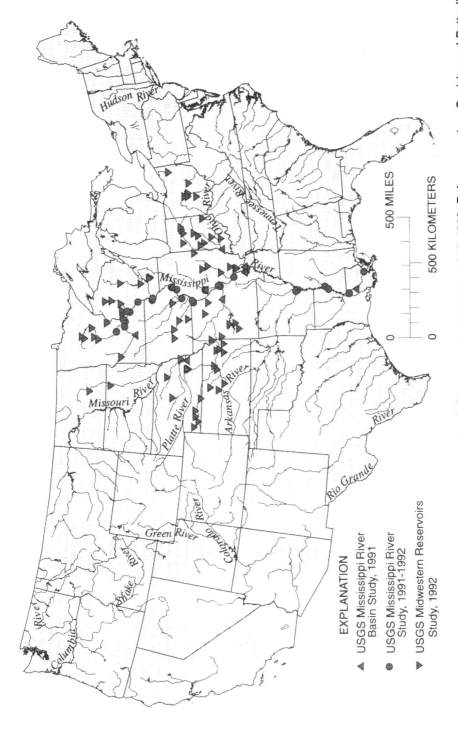

Figure 2.4. Sampling sites of selected national and multistate studies conducted during 1990–1992. References: ▲ – Goolsby and Battaglin (1993); ● – Pereira and Hostettler (1993); ▼ – Goolsby and others (1993).

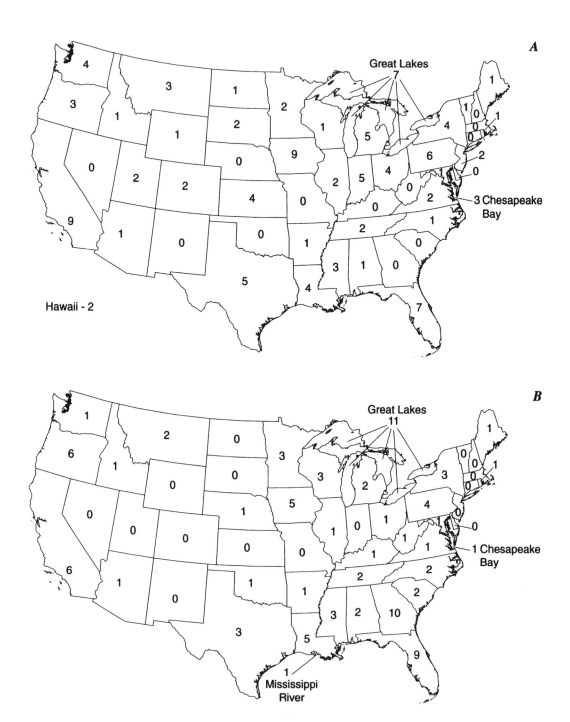

Figure 2.5. Geographic distribution of reviewed (*A*) state and local monitoring studies (Table 2.2) and (*B*) process and matrix distribution studies (Table 2.3).

2.2 GENERAL DESIGN FEATURES

Characteristics of the studies included in Tables 2.1, 2.2, and 2.3 are summarized in Table 2.4. Most of the data are from studies classified as state and local monitoring studies. Studies in all categories generally have been short-term, seldom lasting more than 2 years. Study designs ranged from monitoring a single pesticide at a single site to regional studies of multiple pesticide classes. There was little consistency in sampling methodologies, sampling site selection, timing of sample collection, detection limits, or target analytes (other than the OCs).

2.3 TARGET ANALYTES

Most of the pesticides investigated in the studies tabulated in Tables 2.1, 2.2, and 2.3 can be classified into six major groups: OCs, organophosphorus insecticides (OPs), other insecticides and fungicides, triazine and acetanilide herbicides, phenoxy acid herbicides, and other herbicides. Analytes targeted in the reviewed studies (Tables 2.1 and 2.2) are listed in Table 2.5 (most compounds listed in this table, and throughout this book, are referred to by their common names; chemical names, using standard International Union of Pure and Applied Chemistry (IUPAC) nomenclature, are listed in the Appendix for all pesticides mentioned in the text, tables, and figures of this book). The distribution of sampling effort devoted to each of these six groups, in terms of study years, is plotted as a function of time in Figure 2.6. In compiling the data for Figure 2.6, one study year was assigned for each year in which samples were collected, regardless of starting month. The number of analytes, number of sampling sites, and the sampling intensity were not factored into the compilation, but Figure 2.6 gives a general indication of the trends in monitoring over the last several decades.

Studies in the late 1950's and the 1960's focused on the OCs and a few phenoxy acid herbicides (2,4-D, 2,4,5-T, and silvex [2,4,5-TP]) and OPs (parathion, malathion, methyl parathion, ethion, and diazinon). A great deal of effort has been expended on monitoring residues of OCs since the 1960's (Figure 2.6), even after many of these compounds were banned or their use greatly restricted in the United States. Attention remains focused on the organochlorines for a number of reasons. First, many are listed as priority pollutants by the U.S. Environmental Protection Agency (USEPA), with monitoring required by law in certain cases. Second, they are still detected in the bed sediments of rivers and lakes and in the soil. Third, several have known adverse ecological and human-health effects and can bioaccumulate in fish and other organisms. Finally, they continue to be used in other parts of the world and have the potential for long-range atmospheric transport.

The trend in the 1970's and 1980's was a pronounced increase in the number of different types of pesticides being monitored in surface waters. This trend has been driven by a number of factors. Most of the organochlorines have been replaced with organophosphates or other insecticides. Use of herbicides, particularly the triazines (such as atrazine and cyanazine) and acetanilides (such as alachlor and metolachlor), has increased dramatically since the 1960's. Many of these compounds are much more likely to appear in the water column of surface waters than the organochlorines, due to their greater water solubility and lower tendency to sorb to soil and sediments (Goss, 1992). By the 1980's, approximately the same amount of time was devoted to monitoring triazine and acetanilide herbicides, OPs, and OCs. Insecticides and herbicides in other classes also were targeted in more studies. Increasing environmental regulation and

Table 2.4. General characteristics of studies included in Tables 2.1, 2.2, and 2.3

Study Characteristics	Study Type		
	National and Multistate Monitoring Studies (Table 2.1)	State and Local Monitoring Studies (Table 2.2)	Process and Matrix Distribution Studies (Table 2.3)
Number of Studies	27	109	101
Number of Sites			
Range	6-177	1-142	no data
Median	30	6	no data
Study Duration (months)			
Range	1-150	1-132	1-72
Median	12	12	12
Surface Water Type			
Streams	16	80	52
Lakes and Reservoirs	5	31	20
Estuaries	1	16	10
Forest Streams	0	0	15
Agricultural Runoff	2	9	36
Urban Runoff	1	5	2
Wetlands	0	1	0
Oceans	1	3	2
Drinking Water	1	6	2
Compound Class			
Organochlorine Insecticides	12	88	31
Organophosphorus Insecticides	5	40	12
Other Insecticides & Fungicides	1	12	12
Triazine and Acetanilide Herbicides	8	29	30
Phenoxy Herbicides	4	30	12
Other Herbicides	1	21	22

Table 2.5. Detection frequency of targeted pesticides in surface waters

[Data from studies in Table 2.1 (national and multistate studies) and Table 2.2 (state and local studies). α, alpha; β, beta; γ, gamma; δ, delta. nr, not reported]

Pesticide	Sampling Sites			Samples		
	Total sites	Number of sites with detections	Percent of sites with detections	Total samples	Number of samples with detections	Percent of samples with detections
INSECTICIDES						
Organochlorine Compounds:						
Aldrin	951	65	7	3,910	224	6
Chlordane	838	154	18	3,366	948	28
DDT[1]	1,185	258	22	5,569	945	17
DDT-total (sum of DDT, DDD, DDE)	75	56	75	77	42	55
Dieldrin	1,016	459	45	4,995	1,412	28
Endosulfan	469	9	2	1,614	42	3
Endrin	944	136	14	4,255	359	8
HCH (all isomers)[2]	1,498	462	31	7,144	2,087	29
Heptachlor	948	102	11	3,877	287	7
Kepone	75	nr	nr	750	nr	nr
Methoxychlor	268	33	12	772	33	4
Mirex	212	2	1	512	13	3
Perthane	81	0	0	285	0	0
Toxaphene	215	16	7	1,490	84	6
Organophosphorus Compounds:						
Azinphos-methyl	79	0	0	402	0	0
Chlorpyrifos	108	7	6	987	13	1
Crufomate	33	0	0	33	0	0
DEF	4	2	50	4	2	50
Diazinon	193	36	18	1,836	256	14
Dichlorvos (DDVP)	2	0	0	30	0	0
Dimethoate	33	0	0	33	0	0
Disulfoton	40	0	0	349	0	0
Disyston	4	0	0	4	0	0
Ethion	326	0	0	1,046	0	0
Ethoprop	33	0	0	33	0	0
Fenitrothion	42	0	0	42	0	0
Fensulfothion	9	0	0	9	0	0
Fenthion	232	0	0	538	0	0
Fonofos	94	14	15	945	63	7
Imidan	33	0	0	33	0	0
Malathion	426	16	4	2,415	104	4
Methamidophos	10	0	0	100	0	0
Methidathion	2	2	100	nr	nr	nr
Methyl parathion	387	13	3	2,215	14	1
Methyl trithion	80	0	0	185	0	0
Parathion	326	4	1	1,493	5	0
Phorate	121	0	0	1,008	0	0
Phosphamidon	33	0	0	33	0	0
Ronnel	35	0	0	63	0	0

Table 2.5. Detection frequency of targeted pesticides in surface waters—Continued

Pesticide	Sampling Sites			Samples		
	Total sites	Number of sites with detections	Percent of sites with detections	Total samples	Number of samples with detections	Percent of samples with detections
Sulprofos	33	0	0	33	0	0
Terbufos	94	10	11	945	2	0
Trithion	314	1	0	805	2	0
Other Insecticides[3]:						
Aldicarb	4	0	0	4	0	0
Carbaryl	24	6	25	333	32	10
Carbofuran	84	25	30	396	119	30
Deet	26	22	85	nr	nr	nr
Dibutyltin (DBT)	10	1	10	22	4	18
Fenvalerate	4	4	100	nr	nr	nr
Methomyl	8	0	0	8	0	0
Oxamyl	4	0	0	4	0	0
Permethrin	11	4	36	316	3	1
Propargite	7	3	43	316	3	1
Tributyltin (TBT)	10	8	80	22	15	68
HERBICIDES						
Triazines and Acetanilides:						
Acrolein	17	nr	nr	121	nr	nr
Alachlor	372	272	73	1,549	802	52
Ametryn	123	11	9	947	212	22
Atratone	15	0	0	27	0	0
Atrazine	497	440	89	4,650	3,928	84
Cyanazine	366	242	66	1,473	755	51
Cyprazine	15	0	0	27	0	0
Hexazinone	26	5	19	nr	nr	nr
Metolachlor	362	280	77	1,452	827	57
Metribuzin	349	147	42	1,469	245	17
Prometon	270	74	27	828	140	17
Prometryn	62	9	15	523	4	1
Propachlor	12	7	58	450	48	11
Propazine	244	70	29	827	28	3
Simazine	209	119	57	632	312	49
Simetone	15	0	0	27	0	0
Simetryn	22	0	0	36	0	0
Terbutryn	4	0	0	16	0	0
Phenoxy Acids:						
2,4-D	215	110	51	1,721	359	21
2,4-D (methyl ester)	6	0	0	84	0	0
2,4-DP	50	2	4	141	4	3
2,4,5-T	166	70	42	1,347	214	16
2,4,5-TP (silvex)	196	27	14	1,576	79	5
Other Herbicides:						
Bensulfuron-methyl	3	2	67	54	16	30
Butylate	94	8	9	945	49	5
Chloramben	30	0	0	30	0	0
Dacthal	119	nr	nr	1,074	nr	nr

Table 2.5. Detection frequency of targeted pesticides in surface waters—Continued

Pesticide	Sampling Sites			Samples		
	Total sites	Number of sites with detections	Percent of sites with detections	Total samples	Number of samples with detections	Percent of samples with detections
Dicamba	68	17	25	181	17	9
Dinoseb	4	0	0	16	0	0
Diquat	9	0	0	9	0	0
EPTC	15	7	47	316	63	20
Fluometuron	26	7	27	nr	nr	nr
Linuron	37	9	24	395	2	1
Molinate	27	7	26	16	16	100
Norflurazon	26	5	19	nr	nr	nr
Paraquat	9	0	0	9	0	0
Pendimethalin	15	14	93	316	25	8
Picloram	38	15	39	71	18	25
Propham	8	0	0	8	0	0
Thiobencarb	27	2	7	16	16	100
Trifluralin	104	24	23	1,087	113	10
FUNGICIDES						
Captan	30	0	0	580	0	0
Chlorothalonil	4	0	0	16	0	0
HCB	50	43	86	255	216	85
PCNB	4	0	0	16	0	0
PCP	11	8	73	11	8	73
TRANSFORMATION PRODUCTS						
Azinphos-methyl oxon	6	0	0	20	0	0
Carbofuran phenol	1	1	100	9	9	100
2-Chloro-2′,6′-diethylacetanilide	26	8	31	nr	nr	nr
Cyanazine amide	26	16	62	nr	nr	nr
DDD	876	139	16	3,941	543	14
DDE	1,128	219	19	4,869	939	19
Deethylatrazine	291	254	87	685	559	82
Deisopropylatrazine	242	154	64	685	249	36
Desmethyl norflurazon	26	2	8	nr	nr	nr
Endosulfan sulfate	50	0	0	154	0	0
Endrin aldehyde	50	0	0	154	0	0
ESA (alachlor metabolite)	76	60	79	304	222	73
Heptachlor epoxide	922	181	20	3,714	552	15
2-Hydroxy-2′6′-diethylacetanilide	26	19	73	nr	nr	nr
2-Ketomolinate	1	1	100	nr	nr	nr
4-Ketomolinate	1	1	100	nr	nr	nr
Oxychlordane	14	14	100	14	14	100
Paranitrophenol	1	1	100	9	9	100
Photomirex	14	0	0	14	0	0
Terbufos sulfone	33	0	0	33	0	0

[1]Detection frequencies for DDT, DDD, and DDE include both *p,p′*-, and *o,p′*-isomers, as many studies did not report which isomer was targeted.
[2]HCH data for all isomers, including α, β, γ (lindane), and δ.
[3]Includes compounds used as acaricides, miticides and nematocides.

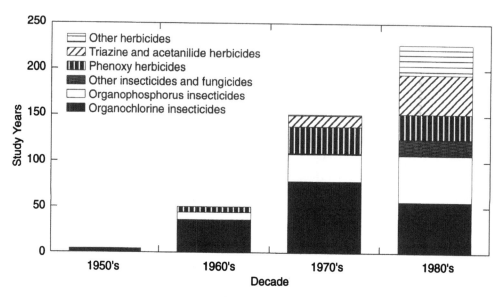

Figure 2.6. Distribution of pesticide study efforts by decade. Each year in which samples were collected in a specific study is defined as one study year, regardless of starting month. Data are from national and multistate studies in Table 2.1 and from state and local studies in Table 2.2.

changing public perceptions of pesticides have resulted in a steady increase in the total effort expended on the monitoring of pesticides in surface waters.

2.4 GEOGRAPHIC DISTRIBUTION

In Figures 2.1 through 2.4, sampling sites are shown for reviewed national and multistate studies conducted during the 1950's–1960's, 1970's, 1980's, and during 1990–1992, respectively. The most extensive data collection efforts have been in the Mississippi River Basin, the Great Lakes, and rivers of the western United States. Figure 2.5a shows that the geographic distribution of reviewed state and local studies is uneven, with no reviewed studies conducted in some states and numerous reviewed studies conducted in others. Iowa, California, Florida, and the Great Lakes had the greatest number of reviewed studies. The reviewed studies span scales from a few hectares (runoff to streams from field plots) to the entire nation.

2.5 TEMPORAL DISTRIBUTION

From the late 1950's through the 1980's, two long-term national-scale studies of pesticide residues in rivers and streams were conducted by the Federal Water Pollution Control Administration (FWPCA), later called the Federal Water Quality Administration, or FWQA (Weaver and others, 1965; Breidenbach and others, 1967; Green and others, 1967; Lichtenberg and others, 1970), and by the U.S. Geological Survey, or USGS (Gilliom, 1985; Gilliom and others, 1985). The USGS also monitored pesticide concentrations in streams throughout the western United States from 1965 to 1971 (Brown and Nishioka, 1967; Manigold and Schulze, 1969; Schulze and others, 1973). In the 1980's and 1990's, the general trend has been toward smaller-scale studies conducted within individual states or specific river basins. No national-scale studies were undertaken during the 1980's, although several large multistate studies were done in the Mississippi River Basin (Pereira and Rostad, 1990; Pereira and others, 1990, 1992; Goolsby and others, 1991a,b; Thurman and others, 1992; Goolsby and Battaglin, 1993; Goolsby and others, 1993). The number of research oriented studies (Table 2.3) rose during the 1980's as well, comprising almost half of the studies reviewed from this period. Along with the trend toward smaller geographical areas, the duration of studies also has decreased. The median duration of sample collection in the state and local studies is 12 months, while the national programs of the 1960's and 1970's sampled the same sites over a multiyear period. A notable exception to this trend is the ongoing program of Baker, Richards, and coworkers (Richards and Baker, 1993), that has been sampling the tributaries of Lake Erie and the drainage basins in Ohio and in parts of Indiana and Michigan continuously since 1981. This data set is probably the most complete and consistent of all of the data reviewed here. Monitoring by Ciba-Geigy Corporation for atrazine throughout the Mississippi River Basin also has provided long-term records at some sites.

2.6 MATRICES SAMPLED

This book includes only studies with research related to pesticides in water-column matrices. A companion review (Nowell, 1996) examines research on pesticides in bed sediments and aquatic macrobiota. Matrices in this review include unfiltered water (whole water), filtered water, suspended solids (biotic or abiotic particles separated from the water by filtration or centrifugation), colloidal/dissolved organic carbon, and the surface microlayer. By far the most common matrix, especially in monitoring studies, was unfiltered water. However, a number of process studies examined other surface water matrices (Table 2.3), and these studies have greatly added to our understanding of the distribution and fate of pesticides in surface waters.

2.7 ANALYTICAL LIMITS OF DETECTION

A major problem in comparing results from different studies is dealing with unknown or variable detection limits. Analytical limits of detection were reported in about 90 percent of the national and multistate monitoring studies reviewed, but in fewer local and state monitoring studies. In some studies, limits of detection for some compounds could be inferred from the reported data when less-than values were given, or from other studies by the same agency in which the detection limits were stated. In other cases, the lowest reported value for a compound or group of similar compounds can be used as an estimate of the detection limit, although this does not necessarily indicate the actual detection limit.

The analytical detection limits for all pesticides in surface water samples are partially determined by the volume or mass of the sample. If a lower detection limit is required, sample size generally can be increased, provided the sampling and extraction efficiencies remain the same. The national studies summarized by Breidenbach and others (1967) used 1,000 L of water to isolate the OCs and had detection limits of 0.001 to 0.002 µg/L. Many other studies used only 1 L of water and had much higher detection limits.

Table 2.6. Example of the effect of detection limits on the frequency of detection of pesticides in surface waters

[Sampling sites for Schulze and others (1973) are shown in Figure 2.1. Sampling sites for Gilliom and others (1985) are shown in Figure 2.2. µg/L, micrograms per liter]

	Study	
	Schulze and others, 1973	Gilliom and others, 1985
Sampling period	1968-71	1975-80
Number of sites	20	186
Number of samples	600	1,764
2,4-D		
Detection limit (µg/L)	0.02	0.5
Percent samples with detections	17	0.2
Percent sites with detections	90	2.4
2,4,5-T		
Detection limit (µg/L)	0.005	0.5
Percent samples with detections	18	0.1
Percent sites with detections	100	0.6

Detection limits can influence the results and, ultimately, the interpretation of a study. As an example, two large-scale studies (Schulze and others, 1973; Gilliom and others, 1985) that both targeted the phenoxy acid herbicides 2,4-D and 2,4,5-T are compared in Table 2.6. In the study by Schulze and others (1973), 20 sites were sampled from 1968 to 1971, and detection limits were 0.02 and 0.005 µg/L for 2,4-D and 2,4,5-T, respectively. In the study by Gilliom and others (1985), 186 sites were sampled from 1975 to 1980, and detection limits were 0.5 µg/L for both compounds. All of the sites from the earlier study were included in the later study (Figures 2.1 and 2.2). As shown in Table 2.6, detection frequencies for both compounds were much higher in the study with the lower detection limits. The large difference in detection limits between these two studies is almost certainly the major reason for the very different results for the detection frequency of 2,4-D, since agricultural use of 2,4-D in the United States was very stable during this period (Eichers and others, 1970; Andrilenas, 1974; Eichers and others, 1978; Gianessi and Puffer, 1991). Use of 2,4,5-T is less well documented during the late 1970's, but substantial quantities were being used as late as 1981 (Gilliom and others, 1985). Furthermore, all of the concentrations reported for these two compounds in the earlier study were lower than the detection limits in the later study. Results from the study with the lower detection limits suggest that 2,4-D and 2,4,5-T were widespread, low-level contaminants in surface waters throughout the western United States for at least part of the year, whereas the other study suggests that these compounds were present in only a few samples at a few sites. The effect of variable detection limits should be kept in mind in reviewing the aggregate statistics and in the discussion of these studies. The national study conducted during 1975–1980 (Gilliom and others, 1985) is not included in the summary statistics of the detection frequencies of pesticides in surface waters (Table 2.5), since the relatively high detection limits and the large number of samples in this study would result in a somewhat misleading picture when combined with other studies from the same period.

2.8 INFLUENCE OF STUDY DESIGN

Interpretation of the results of the reviewed studies can be affected by study design. The choice of analytes, sampling sites, sampling frequency, timing of sampling, matrices sampled, and study duration all have an important influence on the conclusions that can be drawn from a study. Consideration of these study-design components is especially important when comparing the results of different studies. As shown in the preceding section, analytical detection limits can have a large effect on the interpretation of results. Two examples of the effects of other study-design components are described below.

Sampling frequency and the timing of sampling are important considerations when comparing the results of two large-scale studies conducted during the 1960's. The FWQA sampled approximately 70 rivers at over 100 sites throughout the United States from 1964 to 1968 (Table 2.1). These were synoptic studies, in which one sample was taken at each site each year. From 1964 to 1967, samples were collected in September, when most rivers were in a low-flow period. In contrast, in the USGS studies of streams of the western United States conducted from 1965 to 1971 (Table 2.1), samples were collected monthly, so that both low-.and high-flow periods were sampled. Because of the differences in the timing and frequency of sampling, the detection frequencies reported in these two studies cannot be directly compared. Both the FWQA studies and the USGS studies of western streams also are examples of studies in which the choice of the matrix sampled is an important consideration. In these studies, unfiltered water samples were analyzed to include suspended sediment in the samples. The organochlorine compounds targeted in these studies have a tendency to sorb to particles in the water column (see Section 4.2). This has several implications for interpretation of results from these studies. First, much of the variation in detection frequencies and concentrations observed from year to year and between sites in these studies may have reflected differences in suspended sediment concentrations at the time of sampling. Second, the environmental significance of the concentrations of organochlorine compounds observed in these studies is unclear, since the concentrations in the dissolved and sorbed phases were not determined (see Section 6.2). Finally, detection frequencies and concentrations observed in these studies cannot be directly compared with the results of later studies that analyzed filtered water samples.

The viewpoints and purposes of those conducting studies and of those providing funding for studies can also influence the way in which studies are designed and conducted. In many of the studies reviewed, the government agency responsible for managing a resource also conducted or funded studies evaluating the effects of pesticides used in its management program. Other studies were funded, and in some cases conducted, by pesticide manufacturers. Much of the research conducted by pesticide manufacturers is done to satisfy pesticide registration requirements or to demonstrate that a specific pesticide can be used without negative environmental effects. Whether our understanding of problems associated with pesticide use has been influenced by the viewpoints and purposes of those conducting studies is not clear, but it is important that this potential bias is recognized when interpreting the results of the studies. As an example, in the studies conducted by Ciba-Geigy Corporation on atrazine occurrence in streams (Ciba-Geigy, 1992a,b,c,d,f, 1994a), the focus was clearly on comparing observed annual mean concentrations with the USEPA-established maximum contaminant level for atrazine. No transformation products of atrazine were monitored, and concentrations of other herbicides present at the same time were not reported. No effects on aquatic organisms were investigated. The data set resulting from the Ciba-Geigy studies is one of the best available for examining long-term trends in atrazine occurrence and for evaluating the significance of atrazine concentrations with respect to drinking water (see Section 6.1). However, because the studies were designed to focus exclusively on the occurrence of atrazine, no information was obtained on the potential presence of atrazine degradation products or other pesticides in the surface waters sampled.

Table 2.1. National and multistate monitoring studies reviewed

[Matrix: w, whole (unfiltered) water; d, drinking water; f, filtered water; s, suspended sediments. **Bold face type** in compound column indicates a positive detection in one or more samples. Abbreviations used for compounds: Azinphos-m., Azinphos-methyl; DAR, deethylatrazine/atrazine ratio; Deethylatr., deethylatrazine; Deisoatr, deisopropylatrazine; Diethylacetan., diethylacetanilide; Hept. epox., heptachlor epoxide; Methox., methoxychlor; M. parathion, methyl parathion; M. trithion, methyl trithion. tr, trace concentration reported, above detection limit but below reporting level. Technical (following a compound name), a mixture of isomers and related compounds. nr, not reported. α, alpha; β, beta; γ, gamma; δ, delta. FWQA, Federal Water Quality Administration. USEPA, U.S. Environmental Protection Agency. USGS, U.S. Geological Survey. <, less than; >, greater than. ~, number is approximate. µg/L, micrograms per liter. no det., no samples with concentrations above the detection limit]

Reference(s)	Matrix	Sampling dates	Location(s)	Compounds	Detection limit(s) (µg/L)	Number of sites	Percent of sites with detections	Number of samples	Percent of samples with detections	Maximum concentration (µg/L)	Comments
Weaver and others, 1965	w	9/64	United States: 96 sites on rivers throughout the United States and on the Great Lakes	**Dieldrin**	0.002-0.01	96	74	96	74	0.07	FWQA study (then called Public Health Service). Synoptic survey of rivers throughout the United States. One sample taken at each site in September 1964 during low flow. Compounds estimated to constitute >60 percent of chlorinated pesticide use at the time. Data are analyzed in terms of geographic distribution. Generally, higher levels of chlorinated pesticides were observed in the North Atlantic, lower Mississippi, and California basins.
				Endrin	0.002-0.01	96	46	96	46	>0.09	
				DDT	0.002-0.01	96	44	96	44	0.087	
				DDE	0.002-0.01	96	39	96	39	0.018	
				DDD	0.075	96	1	96	1	0.083	
				Aldrin	0.002-0.01	96	10	96	10	0.085	
				Heptachlor	0.002-0.01	96	17	96	17	tr	
				Hept. epox.	0.075	96	0	96	0	no det.	
				HCH	0.025	96	2	96	2	tr	

Table 2.1. National and multistate monitoring studies reviewed—*Continued*

Reference(s)	Matrix	Study Sampling dates	Study Location(s)	Compounds	Detection limit(s) (µg/L)	Sites Number of sites	Sites Percent of sites with detections	Samples Number of samples	Samples Percent of samples with detections	Samples Maximum concentration (µg/L)	Comments
Breidenbach and others, 1967	w	1957–65	United States: 99 sites on rivers throughout the United States and on the Great Lakes	**Dieldrin**	0.002	99	47	99	47	0.1	FWQA study (then called Federal Water Pollution Control Administration). Synoptic survey of rivers throughout the United States in 1965, and summary of data from previous sampling, 1957–65. Concentration data and detection frequency shown for 1965 survey. One sample taken at each site in September 1965 during low flow. Compounds constituted >60 percent of chlorinated pesticide use. Endrin and dieldrin detections decreased from 1964 synoptic survey. DDT group essentially unchanged. Endrin occurrence in lower Mississippi declined after reaching maximum in autumn of 1963.
				Endrin	0.002	99	16	99	16	0.12	
				DDT	0.002	99	25	99	25	0.15	
				DDE	0.002	99	5	99	5	0.009	
				DDD	0.002	99	17	99	17	0.026	
				Aldrin	0.002	99	0	99	0	no det.	
				Heptachlor	0.002	99	24	99	24	0.16	
				Hept. epox.	0.002	99	25	99	25	0.067	
				HCH	0.002	99	5	99	5	0.004	

Table 2.1. National and multistate monitoring studies reviewed—*Continued*

| Reference(s) | Matrix | Study | | | | Sites | | | Samples | | | Comments |
		Sampling dates	Location(s)	Compounds	Detection limit(s) (µg/L)	Number of sites	Percent of sites with detections	Number of samples	Percent of samples with detections	Maximum concentration (µg/L)	
Brown and Nishioka, 1967	w	10/65– 9/66 (monthly)	Western United States: 11 sites on major rivers	**Aldrin**	0.005	11	36	128	3	0.005	USGS western streams study. Data from 10/65 to 9/66 for 11 sites on streams throughout the western United States.
				DDD	0.005	11	55	128	10	0.015	
				DDE	0.005	11	64	128	14	0.02	
				DDT	0.01	11	82	128	11	0.11	
				Dieldrin	0.005	11	91	128	23	0.015	
				Endrin	0.005	11	45	128	6	0.04	
				Heptachlor	0.005	11	73	128	11	0.015	
				Hept. epox.	0.005	11	100	128	16	0.09	
				Lindane	0.005	11	100	128	36	0.02	
				2,4,-D	0.1	11	0	62	0	no det.	
				2,4,5-T	0.005	11	0	62	0	no det.	
				Silvex	0.005	11	0	62	0	no det.	

Table 2.1. National and multistate monitoring studies reviewed—*Continued*

Reference(s)	Matrix	Sampling dates	Location(s)	Compounds	Detection limit(s) (µg/L)	Sites		Samples			Comments
						Number of sites	Percent of sites with detections	Number of samples	Percent of samples with detections	Maximum concentration (µg/L)	
Green and others, 1967	w	9/66	**United States:** 109 sites on rivers throughout the United States and on the Great Lakes	**Dieldrin**	0.002	109	60	109	60	0.17	FWQA study (then called Federal Water Pollution Control Administration). Synoptic survey of rivers throughout the United States. One sample taken at each site in September 1966 during low flow. Compounds constituted >60 percent of chlorinated pesticide use. Dieldrin continues to dominate detections. Endrin levels decreased from 1964 synoptic survey, but increased slightly from 1965 survey. Heptachlor detections down significantly from 1965 survey. Detections most common in the Northeast and Mississippi River Valley. Evidence that impoundments result in lower levels of organochlorines in downstream waters because of sedimentation.
				Endrin	0.002	109	25	109	25	0.069	
				DDT	0.002	109	30	109	30	0.12	
				DDE	0.002	109	10	109	10	0.004	
				DDD	0.002	109	50	109	50	0.013	
				Aldrin	0.002	109	1	109	1	tr	
				Heptachlor	0.002	109	1	109	1	0.004	
				Hept. epox.	0.002	109	20	109	20	0.019	
				HCH	0.002	109	16	109	16	0.13	
				Chlordane	0.001-0.005	109	6	109	6	0.075	

Table 2.1. National and multistate monitoring studies reviewed—*Continued*

		Study			Sites			Samples			
Reference(s)	Matrix	Sampling dates	Location(s)	Compounds	Detection limit(s) (µg/L)	Number of sites	Percent of sites with detections	Number of samples	Percent of samples with detections	Maximum concentration (µg/L)	Comments
Manigold and Schulze, 1969	w	1966–68	Western United States: 20 sites on major rivers and irrigation canals	**Aldrin**	0.005	20	40	~330	3	0.04	USGS western streams study. Summary of data from 10/66 to 9/67.
				DDD	0.005	20	45	~330	10	0.04	
				DDE	0.005	20	45	~330	15	0.06	
				DDT	0.005	20	90	~330	25	0.09	
				Dieldrin	0.005	20	60	~330	7	0.07	
				Endrin	0.005	20	20	~330	1	0.07	
				Heptachlor	0.005	20	55	~330	8	0.04	
				Hept. epox.	0.005	20	10	~330	0.6	0.04	
				Lindane	0.005	20	30	~330	2	0.02	
				2,4-D	0.02	20	70	~330	12	0.35	
				2,4,5-T	0.005	20	45	~330	8	0.07	
				Silvex	0.005	20	25	~330	4	0.21	

Table 2.1. National and multistate monitoring studies reviewed—*Continued*

Reference(s)	Matrix	Sampling dates	Location(s)	Compounds	Detection limit(s) (µg/L)	Number of sites	Percent of sites with detections	Number of samples	Percent of samples with detections	Maximum concentration (µg/L)	Comments
						Sites		**Samples**			
Schafer and others, 1969	w d	1964–67	Mississippi and Missouri Rivers	**Aldrin**	0.06	6	83	67	39	nr	Study to establish methods for organochlorines in surface waters and finished drinking water. Samples of raw and finished drinking water taken at 10 sites along Mississippi and Missouri Rivers. Detection frequencies shown are for raw river water. Detection frequency in finished water samples was zero for toxaphene and methoxychlor; 10 to 25 percent for aldrin, endrin, chlordane, DDE, and DDT; and 40 to 75 percent for dieldrin and the HCHs.
				Endrin	0.06	6	67	67	7	nr	
				Dieldrin	0.1	6	83	67	43	nr	
				HCH (α, β, δ)	0.06	6	100	34	85	nr	
				Lindane	0.06	6	100	34	74	nr	
				Methox.	2.5	6	0	67	0	nr	
				Chlordane	0.12	6	50	46	24	nr	
				Heptachlor	0.1	6	33	67	3	nr	
				DDE (*p,p'*)	0.5	6	100	67	22	nr	
				DDT (*p,p'*)	0.5	6	100	67	22	nr	
				Toxaphene	0.1	6	33	67	7	nr	
					2.5	6	0	67	0	nr	

Table 2.1. National and multistate monitoring studies reviewed—*Continued*

Reference(s)	Matrix	Sampling dates	Location(s)	Compounds	Detection limit(s) (µg/L)	Sites		Samples			Comments
						Number of sites	Percent of sites with detections	Number of samples	Percent of samples with detections	Maximum concentration (µg/L)	
Lichtenberg and others, 1970	w	1964-68	**United States:** ~100 sites throughout the United States, mostly on rivers	**Dieldrin**	0.001-0.002	~100	nr	529	38	0.41	FWQA study. Summary of data for 1964–68. Detailed data reported for 1967–68 only. Percent detections shown reflect 5-year totals. Organophosphates included in 1967–68 data only; some question about applicability of method. Marked decrease in detections of most organochlorines observed after peak in 1966.
				Endrin	0.001-0.002	~100	nr	529	13	0.13	
				DDT	0.001-0.002	~100	nr	529	16	0.32	
				DDE	0.001-0.002	~100	nr	529	5	0.05	
				DDD	0.001-0.002	~100	nr	529	13	0.84	
				Aldrin	0.001-0.002	~100	nr	529	<1	0.09	
				Heptachlor	0.001-0.002	~100	nr	529	3	0.05	
				Hept. epox.	0.001-0.002	~100	nr	529	5	0.07	
				Lindane	0.001-0.002	~100	nr	529	2	0.02	
				HCH	0.001-0.002	~100	nr	529	7	0.11	
				Chlordane	0.005	~100	nr	529	<1	0.17	
				M. parathion	0.01-0.025	~100	0	529	0	no det.	
				Parathion	0.01-0.025	~100	0	529	0	no det.	
				Fenthion	0.01-0.025	~100	0	529	0	no det.	
				Ethion	0.01-0.025	~100	0	529	0	no det.	
				Malathion	0.01-0.025	~100	0	529	0	no det.	
				Carbophenothion	0.01-0.025	~100	0	529	0	no det.	

Table 2.1. National and multistate monitoring studies reviewed—*Continued*

Reference(s)	Matrix	Sampling dates	Location(s)	Compounds	Detection limit(s) (μg/L)	Number of sites	Percent of sites with detections	Number of samples	Percent of samples with detections	Maximum concentration (μg/L)	Comments
Schulze and others, 1973	w	10/68–9/71	**Western United States:** 20 sites on major rivers	**Aldrin**	0.005	20	5	600	0.2	0.01	USGS western streams study. Summary of data for 1968–71. Marked decrease in detections of insecticides between 1968 and 1971. Phenoxy herbicide detections peaked in 1969, then declined by ~50 percent by 1971.
				DDD	0.005	20	55	600	8	0.08	
				DDE	0.005	20	40	600	11	0.1	
				DDT	0.005	20	90	600	16	0.46	
				Dieldrin	0.005	20	45	600	6	0.03	
				Endosulfan	0.005	20	5	600	0.2	0.02	
				Endrin	0.005	20	5	600	0.5	0.03	
				Heptachlor	0.005	20	0	600	0	no det.	
				Hept. epox.	0.005	20	0	600	0	no det.	
				Lindane	0.005	20	55	600	7	0.16	
				Chlordane	0.1	20	5	600	0.02	0.02	
				Toxaphene	0.5–1.0	20	0	600	0	no det.	
				2,4-D	0.02	20	90	600	17	0.99	
				Silvex	0.005	20	55	600	8	0.14	
				2,4,5-T	0.005	20	100	600	18	0.4	
				M. parathion	0.01	20	20	~385	2	1	
				Parathion	0.01	20	15	~385	1	0.16	
				Diazinon	0.01	20	25	~385	2	0.1	
				Malathion	0.01	20	0	~285	0	no det.	

Table 2.1. National and multistate monitoring studies reviewed—*Continued*

Reference(s)	Matrix	Sampling dates	Location(s)	Compounds	Detection limit(s) (µg/L)	Sites: Number of sites	Sites: Percent of sites with detections	Samples: Number of samples	Samples: Percent of samples with detections	Samples: Maximum concentration (µg/L)	Comments
Glooschenko and others, 1976	f s	1974 summer	**Upper Great Lakes:** 17 sites, Lake Superior; 2 sites, North Channel; 5 sites, Georgian Bay; 9 sites, Lake Huron	**Lindane**	0.005	33	100	33	100	no det.	Water, seston, and sediments were analyzed at 33 locations. Water contained no analytes above the indicated detection limits. Seston contained traces of dieldrin at 24 of 31 sites and DDE at 12 of 33 sites. Traces of dieldrin, and measurable amounts of DDD, DDE, and DDT were detected in sediments at 1 to 13 sites. None of the organophosphorus compounds were detected in any of the media.
				Heptachlor	0.005	33	3 (tr)	33	(tr)	no det.	
				Hept. epox.	0.005	33	0	33	3 (tr)	no det.	
				Dieldrin	0.005	33	3 (tr)	33	0	no det.	
				Aldrin	0.01	33	0	33	3 (tr)	no det.	
				Endrin	0.005	33	0	33	0	no det.	
				DDE (p,p')	0.005	33	3 (tr)	33	0	no det.	
				DDD (p,p')	0.005	33	0	33	3 (tr)	no det.	
				DDT (p,p')	0.005	33	0	33	0	no det.	
				DDT (o,p')	0.005	33	0	33	0	no det.	
				Chlordane (α)	0.010	33	0	33	0	no det.	
				Chlordane (γ)	0.010	33	0	33	0	no det.	
				Endosulfan (α)	0.010	33	0	33	0	no det.	
				Endosulfan (β)	0.010	33	0	33	0	no det.	
				Methox. (p,p')	0.010	33	0	33	0	no det.	
				Phorate	0.003	33	0	33	0	no det.	
				Diazinon	0.005	33	0	33	0	no det.	
				Disulfoton	0.003	33	0	33	0	no det.	
				Ronnel	0.005	33	0	33	0	no det.	
				M. parathion	0.005	33	0	33	0	no det.	
				Malathion	0.005	33	0	33	0	no det.	
				Parathion	0.005	33	0	33	0	no det.	
				Crufomate	0.025	33	0	33	0	no det.	
				M. trithion	0.01	33	0	33	0	no det.	
				Ethion	0.005	33	0	33	0	no det.	
				Carbophenothion	0.01	33	0	33	0	no det.	

Table 2.1. National and multistate monitoring studies reviewed—Continued

Reference(s)	Matrix	Sampling dates	Location(s)	Compounds	Detection limit(s) (µg/L)	Number of sites	Percent of sites with detections	Number of samples	Percent of samples with detections	Maximum concentration (µg/L)	Comments
Glooschenko and others, 1976—*Continued*	f s	1974 summer	**Upper Great Lakes—** *Continued*	Imidan	0.05	33	0	33	0	no det.	
				Azinphos-m.	0.05	33	0	33	0	no det.	
				Azinphos-ethyl	0.05	33	0	33	0	no det.	
				Phosphamidon	0.03	33	0	33	0	no det.	
				Dimethoate	0.005	33	0	33	0	no det.	
				Fenitrothion	0.005	33	0	33	0	no det.	
Cole and others, 1983; Cole and others, 1984	w	1980–82 1980–83	**United States:** 21 cities (17 included in report)	Acrolein	nr	17	nr	~121	nr	no det.	Survey of priority pollutant concentrations in urban runoff from cities across the United States. Data shown are from final report (Cole and others, 1984). Detection limits not reported.
				Aldrin	nr	17	18	~121	6	0.1	
				Chlordane	nr	17	24	42	17	10.0	
				DDD	nr	17	0	~121	0	no det.	
				DDE	nr	17	6	~121	6	0.027	
				DDT	nr	17	6	~121	1	0.1	
				Dieldrin	nr	17	12	~121	6	0.1	
				Endosulfan (α)	nr	17	18	49	19	0.2	
				Endosulfan (β)	nr	17	0	~121	0	no det.	
				Endosulfan sulfate	nr	17	0	~121	0	no det.	
				Endrin	nr	17	0	~121	0	no det.	
				Endrin aldehyde	nr	17	0	~121	0	no det.	
				HCH (α)	nr	17	24	106	20	0.1	
				HCH (β)	nr	17	12	~121	5	0.1	
				HCH (δ)	nr	17	12	~121	6	0.1	
				Lindane	nr	17	24	100	15	0.1	
				Heptachlor	nr	17	18	~121	6	0.1	
				Hept. epox.	nr	17	12	~121	2	0.1	
				Toxaphene	nr	17	0	~121	0	no det.	

Table 2.1. National and multistate monitoring studies reviewed—*Continued*

Reference(s)	Matrix	Study		Compounds	Detection limit(s) (µg/L)	Sites		Samples			Comments
		Sampling dates	Location(s)			Number of sites	Percent of sites with detections	Number of samples	Percent of samples with detections	Maximum concentration (µg/L)	
Gilliom and others, 1985	w	1975–80 quarterly (Nov, Feb, May, Aug)	United States: 160 to 180 sites on major rivers throughout the United States	Aldrin	0.01	177	2.3	2,946	0.2	nr	USGS nationwide study of pesticides in major rivers of the United States. Water sampled four times per year and bed sediments two times per year. Less than 10 percent of samples contained detectable levels of any of the analytes. This was partly due to high detection limits in this study. Much lower detection frequencies than in the 1968–71 study (Schulze and others, 1973). Gradual decline in occurrence of organochlorines evident. No clear trends for herbicides or organophosphate insecticides observed.
				Dieldrin	0.03	177	2.3	2,945	0.2	nr	
				Chlordane	0.15	177	0.6	2,943	0	nr	
				DDD	0.05	177	4.0	2,720	0.3	nr	
				DDE	0.03	177	0.6	2,715	0	nr	
				DDT	0.05	177	2.8	2,721	0.4	nr	
				Endrin	0.05	180	1.1	2,950	0.1	nr	
				Hept. epox.	0.01	177	4.5	2,946	0.3	nr	
				Lindane	0.01	177	8.5	2,945	1.1	nr	
				Methox.	0.1	172	0	2,761	0	nr	
				Toxaphene	0.25	177	2.8	2,946	0.4	nr	
				Diazinon	0.1	174	9.8	2,859	1.2	nr	
				Ethion	0.25	174	0.6	2,823	0.1	nr	
				Malathion	0.25	174	0.6	2,859	0.1	nr	
				M. parathion	0.25	174	2.7	2,861	0.1	nr	
				M. trithion	0.5	174	0	2,822	0	nr	
				Parathion	0.25	174	0.6	2,856	0	nr	
				Trithion	0.5	174	1.1	2,819	0.1	nr	
				Atrazine	0.5	144	24	1,363	4.8	nr	
				2,4-D	0.5	186	2.4	1,764	0.2	nr	
				2,4,5-T	0.5	186	0.6	1,765	0.1	nr	
				Silvex	0.5	167	0.6	1,768	0.1	nr	

Table 2.1. National and multistate monitoring studies reviewed—Continued

Reference(s)	Matrix	Sampling dates	Location(s)	Compounds	Detection limit(s) (µg/L)	Sites		Samples			Comments
						Number of sites	Percent of sites with detections	Number of samples	Percent of samples with detections	Maximum concentration (µg/L) (Median)	
Staples and others, 1985	w	1975–82	United States	Acrolein	Variable	nr	nr	798	0.25	<14.0	Retrieval of data on concentrations of priority pollutants in ambient waters from USEPA's STOrage and RETrieval water quality database (STORET) for the years 1975–82. Note that median concentrations are shown, not maximum. Data must be viewed with caution, as samples are not necessarily representative of ambient conditions across the entire United States and seasonality is not taken into account. Levels in biota, sediments, and effluents also are discussed.
				Aldrin	Variable	nr	nr	7,891	40	0.001	
				Chlordane (technical)	Variable	nr	nr	5,250	40	0.02	
				Chlordane (cis)	Variable	nr	nr	827	56	0	
				Chlordane (trans)	Variable	nr	nr	820	56	0	
				DDD	Variable	nr	nr	3,533	45	0	
				DDE	Variable	nr	nr	5,333	48	0.001	
				DDT	Variable	nr	nr	5,718	44	0.001	
				Dieldrin	Variable	nr	nr	7,609	40	0.001	
				Endosulfan (technical)	Variable	nr	nr	2,887	53	0	
				Endosulfan (α)	Variable	nr	nr	864	0.5	<0.1	
				Endosulfan (β)	Variable	nr	nr	871	0.85	<0.1	
				Endosulfan-sulfate	Variable	nr	nr	850	0	<0.2	
				Endrin	Variable	nr	nr	8,789	32	0.001	
				Endrin aldehyde	Variable	nr	nr	770	0	<0.2	
				Heptachlor	Variable	nr	nr	4,650	34	0.001	
				Hept. epox.	Variable	nr	nr	4,632	36	0.001	
				HCH (α)	Variable	nr	nr	1,470	8	>0.18	
				HCH (β)	Variable	nr	nr	1,010	1.3	<0.05	
				HCH (δ)	Variable	nr	nr	880	0.8	<0.1	
				Lindane	Variable	nr	nr	4,505	27	0.02	
				Toxaphene	Variable	nr	nr	7,325	32	0.05	

Table 2.1. National and multistate monitoring studies reviewed—*Continued*

Reference(s)	Matrix	Sampling dates	Location(s)	Compounds	Detection limit(s) (µg/L)	Number of sites	Percent of sites with detections	Number of samples	Percent of samples with detections	Maximum concentration (µg/L)	Comments
						Sites		**Samples**			
DeLeon and others, 1986	w	1984 summer	Mississippi River: 11 sites along entire length	**Atrazine**	nr	11	nr	11	nr	1.1	Highest concentrations (atrazine and alachlor) at site downstream of Memphis, Tennessee. Metals and other organics also monitored. Pesticide data reported for only 4 of 11 sampling sites.
				Propazine	nr	11	nr	11	nr	nr	
				Alachlor	nr	11	nr	11	nr	nr	
				Propachlor							
				Trifluralin	nr	11	nr	11	nr	0.84	
Stevens and Neilson, 1989	w f	1986 spring	Great Lakes: Lakes Huron, Erie, Ontario, and Superior	**HCH** (α)	0.00001	95	100	95	100	0.011	Survey of concentrations of organochlorine compounds in the Great Lakes. Large volume extractor used to achieve low detection limits. Spatial patterns and sources discussed. Whole-water and centrifuged samples were compared. Concentration data and detection frequency shown are for all samples, regardless of lake.
				Lindane	0.00001	95	100	95	100	0.003	
				Chlordane (*cis*)	0.00001	95	70	95	70	1E-04	
				Chlordane (*trans*)	0.00001	95	16.0	95	16.0	1E-04	
				Heptachlor	0.00001	95	0	95	0	no det.	
				Hept. epox.	0.00001	95	100	95	100	3E-04	
				Endosulfan (α)	0.00001	95	0	95	0	no det.	
				Endosulfan (β)	0.00001	95	0	95	0	no det.	
				Aldrin	0.00001	95	0	95	0	no det.	
				Dieldrin	0.00001	95	100	95	100	0.001	
				Endrin	0.00001	95	13	95	13	1E-04	
				DDE (*p,p′*)	0.00001	95	53	95	53	1E-04	
				DDT (*p,p′*)	0.00001	95	0	95	0	no det.	
				DDT (*o,p′*)	0.00001	95	0	95	0	no det.	
				DDD (*p,p′*)	0.00001	95	0	95	0	no det.	
				Methox. (*p,p′*)	0.00001	95	0	95	0	no det.	
				Mirex	0.00001	95	0	95	0	no det.	

Table 2.1. National and multistate monitoring studies reviewed—Continued

Reference(s)	Matrix	Sampling dates	Location(s)	Compounds	Detection limit(s) (µg/L)	Number of sites	Percent of sites with detections	Number of samples	Percent of samples with detections	Maximum concentration (µg/L)	Comments
Pereira and Rostad, 1990; Pereira and others, 1990; Pereira and others, 1992	f, s	7/87–6/89	Mississippi River Basin: Ohio, Mississippi, Illinois, Missouri, and Arkansas Rivers	**Simazine**	0.005	17	nr	85	nr	0.13	Data from five separate sampling cruises are summarized: 7/87–8/87, 11/87–12/87, 5/88–6/88, 3/89–4/89, and 5/89–6/89. Loads at each sampling point and amount entering Gulf of Mexico estimated. Atrazine load entering Gulf of Mexico estimated as 0.4 percent of amount applied in basin in 1987 and 1.7 percent applied in 1989. Load estimates indicate a point source of alachlor, 2,6-diethyl-aniline, and the acetanilides near St. Louis, Missouri; <0.5 percent of total detected in suspended solids. Cross channel mixing downstream from river confluences shown to be slow, implying that samples must be representative of entire river width. Concentration data shown are for the 5/88 to 6/88 sampling trip only.
				Atrazine	0.005	17	nr	85	nr	0.69	
				Deethylatr.	0.005	17	nr	85	nr	0.08	
				Deisoatr.	0.005	17	nr	85	nr	0.05	
				Alachlor	0.005	17	nr	85	nr	0.9	
				2,6-Diethyl-aniline	0.005	17	nr	85	nr	0.9	
				2-Chloro-2',6'-diethylacetan.	0.005	17	nr	85	nr	0.4	
				2-Hydroxy-2',6'-diethylacetan.	0.005	17	nr	85	nr	0.05	
				Metolachlor	0.005	17	nr	85	nr	0.3	
				Cyanazine	0.005	17	nr	85	nr	0.6	

Table 2.1. National and multistate monitoring studies reviewed—*Continued*

Reference(s)	Matrix	Sampling dates	Location(s)	Compounds	Detection limit(s) (µg/L)	Sites		Samples			Comments
						Number of sites	Percent of sites with detections	Number of samples	Percent of samples with detections	Maximum concentration (µg/L)	
Goolsby and others, 1991a,b; Thurman and others, 1991; Goolsby and Battaglin, 1993	f	1989 (spring, summer, autumn)	Midwestern United States: 149 sites on streams in 122 drainage basins	**Alachlor**	0.05	132	86	396	nr	51.0	Samples taken during preplanting, postplanting, and postharvest periods. Concentration data and detection frequencies shown are for post-planting samples (May-June). Concentrations generally low in March and April, higher in May and June, and decreased considerably by October and November. Concentrations of atrazine, simazine, and alachlor frequently exceeded USEPA maximum contaminant levels in May and June. DAR may be used as an indicator of ground water movement into surface waters.
				Atrazine	0.05	132	98	396	nr	108.0	
				Deethylatr.	0.05	132	86	396	nr	404.0	
				Deisoatr.	0.05	132	54	396	nr	3.2	
				Cyanazine	0.20	132	63	396	nr	61.0	
				Metolachlor	0.05	132	83	396	nr	40.0	
				Metribuzin	0.05	132	53	396	nr	7.6	
				Propazine	0.05	132	40	396	nr	1.4	
				Prometon	0.05	132	23	396	nr	0.93	
				Simazine	0.05	132	55	396	nr	7.0	
				Ametryne	0.05	132	0	396	nr	no det.	
				Prometryn	0.05	132	0	396	nr	no det.	
				Terbutryn	0.05	132	0	396	nr	no det.	

Table 2.1. National and multistate monitoring studies reviewed—*Continued*

| Reference(s) | Matrix | Study | | | Sites | | Samples | | | Comments |
		Sampling dates	Location(s)	Compounds	Detection limit(s) (µg/L)	Number of sites	Percent of sites with detections	Number of samples	Percent of samples with detections	Maximum concentration (µg/L)	
Ciba-Geigy, 1992c	w	1975–91	**Midwestern rivers and lakes:** 53 sites on 43 water bodies	**Atrazine**	0.05–0.25	53	98	1,761	87	30.0	Review of monitoring data from Ciba-Geigy Corp., Monsanto Company, USGS, and Topeka, Kansas water utility, 1975–91. Report is from Ciba-Geigy Corp. Duration of monitoring was 1 to 2 years at all sites. Data from 1975–76, 1985–87, or 1990–91, depending on site. Concentration data shown are from all sites and samples. Time-weighted annual means were below 3 µg/L at 94 percent of sites. Eighty-nine percent of individual samples were below 3 µg/L. Maximum concentrations occurred in June (41 percent), May (28 percent), or July (13 percent).

Table 2.1. National and multistate monitoring studies reviewed—*Continued*

Reference(s)	Matrix	Sampling dates	Location(s)	Compounds	Detection limit(s) (µg/L)	Number of sites	Percent of sites with detections	Number of samples	Percent of samples with detections	Maximum concentration (µg/L)	Comments
Ciba-Geigy, 1992d	w	1975–91	Mississippi, Missouri, and Ohio Rivers	Atrazine	0.1–0.2	10	100	~1,120	82-100	17.0	Review of monitoring data from Ciba-Geigy Corp. and Monsanto Company. Report is from Ciba-Geigy Corp. Duration of monitoring varied among sites. Some were monitored in 1975–76 and again in the mid-1980's. One site on the Mississippi River (Vicksburg, Mississippi) was monitored continuously from 1975–89. Three sets of concentration data shown are for the three rivers. Annual mean concentrations for Mississippi River sites ranged from 0.26 to 2.2 µg/L. Annual mean concentrations for Missouri River sites ranged from 0.5 to 3.77 µg/L. Annual mean concentrations for Ohio River sites ranged from 0.38 to 0.84 µg/L.
				Atrazine	0.1–0.2	5	100	83	62-100	14.0	
				Atrazine	0.1–0.2	3	100	145	77-100	4.8	

Table 2.1. National and multistate monitoring studies reviewed—*Continued*

Reference(s)	Matrix	Sampling dates	Location(s)	Compounds	Detection limit(s) (µg/L)	Number of sites	Percent of sites with detections	Number of samples	Percent of samples with detections	Maximum concentration (µg/L)	Comments
Goolsby and Battaglin, 1993	f	4/91–3/92	Mississippi River and tributaries: 3 sites on Mississippi; 6 sites on major tributaries	**Alachlor**	0.002	9	100	381	~85	2.0	Summary of three separate studies that focused on different aspects of pesticide occurrence in mid-western streams and major rivers: a regional reconnaissance study of 122 river basins, a study of the temporal variability of pesticide concentrations in 9 river basins (April-July, 1990), and a study of pesticide occurrence in the Mississippi River and major tributaries. The concentration and frequency of detection data shown are from the Mississippi River study. Concentration data are approximate.
				Ametryne	0.05	8	25	381	<2	nr	
				Atrazine	0.002	9	100	381	~99	11.0	
				Azinphos-m.	0.01	8	0	316	0	no det.	
				Butylate	0.002	8	86	316	~15	0.1	
				Carbaryl	0.002	7	86	316	~10	0.1	
				Carbofuran	0.002	8	100	311	~30	0.11	
				Chlorpyrifos	0.005	8	86	316	~4	0.11	
				Cyanazine	0.01	9	100	381	~85	7.0	
				DDE	0.005	7	86	316	~4	0.02	
				Deethylatr.	0.02	8	100	381	~90	0.8	
				Deisoatr.	0.05	8	100	381	~15	0.6	
				Diazinon	0.002	8	86	316	~45	0.1	
				Dieldrin	0.02	7	71	316	~8	0.03	
				Disulfoton	0.02	7	0	316	0	no det.	
				EPTC	0.002	8	100	316	~20	0.11	
				Ethoprop	0.005	8	0	316	0	no det.	
				Fonofos	0.005	8	100	316	~20	0.03	
				Lindane	0.005	8	0	316	0	no det.	
				Linuron	0.01	8	14	316	<1	nr	
				Malathion	0.005	8	29	316	<2	0.01	
				M. parathion	0.005	8	14	316	<1	0.008	
				Metolachlor	0.002	9	100	381	~90	3.0	
				Metribuzin	0.005	9	100	381	~40	0.03	
				Parathion	0.002	8	0	316	0	no det.	
				Pendimethalin	0.01	8	86	316	~8	0.015	
				Permethrin	0.01	7	28	316	<2	0.018	
				Phorate	0.02	8	0	316	0	no det.	
				Prometon	0.002	8	100	381	~25	0.15	
				Prometryn	0.05	8	63	381	<2	0.08	

Table 2.1. National and multistate monitoring studies reviewed—*Continued*

Reference(s)	Matrix	Sampling dates	Location(s)	Compounds	Detection limit(s) (µg/L)	Number of sites	Percent of sites with detections	Number of samples	Percent of samples with detections	Maximum concentration (µg/L)	Comments
Goolsby and Battaglin, 1993—*Continued*	f	4/91–3/92	Mississippi River and tributaries —*Continued*	Propachlor	0.002	8	86	316	~15	0.1	
				Propargite	0.01	8	43	316	<2	0.08	
				Propazine	0.05	8	63	381	<2	0.07	
				Simazine	0.002	9	100	381	~80	0.4	
				Terbufos	0.02	8	29	416	<1	0.011	
				Trifluralin	0.005	8	100	316	~25	0.015	
Goolsby and others, 1993	w	4/92–11/92	Midwestern United States: 76 reservoirs	Alachlor	0.05	76	48	304	32	~2.0	USGS study of occurrence of herbicides and degradation products in reservoirs throughout the midwestern United States. Seventy-six reservoirs sampled bimonthly from 4/92 to 3/93. Data reported are preliminary results for 4/92 to 11/92. Results indicate that a number of these compounds are present at higher concentrations in reservoirs than in streams at certain times of the year. The ESA metabolite of alachlor appears to be relatively stable in the reservoirs. Concentration data shown are approximate.
				Ametryn	0.05	76	3	304	1	nr	
				Atrazine	0.05	76	92	304	83	~10.0	
				Cyanazine	0.05	76	65	304	54	nr	
				Deethylatr.	0.05	76	78	304	71	~1.50	
				Deisoatr.	0.05	76	70	304	63	~1.0	
				Metolachlor	0.05	76	62	304	53	nr	
				Metribuzin	0.05	76	12	304	6.5	nr	
				Prometon	0.05	76	15	304	12	nr	
				Propazine	0.05	76	10	304	4.5	nr	
				ESA (Alachlor metabolite)	0.1	76	79	304	73	~12.0	

Table 2.1. National and multistate monitoring studies reviewed—*Continued*

Reference(s)	Matrix	Sampling dates	Location(s)	Compounds	Detection limit(s) (µg/L)	Number of sites	Percent of sites with detections	Number of samples	Percent of samples with detections	Maximum concentration (µg/L)	Comments
						Sites		**Samples**			
Pereira and Hostetler, 1993	f	1991–92	Mississippi River Basin: 12 sites on Mississippi River; 14 sites on tributaries	Atrazine	0.005	26	100	~78	nr	4.7	Samples were collected on three sampling cruises. Concentration and detection frequency data shown are from the July-August 1991 cruise. Analytes included pesticides used on major crops (corn/soybean, rice, cotton, forestry) grown in different regions of the basin and several degradation products. Loads from tributaries and in the Mississippi River estimated for a number of the pesticides. Ratios of parent and degradation product concentrations imply that alluvial aquifers serve as storage areas and sources to the rivers.
				Deethylatr.	0.005	26	100	~78	nr	0.86	
				Deisoatr.	0.01	26	85	~78	nr	0.33	
				Ametryn	0.005	26	0	~78	0	no det.	
				Alachlor	0.005	26	96	~78	nr	0.56	
				2-Chloro-2',6'-diethylacetan.	0.005	26	31	~78	nr	0.04	
				2-Hydroxy-2'6'-diethylacetan.	0.005	26	73	~78	nr	0.09	
				Carbofuran	0.005	26	0	~78	nr	no det.	
				Cyanazine	0.025	26	92	~78	0	0.98	
				Cyanazine-amide	0.025	26	62	~78	nr	0.22	
				Deet	0.005	26	85	~78	nr	0.2	
				Diazinon	0.005	26	15	~78	nr	0.02	
				Fluometuron	0.005	26	27	~78	nr	0.41	
				Hexazinone	0.005	26	19	~78	nr	0.07	
				Metolachlor	0.005	26	100	~78	nr	1.9	
				Metribuzin	0.005	26	54	~78	nr	0.08	
				Molinate	0.005	26	23	~78	nr	2.6	
				4-Ketomolinate	0.005	26	27	~78	nr	1.6	
				Norflurazon	0.01	26	19	~78	nr	0.3	
				Desmethyl-norflurazon	0.01	26	8	~78	nr	0.12	
				Prometone	0.005	26	81	~78	nr	0.07	
				Prometryn	0.005	26	15	~78	nr	0.08	
				Simazine	0.005	26	100	~78	nr	0.26	
				Thiobencarb	0.005	26	4	~78	nr	0.06	

Table 2.2. State and local monitoring studies reviewed

[Matrix: w, whole (unfiltered) water; d, drinking water; m, surface microlayer; s, suspended sediments. **Bold face type** in compound column indicates a positive detection in one or more samples. Abbreviations used for compounds: Azinphos-m., Azinphos-methyl; Deethylatr., deethylatrazine; DEA, deethylatrazine; D ethylacetan., diethylacetanilide; Hept. epox., heptachlor epoxide; Methox., methoxychlor; M. parathion, methyl parathion; M. trithion, methyl trithion. PAHs, polycyclic aromatic hydrocarbons; PCBs, polychlorinated biphenyls. max., maximum; nr, data not reported. α, alpha; β, beta; γ, gamma. USEPA, U.S. Environmental Protection Agency. USGS, U.S. Geological Survey. <, less than; >, greater than. μg/kg, microgram(s) per kilogram; μg/L, microgram(s) per liter; kg, kilogram(s); kg/yr, kilogram(s) per year; km, kilometer(s); km², square kilometer(s); L, liter; lb, pound(s); mg/L, milligram(s) per liter; mi, mile; ng/g, nanogram(s) per gram. no det., no samples with concentrations above the detection limit. ?, number is uncertain; ~, number is approximate]

Reference(s)	Matrix	Sampling dates	Location(s)	Compounds	Detection limit(s) (μg/L)	Sites — Number of sites	Sites — Percent of sites with detections	Samples — Number of samples	Samples — Percent of samples with detections	Samples — Maximum concentration (μg/L)	Comments
Mack and others, 1964	w	9/63	**New York:** Four lakes	**DDT (total)**	nr	nr	100	nr	100	0.33	Samples of fish and water of surface waters of New York analyzed for DDT content.
Nicholson and others, 1964	w	1960–62	**Northern Alabama:** Tributaries of Tennessee River	**Toxaphene**	nr	nr	100	84	100	0.41	Organochlorine concentrations in streams in cotton growing area monitored for 4 years. Use estimates for basin included. Detections/concentrations related to use and solubility. Samples of treated and untreated water analyzed. Neither toxaphene nor HCH removed by treatment.
				DDT	nr	nr	0	84	0	no det.	
				HCH	nr	nr	100	84	100	nr	

Table 2.2. State and local monitoring studies reviewed—*Continued*

Reference(s)	Matrix	Sampling dates	Location(s)	Compounds	Detection limit(s) (µg/L)	Number of sites	Percent of sites with detections	Number of samples	Percent of samples with detections	Maximum concentration (µg/L)	Comments
Casper, 1967	w	9/64–10/68	**Texas:** Galveston Bay, Gulf of Mexico	Aldrin	nr	nr	0	nr	0	no det.	Monitoring of insecticide levels in water and oysters of Galveston Bay after increases in insecticide use in the Houston area for mosquito control. No evidence of increased residues in water or oysters. Concentrations reported as <1.0 µg/L represent positive detections below the reporting limit.
				Lindane	nr	nr	33	nr	33	<1.0	
				Chlordane	nr	nr	22	nr	22	<1.0	
				DDE	nr	nr	33	nr	33	<1.0	
				DDT	nr	nr	22	nr	22	<1.0	
				Endrin	nr	nr	0	nr	0	no det.	
				Dieldrin	nr	nr	0	nr	0	no det.	
				Heptachlor	nr	nr	44	nr	44	<1.0	
				Hept. epox.	nr	nr	11	nr	11	<1.0	
				Methox.	nr	nr	11	nr	11	<1.0	
				Trithion	nr	nr	0	nr	0	no det.	
				Malathion	nr	nr	0	nr	0	no det.	
Fahey and others, 1968	w	12/63–3/64	**Michigan:** Battle Creek area, Kalamazoo River, ponds, creeks	Dieldrin	0.1	13	0	22	0	no det.	Monitoring of water, soil, and sediments following application of dieldrin in an urban area for control of Japanese beetles.
Forbes, 1968	w	1967	**Florida:** Farm canals and Lake Apopka, near Zellwood, Florida	DDT	0.01	3	0	nr	0	no det.	Concentration data are for lake. No DDT or parathion detected in lake, but DDT was detected in several samples of canal water. Maximum concentrations were 0.25 and 0.18 µg/L in the canals.
				Parathion	0.01	3	0	nr	0	no det.	

Table 2.2. State and local monitoring studies reviewed—*Continued*

			Study			Sites		Samples			
Reference(s)	Matrix	Sampling dates	Location(s)	Compounds	Detection limit(s) (µg/L)	Number of sites	Percent of sites with detections	Number of samples	Percent of samples with detections	Maximum concentration (µg/L)	Comments
Dupuy and others, 1970	w	1968	**Texas:** Eastern Texas rivers	**Aldrin**	0.01	30	nr	153	0.7	0.01	Survey of organochlorine and phenoxy pesticides in Texas surface waters. Four to five samples taken at most sites throughout 1968.
				DDD	0.01	30	33	153	14	0.09	
				DDE	0.01	30	60	153	29	0.09	
				DDT	0.01	30	67	153	34	0.21	
				Dieldrin	0.01	30	27	153	nr	0.045	
				Endrin	0.01	30	nr	153	nr	0.07	
				Heptachlor	0.01	30	0	153	0	no det.	
				Hept. epox.	0.01	30	0	153	0	no det.	
				Lindane	0.01	30	20	153	25	0.11	
				2,4-D	0.01	30	87	153	34	1.4	
				Silvex	0.01	30	23	153	nr	0.13	
				2,4,5-T	0.01	30	77	153	38	0.15	
Hannon and others, 1970	w	1967–68	**South Dakota:** Lake Poinsett	**Lindane**	0.02	12	nr	12	nr	nr	Study of ecological distribution of organochlorines in lake. Water, bed sediments, and various organisms sampled. Bioconcentration factors calculated at the different trophic levels.
				Heptachlor	0.02	12	nr	12	nr	nr	
				Hept. epox.	0.02	12	nr	12	nr	nr	
				Aldrin	0.02	12	nr	12	nr	nr	
				Dieldrin	0.02	12	nr	12	nr	nr	
				DDT (*p,p'*)	0.08	12	nr	12	nr	nr	
				DDD (*p,p'*)	0.08	12	nr	12	nr	nr	
				DDE (*p,p'*)	0.02	12	nr	12	nr	nr	
				Toxaphene	nr	12	nr	12	nr	nr	
				Endrin	0.40	12	0	12	0	nr	
				Methox.	0.40	12	0	12	0	nr	

Table 2.2. State and local monitoring studies reviewed—*Continued*

Reference(s)	Matrix	Sampling dates	Location(s)	Compounds	Detection limit(s) (µg/L)	Number of sites	Percent of sites with detections	Number of samples	Percent of samples with detections	Maximum concentration (µg/L)	Comments
Johnson and Morris, 1971	w	1968–70	**Iowa:** 10 rivers	**Dieldrin** DDT DDE	0.005 0.005 0.005	10 10 10	90 90 90	179 179 179	40 19 14.5	0.063 0.023 0.017	Report of routine monitoring of organochlorines in Iowa rivers. Differences in concentrations and in seasonal appearance of compounds observed between agricultural and non-agricultural basins. Dieldrin detected most frequently and at highest concentrations.
Knutson and others, 1971	w	1966–69	**Kansas:** Smokey Hill River, Cedar Bluff Reservoir	**Dieldrin** **Endrin** **Heptachlor** **Hept. epox.** **Aldrin** Diazinon Parathion M. parathion	0.1 0.1 0.1 0.1 0.1 0.1 0.1 0.1	nr nr nr nr nr nr nr	67 67 67 33 33 0 0 0	54 54 54 54 54 54 54 54	22 nr nr nr nr 0 0 0	0.051 0.013 0.001 0.006 0.014 no det. no det. no det.	Study related insecticide use and resulting residues in soil, ground water, crops, and surface waters. Detections in surface waters were infrequent and at low levels.
Rowe and others, 1971	w	10/68–5/69	**Louisiana:** 3 estuarine areas: Grand Bayou, Hackberry Bay, Creole Bay	**Dieldrin** **Endrin**	0.1? 0.1?	13 13	nr nr	148 148	nr nr	0.2 0.2	Survey of dieldrin and endrin in water, sediments, and oysters. Dieldrin detected in 70 percent of oysters, endrin detected in 100 percent of oysters. Concentration data shown are from water samples.

Table 2.2. State and local monitoring studies reviewed—*Continued*

Reference(s)	Matrix	Study			Detection limit(s) (µg/L)	Sites			Samples		Comments
		Sampling dates	Location(s)	Compounds		Number of sites	Percent of sites with detections	Number of samples	Percent of samples with detections	Maximum concentration (µg/L)	
Bevenue and others, 1972	w	1970–71	Hawaii: Water from lakes, streams, bays, harbors, canals, and lagoons on four islands	DDE	0.0005?	46	nr	46	nr	0.001	Random sampling of diverse surface waters. Drinking water also sampled, mostly from ground water.
				DDD	0.0005?	46	nr	46	nr	0.018	
				DDT	0.0005?	46	nr	46	100	0.064	
				Dieldrin	0.0005?	46	nr	46	100	0.019	
				Lindane	0.0005?	46	nr	46	nr	0.003	
				Chlordane	0.0005?	nr	nr	nr	nr	0.018	
Bradshaw and others, 1972	w	3/70–7/70 biweekly 7/70–2/71 weekly	Utah: 15 Utah Lake tributaries; one outlet site	Aldrin	nr	16	100	nr	nr	1.2	Temporal patterns of occurrence correlated with seasonal agricultural use.
				Heptachlor and Hept. epox.	nr	16	88	nr	nr	2.9	
				HCH	nr	16	100	nr	nr	nr	
				Methox.	nr	16	94	nr	nr	5.2	
				DDT (total)	nr	16	100	nr	nr	4.1	
Morris and others, 1972	w	1968–71	Iowa: Mississippi, Missouri, Iowa, Cedar, Skunk, Raccoon, Little Sioux, and Nishnabotna Rivers	Dieldrin	nr	10	100	nr	nr	0.065	Maximum concentrations shown are maximum detected in any of the rivers over the 4-year period. Sediments and catfish also analyzed, with much higher concentrations detected.
				DDT	nr	10	90	nr	nr	0.055	
				DDE	nr	10	100	nr	nr	0.03	

Table 2.2. State and local monitoring studies reviewed—*Continued*

Reference(s)	Matrix	Sampling dates	Location(s)	Compounds	Detection limit(s) (µg/L)	Number of sites	Percent of sites with detections	Number of samples	Percent of samples with detections	Maximum concentration (µg/L)	Comments
Schacht, 1974	w	1970–72	Great Lakes: Illinois waters of Lake Michigan and tributaries	Hept. epox.	0.0002	nr	100	13	85	0.005	Concentrations of organochlorine compounds monitored in water, sediments, and fish of Lake Michigan, two tributary rivers, and sewage treatment plant effluents. Values shown are for river samples. Concentrations in open water of Lake Michigan (4 miles offshore) were nearly all below detection limits. Sediment levels in the rivers were much higher. Samples also analyzed for PCBs and phthalates.
				Dieldrin	0.0002	nr	100	13	92	4	
				Methox.	0.0010	nr	100	11	100	0.023	
				DDE (*o,p'*)	0.0003	nr	0	11	0	0.089	
				DDD (*o,p'*)	0.0003	nr	100	11	45	no det.	
				DDT (*o,p'*)	0.0003	nr	100	13	85	0.002	
				DDE (*p,p'*)	0.0003	nr	100	13	100	6	
				DDD (*p,p'*)	0.0003	nr	100	13	100	0.013	
				DDT (*p,p'*)	0.0010	nr	100	13	100	0.02	
				DDT (total)	nr	nr	100	13	100	0.007	
				Heptachlor	0.0003	nr	0	13	0	0.038	
				Aldrin	0.0003	nr	0	13	0	0.058	
				Endrin	0.0003	nr	0	13	0	no det.	
				Lindane	0.0003	nr	0	13	0	no det.	
								13	0	no det.	
Klaassen and Kadoum, 1975	w	1970–71	Kansas: Tuttle Creek Reservoir	Endrin	10	nr	0	nr	0	no det.	Concentrations measured in water, bed sediments, and biota. Limited detections in water or sediments, but most biota had residues. Note relatively high detection limits.
				Aldrin	10	nr	0	nr	0	no det.	
				Dieldrin	10	nr	50	nr	17	10.0	
				Heptachlor	10	nr	0	nr	0	no det.	
				Hept. epox.	10	nr	0	nr	0	no det.	
				DDT (*o,p'*)	10	nr	50	nr	17	10.0	
				DDT (*p,p'*)	10	nr	0	nr	0	no det.	
				DDE (*o,p'*)	10	nr	0	nr	0	no det.	
				DDD (*o,p'*)	10	nr	50	nr	17	10.0	

Table 2.2. State and local monitoring studies reviewed—*Continued*

Reference(s)	Matrix	Sampling dates	Location(s)	Compounds	Detection limit(s) (µg/L)	Number of sites	Percent of sites with detections	Number of samples	Percent of samples with detections	Maximum concentration (µg/L)	Comments
Mattraw, 1975	w	1968–72	Southern Florida: Rivers, lakes, canals (number of sites unknown)	Aldrin	0.005	nr	0	365	0	no det.	Marked decline in frequency of detection of DDT, DDD, and DDE from 1968 to 1972. Percent detections for DDT, DDD, and DDE went from 81, 41, 23, respectively, in 1968 to 1.2, 3.8, 3.1, respectively, in 1972. Majority of detections were at the 0.005 µg/L level.
				Chlordane	0.005	nr	0	188	0	no det.	
				DDT	0.005	nr	nr	369	11	nr	
				DDE	0.005	nr	nr	382	nr	nr	
				DDD	0.005	nr	nr	382	11	nr	
				Dieldrin	0.005	nr	nr	368	11	nr	
				Endrin	0.005	nr	0	368	0	no det.	
				Heptachlor	0.005	nr	0	366	0	no det.	
				Hept. epox.	0.005	nr	0	157	0	no det.	
				Lindane	0.005	nr	nr	367	nr	nr	
				Toxaphene	0.005	nr	0	146	0	no det.	
Palmer and others, 1975[1]	s	1974	Chesapeake Bay	**DDT**	nr	nr	100	20	95	2.0[1]	Study of transport of DDT, chlordane, and PCBs by suspended sediment in Chesapeake Bay. Results indicate that Susquehanna River is the major source of these compounds to the bay. Resuspension of bottom sediments also resulted in elevated concentrations of these compounds.
				Chlordane	nr	nr	100	20	100	2.0[1]	

[1]Detection limits and concentration data for suspended sediment are in µg/kg, not µg/L.

Table 2.2. State and local monitoring studies reviewed—*Continued*

Reference(s)	Matrix	Sampling dates	Study Location(s)	Compounds	Detection limit(s) (µg/L)	Number of sites	Percent of sites with detections	Number of samples	Percent of samples with detections	Maximum concentration (µg/L)	Comments
Brodtmann, 1976	w	1974 weekly	Louisiana: Mississippi River at New Orleans	**Lindane**	nr	nr	nr	52 ?	100	0.0029	Samples taken using a continuous liquid/liquid extraction apparatus, each sample representing a 7-day period. All concentration data shown taken from monthly means. Estimates of loads show that 49.7 lb of the eight pesticides (combined weight) are discharged to the Gulf of Mexico each day (9 tons per year).
				Chlordane (γ)	nr	nr	nr	52 ?	100	0.0012	
				Heptachlor	nr	nr	nr	52 ?	~55	0.0025	
				Hept. epox.	nr	nr	nr	52 ?	100	0.0018	
				Dieldrin	nr	nr	nr	52 ?	100	0.0098	
				Endrin	nr	nr	nr	52 ?	100	0.0072	
				DDD (p,p')	nr	nr	nr	52 ?	100	0.0039	
				DDT (p,p')	nr	nr	nr	52 ?	~90	0.0036	
Junk and others, 1976	w	1974–75	Iowa: Various rivers and treated drinking water	**Atrazine**	nr	nr	100	~40	100	42.0	Concentrations monitored as part of validation study of analytical technique. Data shown are from South Skunk River and Indian Creek during summer and autumn of 1974. Raw water and finished drinking water analyzed for same compounds. Water treatment was not effective in removal.
				DDE	nr	nr	100	~40	100	3.9	
				Dieldrin	nr	nr	100	~40	100	0.08	
Kellogg and Bulkley, 1976	w	1971–73 (April to October)	Iowa: Des Moines River	Dieldrin (1971)	0.01	nr	nr	5	100	0.05	Concentrations in water showed seasonal trend, with highest concentrations in May and June following application of aldrin. Water concentrations were compared to concentrations in catfish.
				Dieldrin (1972)	0.01	nr	nr	14	50	0.04	
				Dieldrin (1973)	0.001	nr	nr	32	100	0.03	

Table 2.2. State and local monitoring studies reviewed—*Continued*

Reference(s)	Matrix	Study Sampling dates	Location(s)	Compounds	Detection limit(s) (µg/L)	Sites Number of sites	Percent of sites with detections	Samples Number of samples	Percent of samples with detections	Maximum concentration (µg/L)	Comments
Tanita and others, 1976	w	4/74–6/74	Hawaii: Marina receiving urban runoff	α-HCH	0.0001	1?	100	12	33	0.0004	Termite control suspected as source of dieldrin and chlordane. Concentration data shown are for unfiltered water. Chlordane not quantified because of interferences. Detection limit not certain.
				Lindane	0.0001	1?	100	12	33	0.0007	
				Aldrin	0.0001	1?	100	12	25	0.0052	
				Heptachlor	0.0001	1?	100	12	25	0.0002	
				DDE	0.0001	1?	100	12	43	0.0047	
				DDD	0.0001	1?	100	12	18	0.224	
				DDT	0.0001	1?	100	12	83	0.0022	
				Dieldrin	0.0001	1?	100	12	100	0.0015	
				Hept. epox.	0.0001	1?	100	12	18	0.13	
				Chlordane (α)	0.0001	1?	100	12	90	nr	
				Chlordane (γ)	0.0001	1?	100	12	68	nr	
Joyce and Sikka, 1977	w	8/75–1/76	Florida: St. Johns River	2,4-D	0.01	nr	100	45	67.0	1.3	2,4-D residues measured along a 312-mi stretch of St. Johns River in coastal Florida. All concentrations were well below criteria limits for drinking water. 2,4-D concentrations measured in Blue Crabs were detected only in May.
Sullivan and Atchison, 1977	w s	3/74–9/74	Michigan: Rouge River	Methox.	nr	nr	0	~60	0	no det.	Monitoring urban watershed (Detroit) for residues from spraying program for control of Dutch Elm disease. No detections in water or suspended sediment samples.

Table 2.2. State and local monitoring studies reviewed—*Continued*

Reference(s)	Matrix	Sampling dates	Study Location(s)	Compounds	Detection limit(s) (µg/L)	Sites Number of sites	Sites Percent of sites with detections	Samples Number of samples	Samples Percent of samples with detections	Samples Maximum concentration (µg/L)	Comments
Ellis, 1978	w	9/80	**Colorado:** Urban storm runoff in Denver area	Aldrin	0.01	nr	0	nr	0	no det.	USGS study of water quality of urban storm runoff. Sites were in residential and residential/commercial areas of suburban Denver. Data collected for use with a storm water management model. Nutrients, major ions, metals, and bacteria also measured.
				Chlordane	0.1	nr	100	nr	100	1.7	
				DDD	0.01	nr	0	nr	0	no det.	
				DDE	0.01	nr	50	nr	33	0.02	
				DDT	0.01	nr	0	nr	0	no det.	
				Diazinon	0.01	nr	100	nr	100	0.78	
				Dieldrin	0.01	nr	50	nr	33	0.01	
				Endrin	0.01	nr	0	nr	0	no det.	
				Ethion	0.01	nr	0	nr	0	no det.	
				Heptachlor	0.01	nr	50	nr	33	0.26	
				Hept. epox.	0.01	nr	50	nr	33	0.04	
				Lindane	0.01	nr	100	nr	100	0.05	
				Malathion	0.01	nr	50	nr	67	3.5	
				M. parathion	0.01	nr	0	nr	0	no det.	
				M. trithion	0.01	nr	0	nr	0	no det.	
				Parathion	0.01	nr	0	nr	0	no det.	
				Toxaphene	nr	nr	0	nr	0	no det.	
				Trithion	0.01	nr	0	nr	0	no det.	
				2,4-D	0.01	nr	100	nr	100	7.5	
				2,4,5-T	0.01	nr	50	nr	33	0.04	
				Silvex	0.01	nr	50	nr	67	3.2	

Table 2.2. State and local monitoring studies reviewed—*Continued*

		Study			Sites			Samples		Comments
Reference(s)	Matrix	Sampling dates	Location(s)	Compounds	Detection limit(s) (μg/L)	Number of sites	Percent of sites with detections	Number of samples	Percent of samples with detections	Maximum concen-tration (μg/L)	
Kurtz, 1978	w d	1974–75	**Pennsylvania:** Nineteen streams; 110 community water supplies; 3 reservoirs	Streams:							Survey of Pennsylvania streams, drinking water supplies, and reservoirs for DDT and PCB contamina-tion. Streams sampled once in 1974 and once in 1975. Reservoirs and water sup-plies sampled once. Three of 19 streams, 4 of 110 water supplies, and 0 of 3 reser-voirs had trace or measurable levels of DDT or PCBs.
				DDE	0.001	19	nr	38	nr	0.42	
				DDD	0.001	19	nr	38	nr	0.09	
				DDT	0.001	19	11	38	nr	0.11	
				DDT (total)		19	11	38	nr	0.62	
				Drinking water:							
				DDE	0.001	110	nr	110	nr	0.006	
				DDD	0.001	110	nr	110	nr	0.009	
				DDT	0.001	110	nr	110	nr	0.06	
				DDT (total)		110	nr	110	nr	0.075	

Table 2.2. State and local monitoring studies reviewed—Continued

Reference(s)	Matrix	Sampling dates	Study Location(s)	Compounds	Detection limit(s) (µg/L)	Sites Number of sites	Sites Percent of sites with detections	Samples Number of samples	Samples Percent of samples with detections	Maximum concentration (µg/L)	Comments
Schacht and others, 1978	w	5/78	Illinois: Rock River	Lindane	0.01	5	0	5	0	no det.	Survey of water quality in Rock River and tributaries. Pesticide samples collected 5/29/78. All pesticides were below detection limits. Sediments and biota also sampled. Concentrations of nutrients, suspended sediments, and metals monitored.
				Heptachlor	0.01	5	0	5	0	no det.	
				Aldrin	0.01	5	0	5	0	no det.	
				Hept. epox.	0.01	5	0	5	0	no det.	
				Chlordane (α)	0.02	5	0	5	0	no det.	
				Chlordane (γ)	0.02	5	0	5	0	no det.	
				Dieldrin	0.02	5	0	5	0	no det.	
				Endrin	0.01	5	0	5	0	no det.	
				Methox.	0.1	5	0	5	0	no det.	
				DDE (o,p')	0.01	5	0	5	0	no det.	
				DDE (p,p')	0.01	5	0	5	0	no det.	
				DDD (o,p')	0.01	5	0	5	0	no det.	
				DDD (p,p')	0.01	5	0	5	0	no det.	
				DDT (o,p')	0.01	5	0	5	0	no det.	
				DDT (p,p')	0.01	5	0	5	0	no det.	
				Toxaphene	nr	5	0	5	0	no det.	
				Silvex	0.02	5	0	5	0	no det.	
				2,4-D	0.02	5	0	5	0	no det.	
				Organophosphates (total)	0.1	5	0	5	0	no det.	
Kent and Johnson, 1979	w	1974	Idaho: American Falls Reservoir	DDD	nr	nr	0	4?	0	no det.	Organochlorines detected in sediments and fish, but not in water.
				DDE	nr	nr	0	4?	0	no det.	
				DDT	nr	nr	0	4?	0	no det.	
				Aldrin	nr	nr	0	4?	0	no det.	
				Dieldrin	nr	nr	0	4?	0	no det.	
				Endrin	nr	nr	0	4?	0	no det.	
				Heptachlor	nr	nr	0	4?	0	no det.	
				Hept. epox.	nr	nr	0	4?	0	no det.	
				Lindane	nr	nr	0	4?	0	no det.	

Table 2.2. State and local monitoring studies reviewed—*Continued*

Reference(s)	Matrix	Sampling dates	Location(s)	Compounds	Detection limit(s) (µg/L)	Sites		Samples			Comments
						Number of sites	Percent of sites with detections	Number of samples	Percent of samples with detections	Maximum concentration (µg/L)	
Rihan and others, 1979	w	1975 (?)	Mississippi: 10 streams and lakes of northern Mississippi	Aldrin	nr	10	100	10	100	0.0005	Study of organochlorine pesticide levels in northern Mississippi rivers and lakes.
				Heptachlor	nr	10	90	10	90	0.0002	
				Lindane	nr	10	100	10	100	0.00016	
				DDT	nr	10	100	10	100	0.012	
Waller and Lee, 1979	w	1972–73	Lake Ontario	DDE	0.0005	nr	100	nr	100	0.045	Summary of data collected during the International Field Year for the Great Lakes. Average total DDT concentration is ~10 times the USEPA objectives. Detection limits not certain.
				DDD	0.0005	nr	75	nr	75	0.014	
				DDT	0.0005	nr	100	nr	100	0.013	
				DDT (total)	0.0005	nr	nr	nr	nr	0.057	
				Dieldrin	0.0005	nr	100	nr	100	0.013	
Wang and others, 1979	w	6/77–7/77	Florida: St. Lucie estuary	DDT	0.01	nr	0	18	0	no det.	River drained Lake Okeechobee. Residues detected in sediments, but not in water samples.
				DDD	0.01	nr	0	18	0	no det.	
				DDE	0.01	nr	0	18	0	no det.	
Dudley and Karr, 1980	w	7/77 and 6/78	Indiana: Black Creek watershed	Dieldrin	0.2	nr	0	nr	0	no det.	Watershed is 80 percent agricultural (row crops). Sampling in 1977 followed 2-week period without storm runoff. 1978 sampling for 2,4,5-T was done after fish kill following 2,4,5-T application to stream banks. Concentrations ranged from 0.2 to 7.7 µg/L. Note high detection limit for atrazine, alachlor, and carbofuran.
				DDE	0.2	nr	0	nr	0	no det.	
				Atrazine	100	nr	0	nr	0	no det.	
				Alachlor	100	nr	0	nr	0	no det.	
				Carbofuran	100	nr	0	nr	0	no det.	
				Malathion	nr	nr	0	nr	0	no det.	
				2,4,5-T	0.2	13	100	13	100	7.7	

Table 2.2. State and local monitoring studies reviewed—*Continued*

Reference(s)	Matrix	Sampling dates	Study Location(s)	Compounds	Detection limit(s) (µg/L)	Sites Number of sites	Sites Percent of sites with detections	Samples Number of samples	Samples Percent of samples with detections	Maximum concentration (µg/L)	Comments
Setmire and Bradford, 1980	w	9/76–5/77	**California:** Urban runoff, suburban San Diego (Tecolote Creek)	Aldrin	0.01	nr	0	nr	0	no det.	Urban runoff from a suburban, single-family residential area monitored. Samples taken in September of 1976 and February and May of 1977. Metals, chemical oxygen demand, nutrients, bacteria, and suspended sediment also monitored.
				Chlordane	0.1	nr	100	nr	100	1.4	
				DDD	0.01	nr	100	nr	50	0.02	
				DDE	0.01	nr	100	nr	25	0.03	
				DDT	0.01	nr	100	nr	50	0.12	
				Diazinon	0.01	nr	100	nr	100	0.51	
				Dieldrin	0.01	nr	100	nr	50	0.07	
				Endrin	0.01	nr	0	nr	0	no det.	
				Ethion	0.01	nr	0	nr	0	no det.	
				Heptachlor	0.01	nr	100	nr	100	0.06	
				Hept. epox.	0.01	nr	0	nr	0	no det.	
				Lindane	0.01	nr	100	nr	100	0.03	
				Malathion	0.01	nr	100	nr	100	2.6	
				M. parathion	0.01	nr	0	nr	0	no det.	
				M. trithion	0.01	nr	0	nr	0	no det.	
				2,4-D	0.01	nr	100	nr	100	0.38	
				2,4,5-T	0.01	nr	100	nr	75	0.14	
				Silvex	0.01	nr	0	nr	0	no det.	
				Toxaphene	1?	nr	0	nr	0	no det.	
Wang and others, 1980	w	1977–78	**Florida:** Indian River	DDT (*o,p'*)	0.01	nr	0	nr	0	no det.	Water and sediment samples collected near major tributaries, sewage plant outfalls, and municipal areas. Compounds were detected in most sediment samples, but not in water column.
				DDT (*p,p'*)	0.01	nr	0	nr	0	no det.	
				DDE (*p,p'*)	0.01	nr	0	nr	0	no det.	
				DDD (*p,p'*)	0.01	nr	0	nr	0	no det.	

Table 2.2. State and local monitoring studies reviewed—*Continued*

		Study				Sites			Samples			Comments
Reference(s)	Matrix	Sampling dates	Location(s)	Compounds	Detection limit(s) (μg/L)	Number of sites	Percent of sites with detections	Number of samples	Percent of samples with detections	Maximum concentration (μg/L)		
Wu and others, 1980	w m	6/77–11/77 6/78–11/78	**Maryland:** Rhode River	**Bulk water: Atrazine** Surficial microlayer: **Atrazine**	0.001 0.001	nr nr	no det. no det.	65 65	100 100	0.19 3.3		Study of atrazine concentrations in bulk water and surface microlayer in an estuarine environment. Enrichment factors (surface concentration/bulk concentration) of 1.1 to 110 were observed. In ~66 percent of the samples, the enrichment factor was 1 to 10.
Zahnow and Riggleman, 1980	w	8/77–9/78	**Chesapeake Bay:** Rhode River, Choptank River, Tuckahoe Creek	Linuron	0.2	22	0	79	0	no det.		Study to determine whether linuron used on area crops is entering Chesapeake Bay. No residues detected.
Erickson and Essig, 1981	w	1980	**Montana:** Lower Flathead River	Diazinon Endosulfan Dicamba Perthane Metham-idophos 2,4-D 2,4-D (methyl ester)	220 40 250 250 200 50 50	nr nr nr nr nr nr nr	0 0 0 0 0 0 0	84 84 84 84 84 84 84	0 0 0 0 0 0 0	no det. no det. no det. no det. no det. no det. no det.		Samples taken over an 11-month period. Note high detection limits for all analytes.

Table 2.2. State and local monitoring studies reviewed—*Continued*

Reference(s)	Matrix	Sampling dates	Location(s)	Compounds	Detection limit(s) (µg/L)	Number of sites	Percent of sites with detections	Number of samples	Percent of samples with detections	Maximum concentration (µg/L)	Comments
Leung and others, 1981	w w	1971–73; 1978	**Iowa:** Des Moines River at Boone, Iowa	**Dieldrin**	0.001	nr	100	~12 0	100	0.05	Study to determine whether dieldrin concentrations decreased after registration withdrawn in 1975. Thirty to 70 percent found in filtered water. Fish also sampled. Concentrations decreased greatly from 1971 to 1973, but not significantly from 1973 to 1978.
Lunsford, 1981	w	1976–78	**Virginia:** James River estuary	**Kepone**	0.01–0.02; 0.05 in 1978	75	nr	~75 0	nr	1.2	Study of kepone distribution in the water column of the estuary. Results are evaluated with respect to spatial and temporal variability. Highest levels observed in summer months, possibly because of increases in phytoplankton containing kepone residues. Water concentrations were 1 to 5 orders of magnitude below reported sediment concentrations in the estuary. Significant correlation between water column concentrations and underlying bed sediment concentrations.

Table 2.2. State and local monitoring studies reviewed—*Continued*

Reference(s)	Matrix	Sampling dates	Study Location(s)	Compounds	Detection limit(s) (µg/L)	Sites Number of sites	Sites Percent of sites with detections	Samples Number of samples	Samples Percent of samples with detections	Samples Maximum concentration (µg/L)	Comments
Murray and others, 1981	w	7/78–5/79	Texas: Galveston Bay	DDT (total)	0.00001	nr	75	29	nr	(mean) 0.00038	Total DDT, PCB, and phthalate concentrations were measured in water and bed sediments at eight sites throughout the bay.
Page, 1981	w	1977–79	New Jersey	HCH (α)	nr	nr	nr	604	39	0.1	Survey of occurrence of 56 organic compounds (13 pesticides) in surface water and ground water throughout New Jersey. Sites were chosen to be representative of all land uses in the state. Timing of sampling of surface waters not given. Authors conclude that ground water is as contaminated as surface waters.
				Lindane	nr	nr	nr	604	34	0.8	
				HCH (β)	nr	nr	nr	604	60	3.1	
				Heptachlor	nr	nr	nr	604	21	5.9	
				Aldrin	nr	nr	nr	604	24	0.6	
				Hept. epox.	nr	nr	nr	604	40	0.5	
				Chlordane	nr	nr	nr	603	56	0.8	
				DDE (o,p')	nr	nr	nr	603	44	0.1	
				Dieldrin	nr	nr	nr	604	39	0.1	
				Endrin	nr	nr	nr	604	14	0.1	
				DDT (o,p')	nr	nr	nr	604	18	0.1	
				DDD (p,p')	nr	nr	nr	604	27	0.1	
				DDT (p,p')	nr	nr	nr	604	17	0.1	

Table 2.2. State and local monitoring studies reviewed—Continued

Reference(s)	Matrix	Sampling dates	Study Location(s)	Compounds	Detection limit(s) (µg/L)[1]	Sites Number of sites	Sites Percent of sites with detections	Samples Number of samples	Samples Percent of samples with detections	Samples Maximum concentration (µg/L)[1]	Comments
Warry and Chan, 1981[1]	s	4/79– 4/80	Great Lakes, New York: Niagara River		(µg/kg)[1]					(µg/kg)[1]	Samples collected every 2 weeks to quantify inputs of organochlorines to Lake Ontario from Niagara River. Concentrations on suspended sediment varied with season. Total DDT highest in May and June. Much of the mirex, DDT, and PCBs associated with suspended sediment appears to emanate from sources between Grand Island and Lake Ontario, rather than from Lake Erie. Concentrations (in micrograms per kilogram) are based on dry weight of sediment.
				Mirex	1.0	nr	50	44	61	258.0	
				Methox.	1.0	nr	50	44	41	91.0	
				Chlordane (α)	1.0	nr	100	44	41	16.0	
				Chlordane (γ)	1.0	nr	100	44	61	58.0	
				Chlordane (total)	1.0	nr	100	44	66	58.0	
				DDT (*p,p'*)	1.0	nr	100	44	80	15.0	
				DDE (*p,p'*)	1.0	nr	100	44	93	51.0	
				DDD (*p,p'*)	1.0	nr	50	44	36	19.0	
				DDT (*o,p'*)	1.0	nr	100	44	5	15.0	
				DDT (total)	1.0	nr	100	44	100	66.0	
				Endosulfan (α)	1.0	nr	50	44	30	14.0	
				Endosulfan (β)	1.0	nr	0	44	0	no det.	
				Dieldrin	1.0	nr	100	44	61	15.0	
				Endrin	1.0	nr	0	44	0	no det.	
				Aldrin	1.0	nr	0	44	0	no det.	
				Heptachlor	1.0	nr	0	44	0	no det.	
				Hept. epox.	1.0	nr	50	44	16	12.0	
				Lindane	0.1	nr	50	44	16	11.0	
Wu, 1981	w s m	5/77– 9/77 5/78– 9/78	Chesapeake Bay: Rhode River	Atrazine	nr	nr	100	~12 0	100	0.1	Eighty percent of atrazine in the dissolved phase. Surface microlayer atrazine concentrations commonly 10 times higher than bulk-water concentrations. Atrazine detected throughout year in estuarine water, as well as in precipitation.

[1]Detection limits and concentration data for suspended sediment are in µg/kg, not µg/L.

Table 2.2. State and local monitoring studies reviewed—*Continued*

			Study			Sites		Samples			Comments
Reference(s)	Matrix	Sampling dates	Location(s)	Compounds	Detection limit(s) (µg/L)	Number of sites	Percent of sites with detections	Number of samples	Percent of samples with detections	Maximum concentration (µg/L)	
Bushway and others, 1982	w	6/81–8/81	**Maine:** Five ponds in coastal area	Azinphos-m.	0.23	nr	0	20	0	no det.	Study of pond and stream waters of a blueberry growing area. Neither compound nor metabolite found in surface waters. Compound found in one ground water site and in effluent of blueberry processing plant, but not metabolite.
				Azinphos-m.-oxon	0.23	nr	0	20	0	no det.	
Leung and others, 1982	w s	9/77–11/78	**Iowa:** Des Moines River, Saylorville Reservoir	**Atrazine**	nr	nr	100	~120	nr	0.22 (mean)	Samples taken upstream, downstream, and in the reservoir. Atrazine, alachlor, and cyanazine were detected in the dissolved phase only. Dieldrin detected in both dissolved and particulate phases. DDE detected almost entirely in particulate phase. Approximately 67 percent of the dieldrin entering the reservoir remained there. Net deposition in the reservoir during the study period estimated as 281 kg (atrazine), 251 kg (alachlor), 26 kg (cyanazine), 16 kg (dieldrin), and 20 kg (DDE). (Deposition defined as total input minus total outfall.)
				Alachlor	nr	nr	100	~120	nr	0.09 (mean)	
				Cyanazine	nr	nr	100	~120	nr	0.09 (mean)	
				Dieldrin	0.001	nr	100	~120	nr	0.03 (max)	
				DDD (p,p')	nr	nr	0	~120	0	no det.	
				DDE (p,p')	0.001	nr	100	~120	nr	0.13 (max, s)	
				DDT (p,p')	nr	nr	0	~120	0	no det.	
				Hept. epox.	nr	nr	0	~120	0	no det.	
				Endrin	nr	nr	0	~120	0	no det.	
				2,4-D	nr	nr	0	~120	0	no det.	
				2,4,5-TP	nr	nr	0	~120	0	no det.	
				Lindane	nr	nr	0	~120	0	no det.	
				Methox.	nr	nr	0	~120	0	no det.	
				Propachlor	nr	nr	0	~120	0	no det.	
				Toxaphene	nr	nr	0	~120	0	no det.	

Table 2.2. State and local monitoring studies reviewed—*Continued*

Reference(s)	Matrix	Study — Sampling dates	Study — Location(s)	Compounds	Detection limit(s) (µg/L)	Sites — Number of sites	Sites — Percent of sites with detections	Samples — Number of samples	Samples — Percent of samples with detections	Samples — Maximum concentration (µg/L)	Comments
Fuhrer and Rinella, 1983	w	5/80–12/80	Oregon, Washington: Columbia River and several other rivers	Aldrin	0.01	15	0	27	0	no det.	USGS study to evaluate potential effects of dredging.
				Ametryn	0.1	15	0	27	0	no det.	
				Atratone	0.1	15	0	27	0	no det.	
				Atrazine	0.1	15	0	27	0	no det.	
				Chlordane	0.1	15	0	27	0	no det.	
				Cyanazine	0.1	15	0	27	0	no det.	
				Cyprazine	0.1	15	0	27	0	no det.	
				DDD	0.01	15	0	27	0	no det.	
				DDE	0.01	15	0	27	0	no det.	
				DDT	0.01	15	0	27	0	no det.	
				Dieldrin	0.01	15	0	27	0	no det.	
				Endosulfan	0.01	15	0	27	0	no det.	
				Endrin	0.01	15	0	27	0	no det.	
				Heptachlor	0.01	15	0	27	0	no det.	
				Hept. epox.	0.01	15	0	27	0	no det.	
				Lindane	0.01	15	0	27	0	no det.	
				Methox.	0.01	15	0	27	0	no det.	
				Mirex	0.01	15	0	27	0	no det.	
				Perthane	0.1	15	0	27	0	no det.	
				Prometone	0.1	15	0	27	0	no det.	
				Prometryn	0.1	15	0	27	0	no det.	
				Propazine	0.1	15	0	27	0	no det.	
				Silvex	0.01	15	0	27	0	no det.	
				Simazine	0.1	15	0	27	0	no det.	
				Simetone	0.01	15	0	27	0	no det.	
				Simetryn	0.1	15	0	27	0	no det.	
				Toxaphene	nr	15	0	27	0	no det.	
				2,4-D	0.01	15	20	27	11	0.04	
				2,4-DP	0.01	15	nr	27	nr	0.04	
				2,4,5-T	0.01	15	0	27	0	no det.	

Table 2.2. State and local monitoring studies reviewed—*Continued*

Reference(s)	Matrix	Sampling dates	Location(s)	Compounds	Detection limit(s) (µg/L)	Number of sites	Percent of sites with detections	Number of samples	Percent of samples with detections	Maximum concentration (µg/L)	Comments
Kauss, 1983	w	1980 (w)	Great Lakes: Niagara River	Aldrin	0.001	3	0	nr	0	no det.	Summary of studies of contaminant levels in Niagara River water, suspended sediments, and bed sediments. Values shown are for water samples. Point sources along river (dumps, chemical manufacturing sites) may be responsible for some of the detections. Pesticides detected in suspended sediments (number of sites and concentration range): dieldrin—5 of 9, 2 to 26 µg/kg; HCH (α)—4 of 9, 6 to 110 µg/kg; HCH (γ)—1 of 9, 38 µg/kg; chlordane (α)—6 of 9, 1 to 193 µg/kg; chlordane (γ)—5 of 9, 5 to 70 µg/kg; DDT (o,p')—2 of 10, 20 to 21 µg/kg; DDT (p,p')—2 of 10, 70 to 74 µg/kg; DDD (p,p')—4 of 10, 2-65 µg/kg; DDE (p,p'): 10 of 10, 1 to 36 µg/kg; DDT (total)—10 of 10, 1 to 190 µg/kg; endrin—3 of 9, 5 to 13 µg/kg; hept. epox.—7 of 10, 1 to 36 µg/kg; HCB—8 of 10, 1 to 250 µg/kg; mirex—6 of 10, 4 to 640 µg/kg; endosulfan (α)—7 of 10, 1 to 15 µg/kg; endosulfan (β)—6 of 10, 1 to 45 µg/kg.
	s	1979 (s)		Dieldrin	0.001	3	nr	nr	4-7	0.002	
				HCH (α)	0.001	3	nr	nr	~70	0.03	
				HCH (β)	0.001	3	nr	nr	9-10	0.005	
				Lindane	0.001	3	nr	nr	13-16	0.005	
				Chlordane (α)	0.001	3	nr	nr	4-7	0.001	
				Chlordane (γ)	0.001	3	nr	nr	4-7	0.015	
				DDT (o,p')	0.005	3	nr	nr	0-2	0.005	
				DDT (p,p')	0.005	3	nr	nr	0-5.	0.01	
				DDD (p,p')	0.005	3	nr	nr	0-4	0.005	
				DDE (p,p')	0.001	3	nr	nr	0-2	0.003	
				DDT (total)	0.005	3	nr	nr	nr	0.01	
				Endrin	0.001	3	nr	nr	0-2	0.012	
				Heptachlor	0.001	3	0	nr	0	no det.	
				Hept. epox.	0.001	3	nr	nr	7-21	0.03	
				HCB	0.001	3	0	nr	0	no det.	
				Mirex	0.005	3	0	nr	0	no det.	
				Endosulfan (α)	0.001	3	nr	nr	0-2	0.005	
				Endosulfan (β)	0.001	3	nr	nr	0-2	0.005	

Table 2.2. State and local monitoring studies reviewed—*Continued*

Reference(s)	Matrix	Sampling dates	Location(s)	Compounds	Detection limit(s) (µg/L)	Sites		Samples			Comments
						Number of sites	Percent of sites with detections	Number of samples	Percent of samples with detections	Maximum concentration (µg/L)	
Kuntz and Warry, 1983	w s	4/79– 12/81, biweekly	**Great Lakes:** Niagara River	**Water**						(means)	Organochlorine pesticide (and PCB) concentrations were measured in water and suspended sediments at Niagara-on-the-Lake (Lake Ontario end of Niagara River). Loading estimates indicate that suspended sediments were responsible for ~40 percent of the loading to Lake Ontario of PCBs, DDT, and HCB, and considerably less for other organochlorines. Comparisons of suspended sediment concentrations with bed sediment concentrations in Lake Erie indicate that Lake Erie was not the major source of PCBs, mirex, and chlorobenzenes to Lake Ontario and that these compounds entered the system along the Niagara River. Detection limits inferred from data.
				Aldrin	0.0001	nr	nr	75	nr	<0.0001	
				Dieldrin	0.0001	nr	nr	75	93	0.0006	
				HCH (α)	0.0001	nr	nr	75	100	0.011	
				Lindane	0.0001	nr	nr	75	99	0.0021	
				Chlordane (α)	0.0001	nr	nr	75	64	0.0003	
				Chlordane (γ)	0.0001	nr	nr	75	68	0.0005	
				DDT (o,p')	0.0001	nr	nr	75	nr	<0.0001	
				DDT (p,p')	0.0001	nr	nr	75	40	0.0002	
				DDD (p,p')	0.0001	nr	nr	75	nr	0.0001	
				DDE (p,p')	0.0001	nr	nr	75	61	0.0003	
				Endrin	0.0001	nr	nr	75	11	<0.0001	
				Heptachlor	0.0001	nr	0	75	0	no det.	
				Hept. epox.	0.0001	nr	nr	75	47	0.0005	
				HCB	0.0001	nr	nr	75	95	0.0008	
				Mirex	0.0001	nr	nr	75	nr	<0.0001	
				Endosulfan (α)	0.0001	nr	nr	75	13	0.0001	
				Endosulfan (β)	0.0001	nr	nr	75	15	0.0001	
				Methox.	0.0001	nr	nr	75	nr	0.0001	

Table 2.2. State and local monitoring studies reviewed—*Continued*

Reference(s)	Matrix	Sampling dates	Location(s)	Compounds	Detection limit(s) (µg/L)[1]	Sites		Samples			Comments
						Number of sites	Percent of sites with detections	Number of samples	Percent of samples with detections	Maximum concentration (µg/L)	
Kuntz and Warry, 1983[1]— *Continued*	w s	4/79– 12/81, biweekly	**Great Lakes:** Niagara River	Suspended sediment:	(µg/kg)[1]					(means, in µg/kg)[1]	*Continued—* Concentrations on suspended sediment (in micrograms per kilogram) are based on dry sediment weight. Detection limits inferred from data.
				Aldrin	1.0	nr	nr	70	nr	2.0	
				Dieldrin	1.0	nr	nr	70	80	4.0	
				HCH (α)	1.0	nr	nr	70	75	12.0	
				Lindane	1.0	nr	nr	70	33	2.0	
				Chlordane (α)	1.0	nr	nr	70	67	3.0	
				Chlordane (γ)	1.0	nr	nr	70	79	6.0	
				DDT (o,p')	1.0	nr	nr	70	30	6.0	
				DDT (p,p')	1.0	nr	nr	70	86	11.0	
				DDD (p,p')	1.0	nr	nr	70	62	4.0	
				DDE (p,p')	1.0	nr	nr	70	92	23.0	
				Endrin	1.0	nr	nr	70	nr	<1.0	
				Heptachlor	1.0	nr	nr	70	nr	1.0	
				Hept. epox.	1.0	nr	nr	70	37	1.0	
				HCB	1.0	nr	nr	70	100	124.0	
				Mirex	1.0	nr	nr	70	76	12.0	
				Endosulfan (α)	1.0	nr	nr	70	36	4.0	
				Endosulfan (β)	1.0	nr	nr	70	nr	<1.0	
				Methox.	1.0	nr	nr	70	51	7.0	

[1]Detection limits and concentration data for suspended sediment are in µg/kg, not µg/L.

Table 2.2. State and local monitoring studies reviewed—*Continued*

Reference(s)	Matrix	Sampling dates	Study Location(s)	Compounds	Detection limit(s) (µg/L)	Sites Number of sites	Percent of sites with detections	Samples Number of samples	Percent of samples with detections	Maximum concentration (µg/L)	Comments
Lurry, 1983	w	1/80–3/81	**Louisiana:** Calcasieu River (11 sites) and effluent samples	Aldrin	0.001	11	0	33	0	no det.	USGS study to evaluate potential effects of dredging. 2,4-D detected in 20 of 33 river samples.
				Chlordane	0.1	11	0	33	0	no det.	
				DDD	0.001	11	0	33	0	no det.	
				DDE	0.001	11	0	33	0	no det.	
				DDT	0.001	11	0	33	0	no det.	
				Diazinon	0.01	11	55	33	18	0.06	
				Dieldrin	0.001	11	nr	33	nr	0.003	
				Endosulfan	0.001	11	0	33	0	no det.	
				Endrin	0.001	11	0	33	0	no det.	
				Ethion	0.01	11	0	33	0	no det.	
				Heptachlor	0.001	11	0	33	0	no det.	
				Hept. epox.	0.001	11	0	33	0	no det.	
				Lindane	0.001	11	0	33	0	no det.	
				Malathion	0.01	11	0	33	0	no det.	
				Methox.	0.01	11	0	33	0	no det.	
				M. parathion	0.01	11	0	33	0	no det.	
				M. trithion	0.01	11	0	33	0	no det.	
				Mirex	0.01	11	0	33	0	no det.	
				Parathion	0.01	11	0	33	0	no det.	
				Perthane	0.01	11	0	33	0	no det.	
				Toxaphene	0.1	11	0	33	0	no det.	
				Trithion	0.01	11	0	33	0	no det.	
				2,4-D	0.01	11	73	33	61	0.37	
				2,4,5-T	0.01	11	0	33	0	no det.	
				Silvex	0.01	11	0	33	0	no det.	

Table 2.2. State and local monitoring studies reviewed—Continued

Reference(s)	Matrix	Study				Sites		Samples			Comments
		Sampling dates	Location(s)	Compounds	Detection limit(s) (µg/L)	Number of sites	Percent of sites with detections	Number of samples	Percent of samples with detections	Maximum concentration (µg/L)	
Ray and others, 1983	w	7/80	**Texas:** Nueces Estuary, near Corpus Christi	**HCB**	0.00001	nr	50	nr	50	0.00035	Survey of water and sediment contamination of estuary by organochlorines. Other organochlorines also included as analytes.
				PCP	nr	nr	100	nr	100	0.068	
				DDT (total)	nr	nr	100	nr	100	0.0031	
				HCH (α)	0.00005	nr	88	nr	88	0.00049	
				Lindane	0.00001	nr	63	nr	63	0.00013	
				Chlordane	nr	nr	100	nr	100	0.00093	
				Dieldrin	0.00001	nr	88	nr	88	0.00013	
Toppin, 1983	w	1978–80	**Vermont:** Lower Black River	Aldrin	0.01	9	0	29	0	no det.	Assessment of water quality before proposed dam construction. Concentrations of all pesticides were below detection limit.
				Chlordane	0.1	9	0	29	0	no det.	
				DDD	0.01	9	0	29	0	no det.	
				DDE	0.01	9	0	29	0	no det.	
				DDT	0.01	9	0	29	0	no det.	
				Dieldrin	0.01	9	0	29	0	no det.	
				Endosulfan	0.01	9	0	29	0	no det.	
				Endrin	0.01	9	0	29	0	no det.	
				Heptachlor	0.01	9	0	29	0	no det.	
				Hept. epox.	0.01	9	0	29	0	no det.	
				Lindane	0.01	9	0	29	0	no det.	
				Methox.	0.01	9	0	29	0	no det.	
				Mirex	0.01	9	0	29	0	no det.	
				Perthane	0.1	9	0	29	0	no det.	
				Toxaphene	nr	9	0	29	0	no det.	

Table 2.2. State and local monitoring studies reviewed—*Continued*

Reference(s)	Matrix	Sampling dates	Study Location(s)	Compounds	Detection limit(s) (µg/L)	Sites Number of sites	Sites Percent of sites with detections	Samples Number of samples	Samples Percent of samples with detections	Samples Maximum concentration (µg/L)	Comments
Wang, 1983	w	1977–78	**Florida:** Indian River Lagoon	DDT	0.01	13	0	~50	0	no det.	Low concentrations of DDT detected in sediments, but not in water. Persistence of malathion and parathion investigated. Malathion degradation attributed to hydrolysis; parathion degradation attributed to biological interaction.
				Malathion	nr	13	0	~50	0	no det.	
				Parathion	nr	13	0	~50	0	no det.	
Wangsness, 1983	w	10/86 12/86	**Indiana:** Eagle Creek watershed, Indianapolis area: 4 sites in October; 3 sites in December	Aldrin	0.01	4	0	nr	0	no det.	USGS report. Diazinon was the only pesticide detected.
				Chlordane	0.1	4	0	nr	0	no det.	
				DDD	0.01	4	0	nr	0	no det.	
				DDE	0.01	4	0	nr	0	no det.	
				DDT	0.01	4	0	nr	0	no det.	
				Dieldrin	0.01	4	0	nr	0	no det.	
				Endosulfan	0.01	4	0	nr	0	no det.	
				Endrin	0.01	4	0	nr	0	no det.	
				Heptachlor	0.01	4	0	nr	0	no det.	
				Hept. epox.	0.01	4	0	nr	0	no det.	
				Lindane	0.01	4	0	nr	0	no det.	
				Methox.	0.01	4	0	nr	0	no det.	
				Mirex	0.01	4	0	nr	0	no det.	
				Perthane	0.10	4	0	nr	0	no det.	
				Toxaphene	nr	4	0	nr	0	no det.	
				Diazinon	0.01	4	25	nr	14	0.09	
				Ethion	0.01	4	0	nr	0	no det.	
				Malathion	0.01	4	0	nr	0	no det.	
				M. parathion	0.01	4	0	nr	0	no det.	
				M. trithion	0.01	4	0	nr	0	no det.	
				Parathion	0.01	4	0	nr	0	no det.	
				Trithion	0.01	4	0	nr	0	no det.	

Table 2.2. State and local monitoring studies reviewed—*Continued*

Reference(s)	Matrix	Study Sampling dates	Location(s)	Compounds	Detection limit(s) (µg/L)	Sites Number of sites	Percent of sites with detections	Samples Number of samples	Percent of samples with detections	Maximum concentration (µg/L)	Comments
Fishel, 1984	w	3/80– 3/81	**Pennsylvania**: Susquehanna River	Aldrin	0.01	nr	0	31	0	no det.	2,4-D concentrations were the most variable of all the pesticides. No direct relationship observed for any of the pesticides with discharge, suspended sediment concentration, or particle size. Atrazine concentrations highest during spring and early summer following application.
				Ametryn	0.1	nr	100	49	nr	0.1	
				Atratone	0.1	nr	100	49	nr	0.1	
				Atrazine	0.1	nr	100	49	nr	3.4	
				Chlordiane	0.1	nr	100	31	nr	0.1	
				Cyanazine	0.1	nr	100	49	nr	0.1	
				Cyprazine	0.1	nr	100	49	nr	0.1	
				DDD	0.01	nr	0	31	0	no det.	
				DDE	0.01	nr	0	31	0	no det.	
				DDT	0.01	nr	0	31	0	no det.	
				Diazinon	0.01	nr	nr	31	nr	0.02	
				Dieldrin	0.01	nr	0	31	0	no det.	
				Endosulfan	0.01	nr	0	30	0	no det.	
				Endrin	0.01	nr	0	31	0	no det.	
				Ethion	0.01	nr	0	31	0	no det.	
				Heptachlor	0.01	nr	0	31	0	no det.	
				Hept. epox.	0.01	nr	0	31	0	no det.	
				Lindane	0.01	nr	0	31	0	no det.	
				Malathion	0.01	nr	0	31	0	no det.	
				Methox.	0.01	nr	0	31	0	no det.	
				M. parathion	0.01	nr	0	31	0	no det.	
				M. trithion	0.01	nr	0	31	0	no det.	
				Mirex	0.01	nr	0	30	0	no det.	
				Parathion	0.01	nr	0	31	0	no det.	
				Perthane	0.01	nr	0	30	0	no det.	
				Prometone	0.1	nr	0	49	0	no det.	
				Prometryn	0.1	nr	0	49	0	no det.	
				Propazine	0.1	nr	0	49	0	no det.	
				Silvex	0.01	nr	0	29	0	no det.	

Table 2.2. State and local monitoring studies reviewed—*Continued*

Reference(s)	Matrix	Sampling dates	Study			Sites		Samples			Comments
			Location(s)	Compounds	Detection limit(s) (µg/L)	Number of sites	Percent of sites with detections	Number of samples	Percent of samples with detections	Maximum concentration (µg/L)	
Fishel, 1984—*Continued*	w	3/80–3/81	**Pennsylvania:** Susquehanna River	**Simazine**	0.1	nr	100	49	nr	0.2	
				Simetone	0.1	nr	0	49	0	no det.	
				Simetryn	0.1	nr	0	49	0	no det.	
				Toxaphene	nr	nr	0	31	0	no det.	
				2,4-D	0.01	nr	100	29	75	0.41	
				2,4-DP	0.01	nr	100	49	nr	0.02	
				2,4,5-T	0.01	nr	0	29	0	no det.	
Fuhrer, 1984	w	4/82	**Oregon:** Chetco and Rogue Rivers, southwest Oregon	Aldrin	0.01	nr	0	nr	0	no det.	Samples taken as part of study to evaluate potential effects of dredging and disposal of dredged material. Concentrations of all pesticides were below detection limit.
				Chlordane	0.1	nr	0	nr	0	no det.	
				DDD	0.01	nr	0	nr	0	no det.	
				DDE	0.01	nr	0	nr	0	no det.	
				DDT	0.01	nr	0	nr	0	no det.	
				Dieldrin	0.01	nr	0	nr	0	no det.	
				Endosulfan	0.01	nr	0	nr	0	no det.	
				Endrin	0.01	nr	0	nr	0	no det.	
				Heptachlor	0.01	nr	0	nr	0	no det.	
				Lindane	0.01	nr	0	nr	0	no det.	
				Mirex	0.01	nr	0	nr	0	no det.	
				Perthane	0.1	nr	0	nr	0	no det.	
				Silvex	0.01	nr	0	nr	0	no det.	
				Toxaphene	nr	nr	0	nr	0	no det.	
				2,4-D	0.01	nr	0	nr	0	no det.	
				2,4-DP	0.01	nr	0	nr	0	no det.	
				2,4,5-T	0.01	nr	0	nr	0	no det.	

Table 2.2. State and local monitoring studies reviewed—*Continued*

Reference(s)	Matrix	Sampling dates	Location(s)	Compounds	Detection limit(s) (µg/L)	Sites		Samples			Comments
						Number of sites	Percent of sites with detections	Number of samples	Percent of samples with detections	Maximum concentration (µg/L)	
Granstrom and others, 1984	w	1979–80	**New Jersey:** Delaware and Raritan Canal	Aldrin	nr	12	0	37	0	no det.	Survey of contamination of water and bed sediments. PCBs and volatile organic compounds also measured.
				Chlordane (γ)	nr	12	0	37	0	no det.	
				DDD (*p,p'*)	nr	12	nr	37	11	0.021	
				DDE (*p,p'*)	nr	12	0	37	0	no det.	
				DDT (*o,p'*)	nr	12	nr	37	14	0.15	
				DDT (*p,p'*)	nr	12	0	37	0	no det.	
				Dieldrin	nr	12	0	37	0	no det.	
				Endrin	nr	12	0	37	0	no det.	
				HCH (α)	nr	12	0	37	0	no det.	
				HCH (β)	nr	12	nr	37	49	0.06	
				Lindane	nr	12	nr	37	nr	0.095	
				Heptachlor	nr	12	0	37	0	no det.	
				Hept. epox.	nr	12	0	37	0	no det.	
				Methox.	nr	12	0	37	0	no det.	
				Mirex	nr	12	0	37	0	no det.	
				Toxaphene	nr	12	0	37	0	no det.	

Table 2.2. State and local monitoring studies reviewed—*Continued*

Reference(s)	Matrix	Sampling dates	Location(s)	Compounds	Detection limit(s) (µg/L)	Number of sites	Percent of sites with detections	Number of samples	Percent of samples with detections	Maximum concentration (µg/L)	Comments
							Sites		**Samples**		
Oliver and Nicol,1984	w	8/81– 9/83 weekly	**Great Lakes, New York:** Niagara River	**HCH** (α)	0.00002	nr	nr	104	100	(medians) 0.01	Large volume (16 L) whole water samples taken weekly over the 2-year period at the mouth of the river (outlet to Lake Ontario) and analyzed for 31 organochlorines. Two samples also collected at source of river (Lake Erie end). Most analytes detected more frequently and at higher concentrations at the river mouth, indicating significant inputs along the Niagara River. Approximate yearly loads were calculated. For pesticides, the loads were: (α-HCH (>2,000 kg/yr), γ-HCH (200 to 2,000 kg/yr), γ- and α-chlordane and DDE (20 to 200 kg/yr), and DDT and mirex (~20 kg/yr).
				Lindane	0.00002	nr	nr	104	99	0.0014	
				Chlordane (γ)	0.00002	nr	nr	104	99	0.0001	
				Chlordane (α)	0.00002	nr	nr	104	98	0.0001	
				DDE (*p,p'*)	0.00002	nr	nr	104	100	0.00017	
				DDT (*p,p'*)	0.00005	nr	nr	104	36	0.00005	
				Mirex	0.00006	nr	nr	104	12	0.00006	
				HCB	0.00001	nr	nr	104	100	0.00061	

Table 2.2. State and local monitoring studies reviewed—*Continued*

Reference(s)	Matrix	Sampling dates	Location(s)	Compounds	Detection limit(s) (µg/L)	Number of sites	Percent of sites with detections	Number of samples	Percent of samples with detections	Maximum concentration (µg/L)	Comments
Rogers, 1984	w	2/81–2/83	**New York:** Saw Mill River	Aldrin	0.01	11	0	11	0	no det.	Basin is primarily urban. Metals, nutrients, and priority pollutants also analyzed. Organochlorine compounds were detected in sediments along entire river. No pesticides detected in water samples.
				Chlordane	0.1	11	0	11	0	no det.	
				DDD	0.01	11	0	11	0	no det.	
				DDE	0.01	11	0	11	0	no det.	
				DDT	0.01	11	0	11	0	no det.	
				Dieldrin	0.01	11	0	11	0	no det.	
				Endosulfan	0.01	11	0	11	0	no det.	
				Endrin	0.01	11	0	11	0	no det.	
				Heptachlor	0.01	11	0	11	0	no det.	
				Hept. epox.	0.01	11	0	11	0	no det.	
				Lindane	0.01	11	0	11	0	no det.	
				Methox.	0.01	11	0	11	0	no det.	
				Mirex	0.01	11	0	11	0	no det.	
				Perthane	0.1	11	0	11	0	no det.	
				Toxaphene	nr	11	0	11	0	no det.	
Takita, 1984	w	6/81–9/81	**Pennsylvania:** Susquehanna River and 15 tributaries; Codorus Creek	**Alachlor**	nr	22	nr	88	42	11.0	Samples collected at or near mouths of 15 tributaries and at 3 sites on Susquehanna River. Three additional sites on Codorus Creek to assess urban runoff in York, Pennsylvania. Monthly sampling. Priority pollutants and metals also analyzed.
				Atrazine	nr	22	nr	88	nr	2.5	
				HCH (α)	nr	22	nr	88	35	5.2	
				Chlordane	nr	22	nr	88	nr	0.29	
				Diazinon	nr	22	nr	88	nr	0.42	
				Lindane	nr	22	nr	88	16	0.16	

Table 2.2. State and local monitoring studies reviewed—*Continued*

Reference(s)	Matrix	Sampling dates	Location(s)	Compounds	Detection limit(s) (µg/L)	Sites			Samples		Comments
						Number of sites	Percent of sites with detections	Number of samples	Percent of samples with detections	Maximum concentration (µg/L)	
Thompson, 1984; Stephens, 1984	w	8/81, 8/82	**Utah:** Jordan River: 5 sites on Jordan River; 3 sites on tributaries	Aldrin	0.01	nr	0	nr	0	no det.	Water from six storm conduits also analyzed for priority pollutants; no pesticides detected.
				Chlordane	0.1	nr	0	nr	0	no det.	
				DDD	0.01	nr	0	nr	0	no det.	
				DDE	0.01	nr	13	nr	nr	0.01	
				DDT	0.01	nr	0	nr	0	no det.	
				Dieldrin	0.01	nr	0	nr	0	no det.	
				Endosulfan	0.01	nr	0	nr	0	no det.	
				Endrin	0.01	nr	0	nr	0	no det.	
				Heptachlor	0.01	nr	0	nr	0	no det.	
				Hept. epox.	0.01	nr	0	nr	0	no det.	
				Lindane	0.01	nr	0	nr	0	no det.	
				Mirex	0.01	nr	0	nr	0	no det.	
				Silvex	nr	nr	38	nr	nr	0.02	
				Perthane	0.1	nr	0	nr	0	no det.	
				2,4-D	nr	nr	38	nr	nr	0.06	
				2,4-DP	nr	nr	0	nr	0	no det.	
				2,4,5-T	nr	nr	0	nr	0	no det.	
Baker, 1985	w	1980–83	**Indiana, Ohio**	**Atrazine**	nr	nr	nr	nr	nr	18.0	Study provides baseline levels for a proposed program encouraging conservation tillage.

Table 2.2. State and local monitoring studies reviewed—*Continued*

Reference(s)	Matrix	Sampling dates	Location(s)	Compounds	Detection limit(s) (µg/L)	Sites — Number of sites	Sites — Percent of sites with detections	Samples — Number of samples	Samples — Percent of samples with detections	Maximum concentration (µg/L)	Comments
Butler and Arruda, 1985	w	1973–84	**Kansas:** Rivers and lakes throughout state	Alachlor	nr	~10	nr	1,035	5.0	nr	Summary of results of monitoring program begun in 1973. Sampling frequency varied, but was semiannual or annual for rivers during most of the 1980's. Lakes were sampled one to six times from 1975 to 1982. Detection frequency data shown are for rivers from 1977 to 1984. Atrazine, alachlor, metolachlor, metribuzin, and 2,4-D were detected most often in rivers and lakes during this period. Compounds listed are apparently those with at least one detection. Compounds targeted but not detected are not reported.
				Aldrin	nr	~10	nr	1,035	0.2	nr	
				HCH (α)	nr	~100	nr	1,035	0.3	nr	
				Atrazine	nr	~100	nr	1,035	17	nr	
				Chlordane	nr	~100	nr	1,035	0.4	nr	
				Dacthal	nr	~100	nr	1,035	1.3	nr	
				DDE	nr	~100	nr	1,035	0.5	nr	
				Diazinon	nr	~100	nr	1,035	0.3	nr	
				Dieldrin	nr	~100	nr	1,035	0.2	nr	
				Metolachlor	nr	~100	nr	1,035	4.0	nr	
				Chlorpyrifos	nr	~100	nr	1,035	0.1	nr	
				HCB	nr	~100	nr	1,035	0.4	nr	
				Lindane	nr	~100	nr	1,035	1.5	nr	
				Malathion	nr	~100	nr	1,035	0.7	nr	
				Metribuzin	nr	~100	nr	1,035	2.4	nr	
				Propazine	nr	~100	nr	1,035	0.5	nr	
				Propachlor	nr	~100	nr	1,035	1.1	nr	
				2,4-D	nr	~100	nr	1,035	4.0	nr	
				2,4,5-T	nr	~100	nr	1,035	0.8	nr	
				Silvex	nr	~100	nr	1,035	0.1	nr	
				1-Hydroxy-chlordene	nr	~100	nr	1,035	1.4	nr	

Table 2.2. State and local monitoring studies reviewed—*Continued*

Reference(s)	Matrix	Sampling dates	Location(s)	Compounds	Detection limit(s) (µg/L)	Number of sites	Percent of sites with detections	Number of samples	Percent of samples with detections	Maximum concentration (µg/L)	Comments
McFall and others, 1985	w	5/80–6/80	Louisiana: Lake Ponchartrain Inner Harbor, Navigation Canal	HCH (α)	nr	nr	nr	10	nr	0.004	Eight samples taken during ebb tide, two during flood tide. Concentrations lower during ebb tide. All maximum concentrations shown occurred in one of the two flood tide samples. None of the positive detections for pesticides (gas chromatography-electron capture detection analysis) were confirmed when analyzed by gas chromotography/mass-spectrometry.
				HCH (β)	nr	nr	nr	10	nr	0.006	
				Lindane	nr	nr	nr	10	nr	0.028	
				Heptachlor	nr	nr	nr	10	nr	0.009	
				Aldrin	nr	nr	nr	10	nr	0.006	
				Hept. epox.	nr	nr	nr	10	nr	0.004	
				Endosulfan (α)	nr	nr	nr	10	nr	0.003	
				Dieldrin	nr	nr	nr	10	nr	0.006	
				DDE (*p,p'*)	nr	nr	nr	10	nr	0.001	
				Endrin	nr	nr	0	10	0	no det.	
				Endosulfan (β)	nr	nr	0	10	0	no det.	
				DDD (*p,p'*)	nr	nr	nr	10	nr	0.006	
Oltmann and others, 1985; Oltmann and Schulters, 1989	w	10/81–4/83	California: Urban runoff, Fresno	Aldrin	0.01	nr	25	86	nr	0.02	Urban runoff monitored from four catchments with different land uses (industrial, single-dwelling residential, multiple-dwelling residential, commercial). Pesticide concentrations also measured in rain and street-surface particulates. Chlordane, diazinon, malathion, and lindane occurrences due in part to urban use. Parathion and 2,4-D used heavily in agriculture in surrounding area. Chlordane detected less frequently in industrial catchment than in others.
				Chlordane	0.1	nr	100	86	69	1.2	
				DDE	0.01	nr	100	86	43	0.06	
				DDT	0.01	nr	50	86	nr	0.01	
				Diazinon	0.01	nr	100	86	100	18.0	
				Dieldrin	0.01	nr	100	86	14	0.02	
				Endosulfan	0.01	nr	75	86	12	0.07	
				Endrin	0.01	nr	25	86	nr	0.01	
				Lindane	0.01	nr	100	86	93	0.27	
				Malathion	0.01	nr	100	86	100	14.0	
				Methox.	0.01	nr	75	86	nr	0.19	
				M. parathion	0.01	nr	50.	86	nr	0.03	
				Parathion	0.01	nr	100	86	59	2.5	
				Silvex	0.01	nr	50	86	nr	0.07	
				Trithion	0.01	nr	25	86	nr	0.1	
				2,4-D	0.01	nr	100	86	74	3.7	

Table 2.2. State and local monitoring studies reviewed—*Continued*

Reference(s)	Matrix	Study Sampling dates	Location(s)	Compounds	Detection limit(s) (µg/L)	Sites Number of sites	Sites Percent of sites with detections	Samples Number of samples	Samples Percent of samples with detections	Samples Maximum concentration (µg/L)	Comments
Pope and others, 1985	w	1983	**Kansas:** 19 water supply lakes, eastern Kansas, 2 to 4 sites per lake	**Alachlor**	0.25	19	11	39	nr	0.63	Synoptic study. Most lakes sampled twice—spring/ early summer and late summer/autumn. Pesticides detected at 8 of the 19 lakes. One detection only for 2,4,5-T, 2,4-D, metolachlor, and DDE. Five lakes with atrazine detections and three lakes with alachlor detections.
				Aldrin	0.025	19	0	39	0	no det.	
				Atrazine	1.2	19	21	39	13	0.63	
				Chlordane	0.25	19	0	39	0	no det.	
				Dacthal	0.05	19	0	39	0	no det.	
				DDE (*o,p'*)	0.1	19	nr	39	2.5	0.12	
				DDE (*p,p'*)	0.1	19	nr	39	2.5	0.12	
				Dieldrin	0.15	19	0	39	0	no det.	
				Metolachlor	0.25	19	nr	39	2.5	0.54	
				Endrin	0.1	19	0	39	0	no det.	
				Lindane	0.03	19	0	39	0	no det.	
				Methox.	0.20	19	0	39	0	no det.	
				Propachlor	0.25	19	0	39	0	no det.	
				Metribuzin	0.1	19	0	39	0	no det.	
				Toxaphene	nr	19	0	39	0	no det.	
				2,4-D	0.4	19	nr	39	2.5	0.5	
				Silvex	0.2	19	0	39	0	no det.	
				2,4,5-T	0.2	19	nr	39	2.5	0.21	

Table 2.2. State and local monitoring studies reviewed—*Continued*

Reference(s)	Matrix	Sampling dates	Study Location(s)	Compounds	Detection limit(s) (µg/L)	Sites Number of sites	Sites Percent of sites with detections	Number of samples	Samples Percent of samples with detections	Samples Maximum concentration (µg/L)	Comments
Yamane and Lum, 1985	w	9/80– 8/88	Hawaii: Storm-water runoff	**Aldrin**	0.01	nr	100	nr	50	0.13	Urban storm drains monitored for pesticides, metals, and nutrients. More than 300 total samples, but pesticides analyzed in only a few. Heptachlor, malathion, and lindane exceeded USEPA criteria for protection of aquatic life in some samples.
				Chlordane	0.1	nr	100	nr	100	2.2	
				DDD	0.01	nr	0	nr	0	no det.	
				DDE	0.01	nr	100	nr	25	0.01	
				DDT	0.01	nr	0	nr	0	no det.	
				Diazinon	0.01	nr	100	nr	100	3.6	
				Dieldrin	0.01	nr	100	nr	63	0.03	
				Endosulfan	0.01	nr	0	nr	0	no det.	
				Endrin	0.01	nr	0	nr	0	no det.	
				Ethion	0.01	nr	0	nr	0	no det.	
				Heptachlor	0.01	nr	100	nr	100	0.58	
				Hept. epox.	0.01	nr	100	nr	25	0.01	
				Lindane	0.01	nr	100	nr	63	0.02	
				Malathion	0.01	nr	100	nr	100	1.7	
				Methox.	0.01	nr	50	nr	13	0.03	
				M. parathion	0.01	nr	0	nr	0	no det.	
				M. trithion	0.01	nr	0	nr	0	no det.	
				Mirex	0.01	nr	0	nr	0	no det.	
				Parathion	0.01	nr	0	nr	0	no det.	
				Perthane	0.1	nr	0	nr	0	no det.	
				Toxaphene	0.1	nr	0	nr	0	no det.	
				Trithion	0.01	nr	0	nr	0	no det.	
				2,4-D	0.01	nr	100	nr	100	0.25	
				2,4-DP	0.01	nr	0	nr	0	no det.	
				2,4,5-T	0.01	nr	50	nr	50	0.04	
				Silvex	0.01	nr	100	nr	100	0.03	

Table 2.2. State and local monitoring studies reviewed—*Continued*

Reference(s)	Matrix	Study Sampling dates	Location(s)	Compounds	Detection limit(s) (µg/L)	Sites Number of sites	Sites Percent of sites with detections	Samples Number of samples	Samples Percent of samples with detections	Samples Maximum concentration (µg/L)	Comments
Yorke and others, 1985	w	1979–80	**Pennsylvania:** Schuylkill River, Pottstown, and Manayunk	Aldrin	0.01	nr	0	26	0	no det.	USGS study of effects of low-level dams on water quality.
				Chlordane	0.1	nr	0	26	0	no det.	
				DDD	0.01	nr	0	26	0	no det.	
				DDE	0.01	nr	0	26	0	no det.	
				DDT	0.01	nr	50	26	nr	0.01	
				Dieldrin	0.01	nr	0	26	0	no det.	
				Endosulfan	0.01	nr	0	26	0	no det.	
				Endrin	0.01	nr	0	26	0	no det.	
				Heptachlor	0.01	nr	0	26	0	no det.	
				Hept. epox.	0.01	nr	0	26	0	no det.	
				Lindane	0.01	nr	0	26	0	no det.	
				Methox.	0.01	nr	0	22	0	no det.	
				Mirex	0.01	nr	0	24	0	no det.	
				Perthane	0.1	nr	0	26	0	no det.	
				Toxaphene	0.1	nr	0	26	0	no det.	

Table 2.2. State and local monitoring studies reviewed—*Continued*

Reference(s)	Matrix	Study Sampling dates	Study Location(s)	Compounds	Detection limit(s) (µg/L)	Sites Number of sites	Sites Percent of sites with detections	Samples Number of samples	Samples Percent of samples with detections	Samples Maximum concentration (µg/L)	Comments
Andrews and Schertz, 1986	w	1973–82 (quarterly)	Texas: Colorado River (4 sites); Concho River (1 site)	Aldrin	0.01	5	40	157	nr	0.01	Concentrations of chlordane, DDT, dieldrin, endrin, ethion, heptachlor, and lindane exceeded recommended criteria for aquatic life in at least one sample. Diazinon and 2,4-D detected most often. Summary data only; no information on trends in concentrations.
				Chlordane	0.1	5	40	157	nr	0.26	
				DDD	0.01	5	60	157	nr	0.01	
				DDE	0.01	5	60	157	nr	0.03	
				DDT	0.01	5	60	158	nr	0.03	
				Diazinon	0.01	5	100	158	39	0.36	
				Dieldrin	0.01	5	40	158	nr	0.01	
				Endrin	0.01	5	40	156	nr	0.01	
				Endosulfan	0.001	5	40	62	nr	0.01	
				Ethion	0.01	5	40	115	nr	0.85	
				Heptachlor	0.01	5	40	157	nr	0.01	
				Hept. epox.	0.01	5	40	157	nr	0.01	
				Lindane	0.01	5	80	157	nr	0.02	
				Malathion	0.01	5	60	157	nr	0.02	
				Methox.	0.01	5	40	56	nr	0.01	
				M. parathion	0.01	5	40	157	nr	0.01	
				M. trithion	0.01	5	40	115	nr	0.01	
				Mirex	0.01	5	40	45	nr	0.01	
				Parathion	0.01	5	40	157	nr	0.01	
				Toxaphene	nr	5	40	135	nr	nr	
				Trithion	0.01	5	40	115	nr	0.01	
				2,4-D	0.01	5	100	131	31	0.01	
				2,4,5-T	0.01	5	100	131	27	0.21	
				Silvex	0.01	5	40	131	nr	0.31	

Table 2.2. State and local monitoring studies reviewed—*Continued*

		Study				Sites		Samples			Comments
Reference(s)	Matrix	Sampling dates	Location(s)	Compounds	Detection limit(s) (µg/L)	Number of sites	Percent of sites with detections	Number of samples	Percent of samples with detections	Maximum concentration (µg/L)	
Briggs and Feiffer, 1986	w	11/78–9/83	**Rhode Island:** Blackstone, Branch, Patuxent, Pawcatuck Rivers	Aldrin	0.01	nr	0	35	0	no det.	Samples taken twice yearly during high and low flow.
				Chlordane	0.1	nr	0	35	0	no det.	
				DDD	0.01	nr	0	35	0	no det.	
				DDE	0.01	nr	0	35	0	no det.	
				DDT	0.01	nr	0	35	0	no det.	
				Dieldrin	0.01	nr	50	35	nr	0.01	
				Endosulfan	0.01	nr	0	34	0	no det.	
				Endrin	0.01	nr	0	35	0	no det.	
				Heptachlor	0.01	nr	0	35	0	no det.	
				Hept. epox.	0.01	nr	0	35	0	no det.	
				Lindane	0.01	nr	17	35	nr	0.02	
				Methox.	0.01	nr	0	28	0	no det.	
				Mirex	0.01	nr	0	28	0	no det.	
				Perthane	0.1	nr	0	34	0	no det.	
				Toxaphene	nr	nr	0	35	0	no det.	

Table 2.2. State and local monitoring studies reviewed—*Continued*

Reference(s)	Matrix	Study			Sites		Samples			Comments
		Sampling dates	Location(s)	Compounds	Detection limit(s) (µg/L)	Number of sites	Percent of sites with detections	Number of samples	Percent of samples with detections	Maximum concentration (µg/L)	
Graczyk, 1986	w	1976 1981	**Wisconsin, Minnesota:** St. Croix River Basin	Aldrin	0.01	3	0	3	0	no det.	1976 sample taken at St. Croix Falls. Two samples taken in 1981 from Namekagon River, above and below drainage from cranberry bogs. All pesticides were below detection limits in all samples.
				Chlordane	0.1	3	0	3	0	no det.	
				DDD	0.01	3	0	3	0	no det.	
				DDE	0.01	3	0	3	0	no det.	
				DDT	0.01	3	0	3	0	no det.	
				Diazinon	0.01	3	0	3	0	no det.	
				Dieldrin	0.01	3	0	3	0	no det.	
				Endrin	0.01	3	0	3	0	no det.	
				Endosulfan	0.01	3	0	3	0	no det.	
				Ethion	0.01	3	0	3	0	no det.	
				Heptachlor	0.01	3	0	3	0	no det.	
				Hept. epox.	0.01	3	0	3	0	no det.	
				Lindane	0.01	3	0	3	0	no det.	
				Malathion	0.01	3	0	3	0	no det.	
				M. trithion	0.01	3	0	3	0	no det.	
				Parathion	0.01	3	0	3	0	no det.	
				Toxaphene	nr	3	0	3	0	no det.	
				Trithion	0.01	3	0	3	0	no det.	
				Methox.	0.01	3	0	3	0	no det.	
				M. parathion	0.01	3	0	3	0	no det.	
				Mirex	0.01	3	0	3	0	no det.	
				Perthane	0.1	3	0	3	0	no det.	
				2,4-D	0.01	3	0	3	0	no det.	
				2,4-DP	0.01	3	0	3	0	no det.	
				2,4,5-T	0.01	3	0	3	0	no det.	
				Silvex	0.01	3	0	3	0	no det.	

Table 2.2. State and local monitoring studies reviewed—*Continued*

Reference(s)	Matrix	Sampling dates	Location(s)	Compounds	Detection limit(s) (µg/L)	Sites		Samples		
						Number of sites	Percent of sites with detections	Number of samples	Percent of samples with detections	Maximum concentration (µg/L)	Comments
Lewis, 1986	w	5/84–10/88	**Ohio:** Little Miami River, near Xenia	Toxaphene	0.1	nr	0	~30	0	no det.	Study of impact of wastewater effluent on water quality. Samples taken above and below municipal wastewater outfall. Diversity and abundance of periphyton and invertebrates also monitored. No positive detections of any pesticides.
				Dichlorvos	0.05	nr	0	~30	0	no det.	
				Phorate	0.05	nr	0	~30	0	no det.	
				Diazinon	0.05	nr	0	~30	0	no det.	
				M. parathion	0.05	nr	0	~30	0	no det.	
				Ronnel	0.05	nr	0	~30	0	no det.	
				Malathion	0.05	nr	0	~30	0	no det.	
				Parathion	0.05	nr	0	~30	0	no det.	
				DDE	0.05	nr	0	~30	0	no det.	
				DDD	0.05	nr	0	~30	0	no det.	
				DDT	0.05	nr	0	~30	0	no det.	
				Dieldrin	0.01	nr	0	~30	0	no det.	
				HCH (α)	0.01	nr	0	~30	0	no det.	
				HCH (β)	0.01	nr	0	~30	0	no det.	
				Lindane	0.01	nr	0	~30	0	no det.	
				HCB	0.01	nr	0	~30	0	no det.	
				Endrin	0.01	nr	0	~30	0	no det.	
				Mirex	0.1	nr	0	~30	0	no det.	
				Methox.	0.1	nr	0	~30	0	no det.	

Table 2.2. State and local monitoring studies reviewed—*Continued*

Reference(s)	Matrix	Sampling dates	Location(s)	Compounds	Detection limit(s) (µg/L)	Sites		Samples			Comments
						Number of sites	Percent of sites with detections	Number of samples	Percent of samples with detections	Maximum concentration (µg/L)	
Oliver and Kaiser, 1986	w s	4/89	**Michigan:** St. Clair River and tributaries	**HCB**	0.0001	21	100	21	100	0.09	The entire length of the St. Clair River was sampled to determine sources of agri-cultural and industrial chem-icals and to investigate sediment/water distribu-tions. Specific industrial discharges were identified as sources. Concentration data given here are for whole water samples.
				HCH (α)	0.0001	21	100	21	100	nr	
				Lindane	0.0001	21	100	21	100	nr	
				Chlordane	0.0001	21	nr	21	nr	nr	
				DDD	0.0001	21	nr	21	nr	nr	
				DDE	0.0001	21	nr	21	nr	nr	
				DDT	0.0001	21	nr	21	nr	nr	

Table 2.2. State and local monitoring studies reviewed—*Continued*

Reference(s)	Matrix	Sampling dates	Location(s)	Compounds	Detection limit (µg/L)	Sites: Number of sites	Percent of sites with detections	Samples: Number of samples	Percent of samples with detections	Maximum concentration (µg/L)	Comments
Biberhofer and Stevens, 1987	w	10/83	**Great Lakes:** Lake Ontario	**HCH-α**	nr	14	100	14	100	0.009	Survey of ambient concentrations of organochlorines in water column of Lake Ontario. Thirty-six liter samples enabled detection of very low levels. Samples taken at one meter depth. Eleven sites were within 10 km of shore. Ratios of parent compounds and degradation products were examined in some cases to determine the source to the lake.
				Lindane	nr	14	100	14	100	0.002	
				Chlordane (α)	nr	14	100	14	100	0.00005	
				Chlordane (γ)	nr	14	100	14	100	0.00006	
				Oxychlordane	nr	14	100	14	100	0.0003	
				Hept. epox.	nr	14	100	14	100	0.0004	
				Dieldrin	nr	14	100	14	100	0.0006	
				Endrin	nr	14	100	14	100	0.0001	
				Mirex	nr	14	0	14	0	no det.	
				Photomirex	nr	14	0	14	0	no det.	
				Methox.	nr	14	79	14	79	0.00009	
				DDT (total)	nr	14	100	14	100	0.0003	
				Toxaphene	nr	14	0	14	0	no det.	
				HCB	nr	14	100	14	100	0.0001	
Butler, 1987	w	1976–78	**Wyoming:** Twelve streams	**2,4,5-T**	0.01	20	15	99	nr	0.02	Sites selected to include most major drainage basins in state. Streams sampled in spring/early summer and autumn each year.
				2,4-D	0.01	20	65	99	29	1.2	
				2,4-DP	0.01	20	nr	99	nr	0.04	
				Aldrin	0.01	20	0	92	0	no det.	
				Chlordane	0.1	20	0	92	0	no det.	
				Chlorpyrifos	0.01	20	0	42	0	no det.	
				DDD	0.01	20	0	92	0	no det.	
				DDE	0.01	20	0	92	0	no det.	
				DDT	0.01	20	0	92	0	no det.	
				Diazinon	0.01	20	nr	93	nr	0.11	
				Dicamba	0.01	20	30	43	14	0.02	
				Dieldrin	0.01	20	0	92	0	no det.	
				Endosulfan	0.01	20	0	70	0	no det.	
				Endrin	0.01	20	0	92	0	no det.	
				Ethion	0.01	20	0	93	0	no det.	
				Heptachlor	0.01	20	0	92	0	no det.	

Table 2.2. State and local monitoring studies reviewed—*Continued*

Study				Compounds	Detection limit (µg/L)	Sites		Samples			Comments
Reference(s)	Matrix	Sampling dates	Location(s)			Number of sites	Percent of sites with detections	Number of samples	Percent of samples with detections	Maximum concentration (µg/L)	
Butler, 1987— *Continued*	w	1976–78	**Wyoming**— *Continued*	Hept. epox.	0.01	20	0	92	nr	no det.	
				Lindane	0.01	20	0	92	0	no det.	
				Malathion	0.01	20	nr	93	0	0.04	
				M. parathion	0.01	20	0	93	0	no det.	
				M. trithion	0.01	20	0	93	nr	no det.	
				Mirex	0.01	20	0	21	0	no det.	
				Parathion	0.01	20	nr	93	23	0.01	
				Perthane	0.1	20	0	75	0	no det.	
				Picloram	0.01	20	40	43	0	0.09	
				Silvex	0.01	20	0	99	0	no det.	
				Toxaphene	nr	20	0	92	0	no det.	
				Trithion	0.01	20	0	93	0	no det.	
Cooper and others, 1987	w	9/76– 9/79 (biweekly)	**Mississippi:** Bear Creek	**DDT**	0.01	nr	100	~200	nr	0.6	Highest concentrations following seasonal winter and spring rains. DDT concentrations have not declined since 1970, 1976 studies.
				DDD	0.01	nr	100	~200	nr	0.2	
				DDE	0.01	nr	100	~200	nr	0.1	
				Toxaphene	0.01	nr	100	~200	nr	1.07	

Table 2.2. State and local monitoring studies reviewed—*Continued*

Reference(s)	Matrix	Sampling dates	Location(s)	Compounds	Detection limit (µg/L)	Sites		Samples			Comments
						Number of sites	Percent of sites with detections	Number of samples	Percent of samples with detections	Maximum concentration (µg/L)	
Funk and others, 1987	w	1984–86	Washington: Lakes and rivers, North Cascades	HCH (α)	0.01	nr	0	nr	0	no det.	Study of water quality of relatively pristine lakes and rivers in the North Cascades. Levels of pesticides, metals, and inorganic constituents measured. Samples of water, sediments, and fish collected. No pesticides above reporting limit detected in water. Very low levels detected in some sediment and fish samples.
				Lindane	0.01	nr	0	nr	0	no det.	
				Heptachlor	0.01	nr	0	nr	0	no det.	
				Aldrin	0.01	nr	0	nr	0	no det.	
				DDE	0.01	nr	0	nr	0	no det.	
				Dieldrin	0.01	nr	0	nr	0	no det.	
				DDD (*o,p'*)	0.01	nr	0	nr	0	no det.	
				Endrin	0.01	nr	0	nr	0	no det.	
				DDD (*p,p'*)	0.01	nr	0	nr	0	no det.	
				Mirex	0.02	nr	0	nr	0	no det.	
				Methox.	0.02	nr	0	nr	0	no det.	
				Toxaphene	0.1	nr	0	nr	0	no det.	
				Chlordane	nr	nr	0	nr	0	no det.	
				Malathion	0.03	nr	0	nr	0	no det.	
				Parathion	0.03	nr	0	nr	0	no det.	
				Diazinon	0.03	nr	0	nr	0	no det.	
				Fenthion	0.05	nr	0	nr	0	no det.	
				M. parathion	0.03	nr	0	nr	0	no det.	
				Fensulfothion	0.05	nr	0	nr	0	no det.	
				Fenitrothion	0.05	nr	0	nr	0	no det.	
				Carbaryl	nr	nr	0	nr	0	no det.	
				Carbofuran	nr	nr	0	nr	0	no det.	
				Paraquat	10	nr	0	nr	0	no det.	
				Diquat	10	nr	0	nr	0	no det.	
				Atrazine	0.50	nr	0	nr	0	no det.	
				2,4-D	0.10	nr	0	nr	0	no det.	
				Silvex	0.10	nr	0	nr	0	no det.	
				2,4,5-T	0.10	nr	0	nr	0	no det.	

Table 2.2. State and local monitoring studies reviewed—*Continued*

Reference(s)	Matrix	Study Sampling dates	Study Location(s)	Compounds	Detection limit (µg/L)	Sites Number of sites	Sites Percent of sites with detections	Samples Number of samples	Samples Percent of samples with detections	Samples Maximum concentration (µg/L)	Comments
Lym and Messersmith, 1987	w	1985–86	North Dakota: Des Lacs, Souris, and Heart Rivers	Picloram	0.10	10	50	20	30	5.0	Survey of streams and ground water in areas where picloram is used. Samples collected in June and September both years.
Ward, 1987	w	2/77– 3/79	Pennsylvania: Seven sites in Pequea Creek Basin (tributary of Susquehanna River)	Aldrin	0.01	nr	42	100	nr	0.01	Data tabulated for baseflow and stormflow samples. Maximum concentrations refer to data from site at mouth of Pequea River only, with baseflow and stormflow data combined. Occurrence data from all seven sites, in baseflow and stormflow.
				Chlordane	0.1	nr	86	100	nr	0.2	
				DDD	0.01	nr	86	100	nr	0.01	
				DDE	0.01	nr	86	100	nr	0.01	
				DDT	0.01	nr	100	100	nr	0.04	
				Diazinon	0.01	nr	57	100	nr	0.08	
				Dieldrin	0.01	nr	86	100	nr	0.03	
				Endrin	0.01	nr	0	100	0	no det.	
				Endosulfan	0.01	nr	0	100	0	no det.	
				Ethion	0.01	nr	0	100	0	no det.	
				Heptachlor	0.01	nr	14	100	nr	0.01	
				Hept. epox.	0.01	nr	86	100	nr	0.02	
				Lindane	0.01	nr	86	100	nr	0.03	
				Malathion	0.01	nr	0	100	0	no det.	
				Methox.	0.01	nr	0	100	0	no det.	
				M. parathion	0.01	nr	14	100	0	no det.	
				M. trithion	0.01	nr	0	100	0	no det.	
				Mirex	0.01	nr	0	100	0	no det.	
				Parathion	0.01	nr	0	100	0	no det.	
				Perthane	0.1	nr	0	100	0	no det.	
				Toxaphene	nr	nr	0	100	0	no det.	
				Trithion	0.01	nr	0	100	0	no det.	
				2,4-D	0.01	nr	86	100	nr	1.2	
				2,4-DP	0.01	nr	0	100	0	no det.	
				2,4,5-T	0.01	nr	57	100	nr	0.04	
				Silvex	0.01	nr	43	100	nr	0.09	

Table 2.2. State and local monitoring studies reviewed—*Continued*

Reference(s)	Matrix	Sampling dates	Location(s)	Compounds	Detection limit (µg/L)	Number of sites	Percent of sites with detections	Number of samples	Percent of samples with detections	Maximum concentration (µg/L)	Comments
						Sites		**Samples**			
Ward, 1987— *Continued*	w	2/77– 3/79	Pennsylvania— *Continued*	Ametryn	0.1	nr	0	100	0	no det.	
				Atratone	0.1	nr	0	100	0	no det.	
				Atrazine	0.01	nr	100	100	nr	12	
				Cyanazine	0.1	nr	0	100	0	no det.	
				Cyprazine	0.1	nr	0	100	0	no det.	
				Prometone	0.1	nr	29	100	nr	0.7	
				Prometryn	0.1	nr	0	100	0	no det.	
				Propazine	0.1	nr	0	100	0	no det.	
				Simazine	nr	nr	86	100	nr	5.4	
				Simetone	0.1	nr	0	100	0	no det.	
				Simetryn	0.1	nr	0	100	0	no det.	
				Alachlor	0.1	nr	0	100	0	no det.	
Wnuk and others, 1987	w d	5/86– 7/86	**Iowa** 33 public water supplies using surface water sources; 14 of the surface water sources used for these public water supplies	Aldrin	0.04	33	0	33	0	no det.	Samples of treated water were collected after rain events to show pesticide residue levels affected by agricultural runoff. Concentration and occurrence data shown are for samples of treated water. In the samples collected from 14 surface water sources, the same compounds were present with similar concentration ranges, except for trifluralin and butylate, which were not observed.
				HCH (α)	0.04	33	0	33	0	no det.	
				HCH (β)	0.04	33	0	33	0	no det.	
				Lindane	0.04	33	0	33	0	no det.	
				Chlordane	0.02	33	0	33	0	no det.	
				DDD	0.04	33	0	33	0	no det.	
				DDE	0.04	33	0	33	0	no det.	
				DDT	0.04	33	0	33	0	no det.	
				Dieldrin	0.04	33	0	33	0	no det.	
				Endosulfan I	0.04	33	0	33	0	no det.	
				Endosulfan II	0.04	33	0	33	0	no det.	
				Endosulfan-sulfate	0.04	33	0	33	0	no det.	
				Endrin	0.04	33	0	33	0	no det.	
				Endrin-aldehyde	0.04	33	0	33	0	no det.	
				Heptachlor	0.04	33	0	33	0	no det.	
				Hept. epox.	0.04	33	0	33	0	no det.	
				Toxaphene	0.5	33	0	33	0	no det.	

Table 2.2. State and local monitoring studies reviewed—Continued

Reference(s)	Matrix	Sampling dates	Location(s)	Compounds	Detection limit (µg/L)	Sites		Samples			Comments
						Number of sites	Percent of sites with detections	Number of samples	Percent of samples with detections	Maximum concentration (µg/L)	
Wnuk and others, 1987—*Continued*	w d	5/86–7/86	**Iowa** 33 public water supplies using surface water sources; 14 of the surface water sources used for these public water supplies	Fonofos	0.1	33	0	33	0	no det.	
				Terbufos	0.1	33	0	33	0	no det.	
				Chlorpyrifos	0.1	33	0	33	0	no det.	
				Phorate	0.1	33	0	33	0	no det.	
				Ethoprop	0.1	33	0	33	0	no det.	
				Atrazine	0.1	33	91	33	91	24	
				Cyanazine	0.1	33	78	33	78	17	
				Alachlor	0.1	33	52	33	52	8.8	
				Trifluralin	0.1	33	nr	33	nr	0.13	
				Metribuzin	0.1	33	12	33	12	0.45	
				Metolachlor	0.1	33	64	33	64	21	
				Chloramben	0.1	30	0	30	0	no det.	
				Dicamba	0.1	30	nr	30	nr	1.4	
				2,4-D	0.1	30	nr	30	nr	0.3	
				Silvex	0.1	30	0	30	0	no det.	
				Butylate	0.1	33	nr	33	nr	0.27	
				Carbofuran	0.1	33	27	33	27	14	
				Sulprofos	0.1	33	0	33	0	no det.	
				Terbufos-sulfone	0.1	33	0	33	0	no det.	
Arruda and others, 1988	w	4/85–10/85 monthly	**Kansas:** Tuttle Creek Lake (on Big Blue River)	**Atrazine**	nr	nr	100	14	79	nr	Agricultural watershed in northeastern Kansas (249,000 km^2). Organochlorines detected in fish samples.
				Alachlor	nr	nr	100	14	64	nr	
				Metolachlor	nr	nr	100	14	57	nr	
				Propachlor	nr	nr	50	14	nr	nr	
				Metribuzin	nr	nr	50	14	nr	nr	
Dietrich and others, 1988	w	6/87	**North Carolina:** Haw River	**Atrazine**	nr	2	100	nr	100	2.8	Many other industrial chemicals measured. Primarily an analytical study.

Table 2.2. State and local monitoring studies reviewed—*Continued*

Reference(s)	Matrix	Location(s)	Sampling dates	Compounds	Detection limit (µg/L)	Number of sites	Percent of sites with detections	Number of samples	Percent of samples with detections	Maximum concentration (µg/L)	Comments
Fujii, 1988	w	California: Tulare Lake area agricultural drainage	3/85–9/85	Aldrin	0.01	1	0	2	0	no det.	USGS study evaluated water quality of agricultural drainage water proposed for diversion to Kern National Wildlife Refuge. Two to three samples analyzed for triazines (March, May, June). Two samples analyzed for organochlorines (August, September). Two to four samples analyzed for organophosphates (March-September). Samples taken at pumping station where water would be diverted.
				Chlordane	0.1	1	0	2	0	no det.	
				DDD	0.01	1	0	2	0	no det.	
				DDE	0.01	1	0	2	0	no det.	
				DDT	0.01	1	0	2	0	no det.	
				Dieldrin	0.01	1	0	2	0	no det.	
				Endosulfan	0.01	1	0	2	0	no det.	
				Endrin	0.01	1	0	2	0	no det.	
				Heptachlor	0.01	1	0	2	0	no det.	
				Hept. epox.	0.01	1	0	2	0	no det.	
				Lindane	0.01	1	0	2	0	no det.	
				Methox.	0.01	1	0	2	0	no det.	
				Mirex	0.01	1	0	2	0	no det.	
				Perthane	0.1	1	0	2	0	no det.	
				Toxaphene	nr	1	0	2	0	no det.	
				Diazinon	0.01	1	0	2-4	0	no det.	
				Ethion	0.01	1	0	2-4	0	no det.	
				Malathion	0.01	1	0	2-4	0	no det.	
				M. parathion	0.01	1	0	2-4	0	no det.	
				M. trithion	0.01	1	0	2-4	0	no det.	
				Parathion	0.01	1	0	2-4	0	no det.	
				Trithion	0.01	1	0	2-4	0	no det.	
				Ametryn	0.1	1	0	2-3	0	no det.	
				Atrazine	0.1	1	100	2-3	100	nr	
				Cyanazine	0.1	1	0	2-3	0	no det.	
				Prometone	0.1	1	100	2-3	100	0.2	
				Prometryn	0.1	1	0	2-3	0	no det.	
				Propazine	0.1	1	100	2-3	100	0.1	
				Simetryn	0.1	1	0	2-3	0	no det.	
				Simazine	0.1	1	100	2-3	100	0.1	

Table 2.2. State and local monitoring studies reviewed—Continued

Reference(s)	Matrix	Sampling dates	Location(s)	Compounds	Detection limit (µg/L)	Number of sites	Percent of sites with detections	Number of samples	Percent of samples with detections	Maximum concentration (µg/L)	Comments
Johnson and others, 1988	w	5/85– 10/85	Washington: Yakima River Basin	**DDT**	0.01– 0.02	14	21	56	18	0.03	Elevated concentrations seen during irrigation season.
				DDE	0.005– 0.01	14	29	56	20	0.03	
				DDD	0.01– 0.02	14	nr	56	nr	0.01	
Knapton and others, 1988	w	7/86– 8/86	Montana: Irrigation drainage, Sun River area (Benton National Wildlife Refuge)	**2,4-D**	0.01	3	67	3	67	0.22	USGS reconnaissance study. Pesticide samples taken at two creek sites and one lake site. Study focused primarily on sampling of sediments and biota.
				2,4-DP	0.01	3	0	3	0	no det.	
				2,4,5-T	0.01	3	0	3	0	no det.	
				Dicamba	0.01	3	67	3	67	0.08	
				Picloram	0.01	3	33	3	33	0.01	
				Silvex	0.01	3	0	3	0	no det.	
Lambing and others, 1988	w	6/86– 8/86	Montana: Lake Bowdoin National Wildlife Refuge	**2,4-D**	0.01	3	100	nr	100	0.08	USGS reconnaissance study. Study focused primarily on sampling of sediments and biota. Canal supplying irrigation water, irrigation return flow, and Lake Bowdoin were sampled. No substantial differences in pesticide concentration were noted between the supply water and the other two sites.
				2,4-DP	0.01	3	0	nr	0	no det.	
				2,4,5-T	0.01	3	0	nr	0	no det.	
				Dicamba	0.01	3	100	nr	100	0.03	
				Picloram	0.01	3	33	nr	33	0.01	
				Silvex	0.01	3	0	nr	0	no det.	

Table 2.2. State and local monitoring studies reviewed—*Continued*

Reference(s)	Matrix	Study		Sites			Samples			Comments
		Location(s)	Sampling dates	Detection limit (µg/L)	Number of sites	Percent of sites with detections	Number of samples	Percent of samples with detections	Maximum concentration (µg/L)	
Merriman, 1988	w f s	**Minnesota:** Rainy River	1986							Monitoring of a wide variety of hydrophobic organic compounds (PCBs, organochlorine pesticides, PAHs, chlorophenols, chlorobenzenes, dioxins) at three sites on the river and in two paper mill effluents. Data shown are for filtered water samples from the river. All pesticides were below detection limits (DL) in suspended sediments (DL=10 ng/g for pentachlorophenol and 4 ng/g for remainder.)
		Aldrin		0.0001	3	0	3	0	no det.	
		Chlordane (α)		0.0001	3	0	3	0	no det.	
		Chlordane (γ)		0.0001	3	0	3	0	no det.	
		DDD (p,p')		0.0001	3	0	3	0	no det.	
		DDE (p,p')		0.0001	3	0	3	0	no det.	
		DDT (o,p')		0.0001	3	0	3	0	no det.	
		DDT (p,p')		0.0001	3	0	3	0	no det.	
		Dieldrin		0.0001	3	0	3	0	no det.	
		Endrin		0.0001	3	0	3	0	no det.	
		HCH (α)		0.0001	3	0	3	0	no det.	
		HCH (β)		0.0001	3	0	3	0	no det.	
		Lindane		0.0001	3	0	3	0	no det.	
		Heptachlor		0.0001	3	0	3	0	no det.	
		Hept. epox.		0.0001	3	0	3	0	no det.	
		Methox.		0.0001	3	0	3	0	no det.	
		Mirex		0.0001	3	0	3	0	no det.	
		Endosulfan (α)		0.0001	3	0	3	0	no det.	
		Endosulfan (β)		0.0001	3	0	3	0	no det.	
		HCB		0.0001	3	67	3	67	0.0002	
		PCP		0.1000	3	0	3	0	no det.	

Table 2.2. State and local monitoring studies reviewed—Continued

Reference(s)	Matrix	Sampling dates	Location(s)	Compounds	Detection limit (µg/L)	Number of sites	Percent of sites with detections	Number of samples	Percent of samples with detections	Maximum concentration (µg/L)	Comments
Radtke and others, 1988	w	8/86	**Arizona, California, Nevada:** Irrigation drainage, lower Colorado River valley	Aldrin	0.01	11	0	14	0	no det.	USGS reconnaissance study to assess the impact of irrigation drainage. Samples of water, sediment, and biota were analyzed for metals, nutrients, pesticides, and other organic compounds.
				Chlordane	0.1	11	0	14	0	no det.	
				DDD	0.01	11	0	14	0	no det.	
				DDE	0.01	11	0	14	0	no det.	
				DDT	0.01	11	0	14	0	no det.	
				Dieldrin	0.01	11	0	14	0	no det.	
				Endosulfan	0.01	11	0	14	0	no det.	
				Endrin	0.01	11	0	14	0	no det.	
				Heptachlor	0.01	11	0	14	0	no det.	
				Hept. epox.	0.01	11	0	14	0	no det.	
				Lindane	0.01	11	0	14	0	no det.	
				Methox.	0.01	11	0	14	0	no det.	
				Mirex	0.01	11	0	14	0	no det.	
				Perthane	0.1	11	0	14	0	no det.	
				Toxaphene	nr	11	nr	14	0	no det.	
				Diazinon	nr	11	nr	14	nr	0.01	
				M. parathion	nr	11	nr	14	nr	0.05	
				Parathion	nr	11	nr	14	nr	0.11	
				Chlorpyrifos	nr	11	nr	14	nr	0.15	

Table 2.2. State and local monitoring studies reviewed—*Continued*

Reference(s)	Matrix	Sampling dates	Location(s)	Compounds	Detection limit (µg/L)	Number of sites	Percent of sites with detections	Number of samples	Percent of samples with detections	Maximum concentration (µg/L)	Comments
						Sites		**Samples**			
Schroeder and others, 1988	w	8/86	**California:** Irrigation drainage, Tulare Lake Bed area (Kern and Pixley National Wildlife Refuges)	Methomyl	nr	nr	0	nr	0	no det.	USGS reconnaissance study. Pesticides analyzed at 1 to 7 sites, depending on analyte. Sites included creeks, canals, and irrigation water evaporation ponds.
				Propham	nr	nr	0	nr	0	no det.	
				Sevin	nr	nr	0	nr	0	no det.	
				2,4-D	0.01	nr	20	nr	20	0.04	
				2,4-DP	0.01	nr	0	nr	0	no det.	
				Silvex	0.01	nr	0	nr	0	no det.	
				2,4,5-T	0.01	nr	0	nr	0	no det.	
				Diazinon	0.01	nr	57	nr	57	0.01	
				Disyston	0.01	nr	0	nr	0	no det.	
				Ethion	0.01	nr	0	nr	0	no det.	
				Azinphos-m.	0.1	nr	0	nr	0	no det.	
				Malathion	0.01	nr	0	nr	0	no det.	
				M. parathion	0.01	nr	0	nr	0	no det.	
				M. trithion	0.01	nr	0	nr	0	no det.	
				Parathion	0.01	nr	0	nr	0	no det.	
				Phorate	0.01	nr	0	nr	0	no det.	
				Trithion	0.01	nr	0	nr	0	no det.	
				Ametryn	0.1	nr	0	nr	0	no det.	
				Atrazine	0.1	nr	17	nr	17	0.4	
				Cyanazine	0.1	nr	0	nr	0	no det.	
				Cyprazine	0.1	nr	0	nr	0	no det.	
				Prometone	0.1	nr	0	nr	0	no det.	
				Prometryn	0.1	nr	67	nr	67	0.4	
				Propazine	0.1	nr	0	nr	0	no det.	
				Simazine	0.1	nr	0	nr	0	no det.	
				Simetone	0.1	nr	0	nr	0	no det.	
				Simetryn	0.1	nr	0	nr	0	no det.	

Table 2.2. State and local monitoring studies reviewed—Continued

Reference(s)	Matrix	Sampling dates	Location(s)	Compounds	Detection limit (µg/L)	Number of sites	Percent of sites with detections	Number of samples	Percent of samples with detections	Maximum concentration (µg/L)	Comments
Squillace and Engberg, 1988	w	5/84– 11/85	Iowa: Cedar River Basin	Alachlor	0.05– 0.1	6	100	17	30	21	Herbicide concentrations monitored at six sites in basin. Intensive corn growing region. Highest concentrations in spring/ early summer. Atrazine detected year round in snowmelt and ground water inputs. Herbicides primarily in dissolved phase.
				Atrazine	0.05– 0.1	6	100	17	94	16	
				Cyanazine	0.05– 0.1	6	100	17	34	8.7	
				Metolachlor	0.05– 0.1	6	100	17	41	11	
				Metribuzin	0.05– 0.1	6	100	17	12	nr	
				Trifluralin	0.05– 0.1	6	0	17	0	no det.	
Yurewicz and others, 1988	w	11/87– 3/88	Tennessee: Three tributaries of Reelfoot Lake	Alachlor	0.1	3	0	14	0	no det.	Atrazine detected at two of the three sites. Sampling period covers nongrowing season and period of low discharge.
				Ametryn	0.1	3	0	14	0	no det.	
				Atrazine	0.1	3	67	14	64	1.6	
				Cyanazine	0.1	3	0	14	0	no det.	
				Metolachlor	0.1	3	0	14	0	no det.	
				Metribuzin	0.1	3	0	14	0	no det.	
				Prometone	0.1	3	0	14	0	no det.	
				Prometryn	0.1	3	0	14	0	no det.	
				Propazine	0.1	3	0	14	0	no det.	
				Simazine	0.1	3	0	14	0	no det.	
				Trifluralin	0.1	3	0	14	0	no det.	

Table 2.2. State and local monitoring studies reviewed—*Continued*

Reference(s)	Matrix	Sampling dates	Location(s)	Compounds	Detection limit (µg/L)	Number of sites	Percent of sites with detections	Number of samples	Percent of samples with detections	Maximum concentration (µg/L)	Comments
						Sites		**Samples**			
Gonzalez and others, 1989	w	1987	**California:** Ten marinas on six freshwater lakes	**Tributyltin** **Dibutyltin**	0.017 0.033	10 10	80 10	22 22	68 18	1.22 0.69	Study to determine extent of tributyltin contamination in freshwater lakes. Marinas selected for study were the 10 largest in the state. Tributyltin residues detected in 4 of the 6 lakes (8 of the 10 marinas). Highest levels found in Lake Tahoe marina, which was subject of further study. Residues primarily found in marina; open water had nondetectable levels. Levels in the marina water exceeded chronic toxicity levels for aquatic organisms. Fish in open waters of Lake Tahoe had tributyltin levels of concern.

Table 2.2. State and local monitoring studies reviewed—*Continued*

Reference(s)	Matrix	Study Sampling dates	Location(s)	Compounds	Detection limit (µg/L)	Sites Number of sites	Sites Percent of sites with detections	Number of samples	Samples Percent of samples with detections	Samples Maximum concentration (µg/L)	Comments
Smith, 1989	w d	1987	**Virginia, Pennsylvania:** Roanoke River (5 sites); Lower Susquehanna River (4 sites)	**Alachlor**	0.005	9	nr	nr	nr	nr	Raw river water and treated drinking water analyzed for the presence of chemicals with possible health risks; 187 chemicals (metals, pesticides, volatile hydrocarbons, PAHs) were included in the study. Results are presented as assessments of the upper-bound lifetime cancer risk. No contaminants other than trihalomethanes contributed significantly to total cancer risks or hazard indices.
				Atrazine	0.001	9	nr	nr	nr	nr	
				Butylate	0.005	9	0	nr	0	no det.	
				Carbaryl	0.005	9	0	nr	0	no det.	
				Carbofuran	0.005	9	nr	nr	nr	nr	
				Cyanazine	0.001	9	nr	nr	nr	nr	
				Dinoseb	0.01	9	0	nr	0	no det.	
				HCB	10	9	0	nr	0	no det.	
				Metolachlor	0.001	9	nr	nr	nr	nr	
				Metribuzin	0.001	9	nr	nr	nr	nr	
				PCP	50	9	0	nr	0	no det.	
				Simazine	0.001	9	nr	nr	nr	nr	
				Trifluralin	0.001	9	nr	nr	nr	nr	

Table 2.2. State and local monitoring studies reviewed—*Continued*

Reference(s)	Matrix	Sampling dates	Location(s)	Compounds	Detection limit (µg/L)	Number of sites	Percent of sites with detections	Number of samples	Percent of samples with detections	Maximum concentration (µg/L)	Comments
Mayer and Elkins, 1990	w	1987–88	**Washington:** Padilla Bay in Puget Sound	Atrazine	0.49	4	0	16	0	no det.	Study of pesticide levels in Padilla Bay and in agricultural drainage sloughs discharging to bay. Water and sediment samples taken before and after application of pesticides on nearby agricultural fields in 1987 and 1988. Only dicamba and 2,4-D were detected, and only in 1987 after major rain event and after application. Concentrations of dicamba also detected in sloughs in summer 1987 sampling, ranging from 10 to 160 µg/L. Authors concluded that pesticides were not adversely affecting ecology of the bay.
				Chlorothalonil	0.15	4	0	16	0	no det.	
				Diazinon	0.18	4	0	16	nr	no det.	
				Dicamba	6.10	4	100	16	0	90	
				Dinoseb	0.11	4	0	16	0	no det.	
				Methamido-phos	10.80	4	0	16	0	no det.	
				M. parathion	0.03	4	0	16	0	no det.	
				Metribuzin	0.01	4	0	16	0	no det.	
				Parathion	0.06	4	0	16	0	no det.	
				PCNB	0.01	4	0	16	0	no det.	
				Simazine	0.63	4	0	16	0	no det.	
				Terbutryn	5.76	4	0	16	0	no det.	
				Trifluralin	0.02	4	0	16	0	no det.	
				2,4-D	0.1	4	25	16	nr	0.1	

Table 2.2. State and local monitoring studies reviewed—*Continued*

Reference(s)	Matrix	Sampling dates	Location(s)	Compounds	Detection limit (µg/L)	Number of sites	Percent of sites with detections	Number of samples	Percent of samples with detections	Maximum concentration (µg/L)	Comments
Moyer and Cross, 1990	w	10/85– 10/88	**Illinois:** Thirty sites on rivers throughout state	**Atrazine**	0.05	30	100	580	77	39	Summary of Illinois USEPA pesticide monitoring program results for 1985-88. Thirty stations on rivers in predominantly agricultural areas. Intensive monitoring of storm events at two stations. Detection rates for herbicides generally reflected use in the state. No insecticides were detected at any of the sites.
				Cyanazine	0.05	30	100	580	38	38	
				Metolachlor	0.05	30	100	580	46	17	
				Alachlor	0.02	30	100	580	46	18	
				Metribuzin	0.05	30	83	580	10	3.7	
				Butylate	0.05	30	nr	580	0.1	0.39	
				Trifluralin	0.01	30	43	580	nr	0.73	
				Terbufos	0.05	30	0	580	0	no det.	
				Chlorpyrifos	0.05	30	0	580	0	no det.	
				Fonofos	0.05	30	0	580	0	no det.	
				Malathion	0.05	30	0	580	0	no det.	
				M. parathion	0.05	30	0	580	0	no det.	
				Diazinon	0.05	30	0	580	0	no det.	
				Phorate	0.05	30	0	580	0	no det.	
				Captan	0.05	30	0	580	0	no det.	

Characteristics of Studies Reviewed 105

Table 2.2. State and local monitoring studies reviewed—*Continued*

| | | Study | | Compounds | Detection limit (µg/L) | Sites | | Number of samples | Samples | | Comments |
| | | | | | | | | | | | |
Reference(s)	Matrix	Sampling dates	Location(s)			Number of sites	Percent of sites with detections		Percent of samples with detections	Maximum concentration (µg/L)	
Petersen, 1990	w	1974–85	Arkansas: ~15 rivers and streams in northeastern Arkansas	**Aldrin**	0.002–0.01	38?	nr	nr	nr	0.01	Summary of data collected by several agencies on water quality of ~15 rivers, creeks, bayous, and bays in northeastern Arkansas. Analysis of trends for many constituents, but not for pesticides. Of the 38 sites where pesticides were targeted, dieldrin, endrin, DDT, methyl parathion, 2,4-D, and 2,4,5-T were detected at the most sites—22, 18, 13, 11, 8, and 6 sites, respectively. Maximum concentrations shown represent values from the entire data set, including all sites, and at any time during the 11-year period.
				Chlordane	0.1	38?	0	nr	0	no det.	
				DDD	0.01	38?	nr	nr	nr	0.01	
				DDE	0.002–0.01	38?	nr	nr	nr	0.02	
				DDT	0.004–0.01	38?	nr	nr	nr	0.08	
				Diazinon	0.01	38?	nr	nr	nr	0.02	
				Dieldrin	0.002–0.01	38?	nr	nr	nr	0.03	
				Endosulfan	0.01	38?	0	nr	0	no det.	
				Endrin	0.002–0.01	38?	nr	nr	nr	0.05	
				Ethion	0.01	38?	nr	nr	nr	0.01	
				Heptachlor	0.01	38?	nr	nr	nr	0.01	
				Hept. epox.	0.01	38?	nr	nr	nr	0.01	
				Lindane	?–0.01	38?	nr	nr	nr	0.01	
				Malathion	0.01–0.05	38?	0	nr	0	no det.	
				Methox.	0.01	38?	0	nr	0	no det.	
				M. parathion	0.01–0.04	38?	nr	nr	nr	0.52	
				M. trithion	0.01	38?	0	nr	0	no det.	
				Mirex	0.01	38?	0	nr	0	no det.	
				Parathion	0.01	38?	0	nr	0	no det.	
				Perthane	0.10	38?	nr	nr	nr	0.1	
				Toxaphene	1.0–2.0	38?	nr	nr	nr	nr	
				Trithion	0.01	38?	0	nr	0	no det.	
				2,4-D	?–0.01	38?	nr	nr	nr	0.59	
				2,4,5-T	0.01	38?	nr	nr	nr	2.8	
				Silvex	0.01	38?	nr	nr	nr	0.08	

Table 2.2. State and local monitoring studies reviewed—*Continued*

Reference(s)	Matrix	Study Sampling dates	Study Location(s)	Compounds	Sites Detection limit (μg/L)	Sites Number of sites	Sites Percent of sites with detections	Samples Number of samples	Samples Percent of samples with detections	Samples Maximum concentration (μg/L)	Comments
Setmire and others, 1990	w	10/86	**California:** Agricultural drainage in the Salton Sea area	2,4,5-T	0.01	nr	0	nr	0	no det.	USGS reconnaissance study. Sediments and biota also sampled.
				2,4-D	0.01	nr	75	nr	75	2.6	
				2,4-DP	0.01	nr	0	nr	0	no det.	
				Ametryn	0.01	nr	0	nr	0	no det.	
				Atrazine	0.01	nr	0	nr	0	no det.	
				Carbaryl	0.01	nr	0	nr	0	no det.	
				Cyanazine	0.01	nr	0	nr	0	no det.	
				DEF	0.01	nr	50	nr	50	0.06	
				Diazinon	0.01	nr	0	nr	0	no det.	
				Ethion	0.01	nr	0	nr	0	no det.	
				Malathion	0.01	nr	0	nr	0	no det.	
				Methomyl	0.01	nr	0	nr	0	no det.	
				M. parathion	0.01	nr	0	nr	0	no det.	
				M. trithion	0.01	nr	0	nr	0	no det.	
				Parathion	0.01	nr	0	nr	0	no det.	
				Prometone	0.01	nr	0	nr	0	no det.	
				Prometryn	0.01	nr	0	nr	0	no det.	
				Propazine	0.01	nr	0	nr	0	no det.	
				Propham	0.01	nr	0	nr	0	no det.	
				Silvex	0.01	nr	25	nr	25	0.04	
				Simazine	0.01	nr	0	nr	0	no det.	
				Simetryn	0.01	nr	0	nr	0	no det.	
				Trithion	0.01	nr	0	nr	0	no det.	

Table 2.2. State and local monitoring studies reviewed—*Continued*

Reference(s)	Matrix	Sampling dates	Location(s)	Compounds	Detection limit (µg/L)	Number of sites	Percent of sites with detections	Number of samples	Percent of samples with detections	Maximum concentration (µg/L)	Comments
Butler and others, 1991	w	7/88	**Colorado:** Gunnison River, Uncompahgre River, Sweitzer Lake	**2,4-D**	0.01	4	100	4	100	0.13	USGS reconnaissance study evaluating effects of irrigation drainage.
				2,4-DP	0.01	4	0	4	0	no det.	
				Silvex	0.01	4	0	4	0	no det.	
				2,4,5-T	0.01	4	0	4	0	no det.	
				Diazinon	0.01	4	0	4	0	no det.	
				Disyston	0.01	4	0	4	0	no det.	
				Ethion	0.01	4	0	4	0	no det.	
				Malathion	0.01	4	25	4	25	nr	
				M. parathion	0.01	4	50	4	50	nr	
				M. trithion	0.01	4	0	4	0	no det.	
				Parathion	0.01	4	100	4	100	0.33	
				Trithion	0.01	4	0	4	0	no det.	
				Aldicarb	0.5	4	0	4	0	no det.	
				Carbofuran	0.5	4	25	4	25	nr	
				Methomyl	0.5	4	0	4	0	no det.	
				Oxamyl	0.5	4	0	4	0	no det.	
				Propham	0.5	4	0	4	0	no det.	
				Carbaryl	0.5	4	0	4	0	no det.	
Cooper, 1991	w	1982–85	**Mississippi:** Moon Lake area	**DDT (total)**	nr	nr	100	~200	nr	nr	Samples collected biweekly for 2 years. Soils, sediments, and fish also analyzed. Currently used pesticides detected sporadically during study (27 total detections). DDT concentrations significantly greater during wet seasons, indicating that contaminated soil in runoff acts as a source of DDT to the lake.
				Toxaphene	nr	nr	100	~200	nr	nr	
				Fenvalerate	nr	nr	100	~200	nr	nr	
				Permethrin	nr	nr	50	~200	nr	0.13	
				M. parathion	nr	nr	100	~200	nr	0.49	

Table 2.2. State and local monitoring studies reviewed—*Continued*

Reference(s)	Matrix	Study			Sites			Samples			Comments
		Sampling dates	Location(s)	Compounds	Detection limit (µg/L)	Number of sites	Percent of sites with detections	Number of samples	Percent of samples with detections	Maximum concentration (µg/L)	
Domagalski and Kuivila, 1991	w	6/90	California: Sacramento River	**Carbofuran**	0.05	1	100	16	100	0.35	Concentrations of rice pesticides monitored in 45-mi stretch of river. Samples taken after release of irrigation water from rice fields. Rates of dissipation monitored by using Lagrangian method of sampling. Carbofuran and thiobencarb concentrations remained constant over the 96-hour period. Molinate concentrations decreased slightly. Methyl parathion was not detected and apparently dissipated before release of irrigation water from rice fields. Concentrations of transformation products of carbofuran and molinate remained fairly constant over sampling period.
				Molinate	0.1	1	100	16	100	13	
				Thiobencarb	0.025	1	100	16	100	0.07	
				M. parathion	0.1	1	0	16	0	no det.	
				Carbofuran-phenol	0.025	1	100	9	100	0.08	
				Paranitro-phenol (M. parathion degradate)	0.05	1	100	9	100	0.3	
				4-Keto-molinate	nr	1	100	16	nr	nr	
				2-Keto-molinate	nr	1	100	16	nr	nr	
Nicosia and others, 1991b	w	5/89–7/89	California: Sacramento River, Colusa Basin Drain, Sacramento Slough Drain	**Bensulfuron-methyl**	0.5	3	67	54	30	2.3	Study of dissipation of rice herbicide and levels discharged in paddy water and receiving waters. Compound detected in drains in late May and June, but never in the river, primarily because of dilution.

Table 2.2. State and local monitoring studies reviewed—Continued

Reference(s)	Matrix	Sampling dates	Location(s)	Compounds	Detection limit (µg/L)	Number of sites	Percent of sites with detections	Number of samples	Percent of samples with detections	Maximum concentration (µg/L)	Comments
Ruelle, 1991	w	6/90	**South Dakota:** Wetlands on five national wildlife refuges	**Atrazine**	0.1	16	50	16	50	2.1	Wetlands sampled at time of peak duckling populations. Each sample was a composite of subsamples from various points in wetland. Sediments also sampled. Authors conclude that there is no evidence that agricultural pesticides were having adverse impacts on waterfowl or water quality in these wetlands.
				Cyanazine	0.1	16	13	16	13	0.29	
				Metolachlor	0.1	16	13	16	13	0.65	
				Alachlor	0.1	16	19	16	19	1.3	
				Metribuzin	0.1	16	0	16	0	no det.	
				Butylate	0.1	16	0	16	0	no det.	
				Trifluralin	0.1	16	0	16	0	no det.	
				Terbufos	0.1	16	0	16	0	no det.	
				Fonofos	0.1	16	0	16	0	no det.	
				Chlorpyrifos	0.1	16	0	16	0	no det.	
				Ethoprop	0.1	16	0	16	0	no det.	
				Phorate	0.1	16	0	16	0	no det.	
				Carbofuran	0.1	16	0	16	0	no det.	
Lewis and others, 1992	w	1988–89	**Tennessee:** Three tributaries of Reelfoot Lake	**Alachlor**	0.1	3	100	93	46	45	Study of nutrient, suspended sediment, and pesticide inputs to Reelfoot Lake. High atrazine concentrations in some samples were believed to be short-term responses to storms immediately preceding sampling. Median concentrations of atrazine were 0.01 to 0.03 µg/L for the three sites. Most detections occurred during summer months.
				Atrazine	0.1	3	100	93	94	58	
				Cyanazine	0.1	3	100	92	14	1.2	
				Metolachlor	0.1	3	100	91	20	0.7	
				Metribuzin	0.1	3	67	92	9	0.7	
				Prometon	0.1	3	100	93	5	0.2	
				Prometryn	0.1	3	0	93	0	no det.	
				Propazine	0.1	3	100	93	10	0.6	
				Simatryn	0.1	3	0	93	0	no det.	
				Simazine	0.1	3	100	93	5	0.2	
				Trifluralin	0.1	3	100	93	17	1.6	

Table 2.2. State and local monitoring studies reviewed—*Continued*

Reference(s)	Matrix	Sampling dates	Location(s)	Compounds	Detection limit (μg/L)	Number of sites	Percent of sites with detections	Number of samples	Percent of samples with detections	Maximum concentration (μg/L)	Comments
Rinella and Schuler, 1992	w	7/88	**Oregon:** Malheur, Harney Lakes (Malheur National Wildlife Refuge) Irrigation drainage area	Alachlor	0.1	2	0	nr	0	no det.	USGS reconnaissance study. Water samples analyzed for pesticides at only two sites, representing the terminus of the hydrologic system.
				Ametryn	0.1	2	0	nr	0	no det.	
				Atrazine	0.1	2	0	nr	0	no det.	
				Cyanazine	0.1	2	0	nr	0	no det.	
				Metolachlor	0.1	2	0	nr	0	no det.	
				Metribuzin	0.1	2	0	nr	0	no det.	
				Prometone	0.1	2	0	nr	0	no det.	
				Prometryn	0.1	2	0	nr	0	no det.	
				Propazine	0.1	2	0	nr	0	no det.	
				Simazine	0.1	2	0	nr	0	no det.	
				Simetryn	0.1	2	0	nr	0	no det.	
				Trifluralin	0.1	2	0	nr	0	no det.	
				2,4-D	0.01	2	0	nr	0	no det.	
				2,4-DP	0.01	2	0	nr	0	no det.	
				2,4,5-T	0.01	2	0	nr	0	no det.	
				Silvex	0.01	2	0	nr	0	no det.	
				Dicamba	0.02	2	50	nr	nr	0.1	
				Picloram	0.01	2	0	nr	0	no det.	
Wang and others, 1992	w	3/90– 10/90	**Florida:** Indian River Lagoon	Lindane	0.01	nr	0	88	0	no det.	Study of organochlorine pesticide residues in 18 impoundments along the Indian River Lagoon. Residues detected in sediments and sediment interstitial water, but not in water column.
				Dieldrin	0.01	nr	0	88	0	no det.	
				DDE	0.01	nr	0	88	0	no det.	
				DDT	0.01	nr	0	88	0	no det.	

Table 2.2. State and local monitoring studies reviewed—*Continued*

Reference(s)	Matrix	Sampling dates	Location(s)	Compounds	Detection limit (µg/L)	Number of sites	Percent of sites with detections	Number of samples	Percent of samples with detections	Maximum concentration (µg/L)	Comments
Richards and Baker, 1993; Baker, 1988b	w	1983–91 1982–85	Indiana, Ohio, Michigan: Lake Erie tributaries	Atrazine	0.05	nr	100	~4,000	>50	226	Summaries of data from Lake Erie tributaries from 1982 to 1985, and 1983 to 1991. Mean concentrations and loads for each river are calculated for each year of the study. Peak concentrations seen in spring/early summer, shortly after application. Concentrations generally much higher for herbicides than insecticides. Atrazine residues were present for the longest time in all tributaries. Peak concentrations higher in smaller basins, but drop off more quickly. Maximum concentrations shown are primarily from rivers with very small watersheds and discharge. Detection frequency data from 1983 to 1991 summary. Range covers detection frequency at all sites. Phorate, EPTC, and butylate not quantified before 1986. Carbofuran, pendimethalin data from 1982 to 1985 summary. High maximum linuron value (160 µg/L) may have been due to spill.
				Alachlor	0.05	nr	100	~4,000	nr	75	
				Metolachlor	0.05	nr	100	~4,000	nr	96	
				Cyanazine	0.05	nr	100	~4,000	nr	86	
				Simazine	0.05	nr	100	~4,000	30–58	11	
				Carbofuran	0.2	nr	100	~4,000	nr	nr	
				Terbufos	0.01	nr	100	~4,000	0–10	2.3	
				Chlorpyrifos	0.02	nr	100	~4,000	<0.01–1	3.8	
				Phorate	0.01	nr	100	~4,000	<0.01–5	0.94	
				Linuron	0.2	nr	100	~4,000	nr	160	
				Metribuzin	0.1	nr	100	~4,000	nr	25	
				EPTC	0.05	nr	100	~4,000	6–24	21	
				Butylate	0.05	nr	100	~4,000	3–22	5.7	
				Fonofos	0.01	nr	88	~4,000	11–28	11.9	
				Pendimethalin	0.05	nr	100	~4,000	nr	3.7	

Table 2.2. State and local monitoring studies reviewed—*Continued*

Study					Sites			Samples		
Reference(s)	Matrix	Sampling dates	Location(s)	Compounds	Detection limit (µg/L)	Number of sites	Percent of sites with detections	Number of samples	Percent of samples with detections	Maximum concentration (µg/L)	Comments
Squillace and others, 1993	w	9/89– 10/89	**Iowa:** Cedar and Iowa Rivers and tributaries	**Atrazine**	0.05	32	100	nr	nr	0.58 (tributary)	Examination of the concentrations and sources of atrazine and DEA during low-flow periods. Tributaries contributed 25 percent, and alluvial ground water aquifers contributed 75 percent of both compounds to the mainstem river.
				Deethylatr.	0.05	32	100	nr	nr	0.54 (tributary)	
Kuivila and Foe, 1995	w	1/93– 2/93	**California:** Sacramento River, San Joaquin River, San Francisco Estuary	**Diazinon**	0.030	2	100	35	nr	1.07	USGS study of the transport and biological effects of dormant spray pesticides used on orchards. Data shown are from two sites on the Sacramento and San Joaquin Rivers. Distinct pulses of diazinon and methidathion observed in January and(or) February in the two rivers. Elevated concentrations detected in eastern San Francisco estuary. River water with highest diazinon concentrations was toxic to daphnia, with 100 percent mortality in 7-day bioassay tests.
				Methidathion	0.035	2	100	35	nr	0.21	
				Chlorpyrifos	0.040	2	50	35	nr	0.04	
				Malathion	0.035	2	0	35	0	no det.	

Table 2.3. Process and matrix distribution studies reviewed

[Matrix: w, whole (unfiltered) water; d, drinking water; f, filtered water; s, suspended sediments; m, surface microlayer. Abbreviations used in pesticide names: M. Azinphos-m., Azinphos-methyl; BSM, bensulfuron-methyl; Deethylatr., Deethylatrazine; Deisoatr., Deisopropylatrazine; Hept. epox., Heptachlor epoxide; M. parathion, Methyl parathion; NMF, N-methylformamide; OCs, organochlorine insecticides; PCB, polychlorinated biphenyl. α, alpha; β, beta; γ, gamma; δ, delta. USGS, U.S. Geological Survey. ppb, parts per billion; ppm, parts per million. ha, hectare; h, hour (s); kg, kilogram(s); kg/ha, kilogram(s) per hectare; km, kilometer(s); L, liter(s); mi, mile; min, minutes(s); pg/L, picogram(s) per liter; μg/L, microgram(s) per liter; <, less than; >, greater than; ?, number is uncertain; ~, number is approximate]

Reference(s)	Matrix	Sampling dates	Location(s)	Compounds	Comments
Johnson and others, 1966	w	7/65	**Wisconsin:** 8 lakes	Toxaphene	Study of toxaphene residues in lakes 3 to 9 years after treatment for rough fish control. Water concentrations ranged from 1 to 4 μg/L. Toxic toxaphene components degraded faster than other components. Toxaphene levels in lakes treated 3 to 9 years earlier were 1-4 ppb (water), 0.2-1 ppm (sediments), and 0.05-0.4 ppm (plants).
Sparr and others, 1966	w	1964–65	**Arkansas:** Field plots near Almyra, in White River watershed	Aldrin Dieldrin Endrin	Study of runoff from rice fields. Standing water, runoff water, river water, bed sediment, and fish were analyzed. Only trace levels detected in river water.
Hendrick and others, 1966	w	1963–64	**Louisiana:** Field plots near Crowley	Aldrin Dieldrin	Study of dissipation in rice paddy water, soils, and crawfish.
Cole and others, 1967	w	1965–66	**Pennsylvania:** Northern forest streams	DDT	Study of environmental distribution of DDT after aerial application to forest. Concentrations in water, sediments, fish, and soil were monitored for 380 days after application. Water concentrations up to 24 μg/L observed immediately after application, declining to 0.12 μg/L 14 days after application.
Marston and others, 1968	w	7/66	**Oregon:** Astoria area watershed	Amitrole	Study examined the environmental fate of amitrole after aerial application to clear-cut forested watershed. Amitrole concentrations in stream water decreased to below detectable levels within 6 days.
Trichell and others, 1968	w	1967 (year is uncertain)	**Texas:** Small agricultural plots	Dicamba 2,4,5-T Picloram	Runoff from agricultural plots studied. Effects of slope, rate of application, and movement over untreated soil or sod examined. Amounts in runoff from fallow and sod plots compared.
Marston and others, 1969	w	1/67 (1/23–1/29)	**Oregon:** Alsea River watershed	Endrin	Concentrations of endrin monitored in streamwater after aerial application of endrin-coated Douglas Fir seeds following clear-cutting. Total amount entering the stream during the 6-day period estimated as 0.12 percent of amount applied to watershed. Maximum concentration measured was 0.1 μg/L.

Table 2.3. Process and matrix distribution studies reviewed—*Continued*

Reference(s)	Matrix	Sampling dates	Location(s)	Compounds	Comments
Wojtalik and others, 1971	w	3/69–4/70	**Tennessee:** Guntersville Reservoir	2,4-D (Dimethylamine salt)	Water, sediment, biota concentration monitored after extensive spraying for Eurasian watermilfoil. Concentrations peaked ~8 h after spraying; declined to pretreatment levels within 1 month at three of four sites. No adverse effects noted. Detection limit was 1 µg/L.
Hall and others, 1972	f s	5/67–11/67 5/68–10/68	**Pennsylvania:** Agricultural plots	Atrazine	Study of runoff losses of atrazine applied at seven different rates. Losses in runoff water ranged from 1.7 to 3.6 percent (2.4 percent mean) of the amount applied for the different application rates. No correlation was seen between application rate and percentage lost in runoff water. Losses in runoff suspended sediment ranged from 0.03 to 0.28 percent (0.16 percent mean) of the amount applied with higher percentages lost at the higher application rates. Composite loss at the recommended rate (2.2 kg/ha) was 2.5 percent of the amount applied. First runoff event occurred 23 days after atrazine application, with runoff water concentrations ranging from 0.39 to 4.7 ppm. Study design maximized potential for runoff losses.
Borthwick and others, 1973	w	1969–71	**South Carolina:** Estuaries	Mirex	Mirex aerially applied to coastal areas to control fire ants. Study examines movement up the food chain and in water and bed sediment samples. No mirex was detected in 4-L water samples.
Ginn and Fisher, 1974	w	1971	**Texas:** Drainage from coastal prairie and marshland entering Galveston Bay	Aldrin Dieldrin	Distribution and fate of aldrin studied after application to rice. Aldrin quickly converted to dieldrin, both in water and aquatic organisms. Dieldrin concentrations in estuarine water declined from 0.17 µg/L after 4 days to 0.03 µg/L after 15 days. Dieldrin concentrations in biota persisted at low levels for duration of study (several months).
Moore and others, 1974	w	1967–68	**Oregon:** Forest streams	Endrin	Study of movement of endrin used as seed coating after aerial seeding of forested watersheds. Endrin detected in stream at low levels (~0.005-0.01 µg/L) up to 20 days after application. Authors conclude that aerial application of endrin-coated seed did not constitute a hazard to aquatic habitats. Use of buffer strips to avoid direct application to streams was stressed.

Table 2.3. Process and matrix distribution studies reviewed—*Continued*

Reference(s)	Matrix	Sampling dates	Location(s)	Compounds	Comments
Ritter and others, 1974	w s	1967–70	**Iowa:** Agricultural plots	Atrazine Propachlor Diazinon	Study of runoff losses of pesticides from surface-contoured and ridged watersheds (field plots). Ridge planting of corn greatly reduced losses of atrazine and propachlor. No diazinon was detected in runoff water or sediment owing to incorporation in soil and rapid degradation. Up to 15 percent of the applied atrazine and 2.5 percent of the applied propachlor were lost in runoff water and sediment in a runoff event 7 to 8 days after application. Possible interference from hydroxyatrazine in the analysis of atrazine noted.
Schultz and Whitney, 1974	w	4/71–10/71	**Florida:** Canal in Loxahatchee National Wildlife Refuge	2,4-D (Dodecyl-tetradecyl-amine and dimethyl-amine salts)	Monitored effects of spraying program to control water hyacinth. Water concentrations ranged from <1 to 16 µg/L. Concentrations decreased over 6-month period. Very low levels in fish and sediment samples.
Norris and Montgomery, 1975	w	1971–72	**Oregon:** Farmer Creek watershed	Dicamba	Study of movement of dicamba in forest streams after aerial application. Concentrations up to 38 µg/L observed at three sites within 35 h of application. Less than 0.1 percent of the amount applied appeared in stream.
Mauck and others, 1976	w	1971	**Missouri:** Five small ponds near Columbia	Simazine	Study of persistence of simazine used as a preemergent aquatic herbicide in experimental ponds. Detectable levels were present in water, bed sediments, and biota 456 days after application. No adverse effects on invertebrates or fish were noted.
Schultz and Gangstad, 1976	w	1971	**Florida, Georgia:** Seven ponds	2,4-D (Dimethyl-amine salt)	Study of dissipation and biotic uptake of 2,4-D when applied as an aquatic herbicide for control of Eurasian watermilfoil and water hyacinth. Water concentrations declined to near or below the detection limit within 14 days.
Truhlar and Reed, 1976	w	2/69–4/71	**Pennsylvania:** Four small drainage basins with differing land uses	DDD DDE DDT Dieldrin Endrin Lindane 2,4-D 2,4,5-T Silvex	Study of effects of differing land uses (residential, forest, general farming, orchard) on water quality of streams receiving runoff. More than 80 water samples were collected, 20 percent during base flow and 80 percent during stormflow periods. DDT and metabolites were the most commonly detected pesticides; highest levels occurred in the residential and orchard basins, even though DDT had not been used in the residential and orchard basins for several years. Residues in the forest and farming basins were infrequent and low.

Table 2.3. Process and matrix distribution studies reviewed—*Continued*

Reference(s)	Matrix	Sampling dates	Location(s)	Compounds	Comments
Pierce and others, 1977	w	1975–76	**Mississippi:** Small lake near Hattiesburg	PCP	Effects of a PCP spill into the lake monitored over a 15-month period. PCP accumulated in sediments, but measurable concentrations of PCP remained in water column throughout the study period. Long-term, low-level contamination attributed to influx from contaminated areas in watershed.
Eisenreich and others, 1978	w m	10/73–12/74	**Wisconsin:** Lake Mendota lake foam	Dieldrin DDE DDD DDT	Study of concentration of organochlorines and other compounds in wind-generated lake foam. Concentrations of these compounds were much higher in the foam than in underlying water. Average concentrations in foam (4 samples) were 0.044 µg/L for dieldrin and ~0.2 µg/L for DDT compounds. Lake water concentrations were <0.001 µg/L.
Baker and others, 1979	w s	5/76–6/76	**Iowa:** Four Mile Creek basin	Alachlor Metribuzin Propachlor Cyanazine Paraquat	Study of pesticide losses in runoff from agricultural fields. Paraquat transported in adsorbed phase. Others primarily in dissolved phase. Flux calculations imply attenuation occurring between field and stream for all compounds. Loss as percent of applied compound from plots: alachlor–0.48 percent, metribuzin–0.72 percent, propachlor–0.21 percent, cyanazine–0.96 percent, paraquat–3.34 percent (all applied to bare ground, nonincorporated).
Leonard and others, 1979	f s	7/72–10/75	**Georgia:** Agricultural plots	Atrazine Cyanazine Diphenamid Paraquat Propazine Trifluralin	Study of runoff losses and persistence of herbicides applied to field plots (sandy loam soils). Paraquat used primarily as a tracer of sediment transport; application was atypical. Total losses were usually <2 percent of the amount applied. Large losses (7.2 and 6.7 percent) for diphenamid and propazine were observed when application was closely followed by intense rainfall. Incorporated trifluralin losses ranged from 0.1 to 0.3 percent of the amount applied. All compounds (except paraquat) were transported primarily in the solution phase (83–100 percent). Herbicide concentrations in runoff were related to soil surface concentrations by a nonlinear power function.
Neary and others, 1979	w	1978	**North Carolina:** Forest stream	Picloram	Residues in streamflow monitored after application (5 kg active ingredient/ha). Two pulses (2–8 µg/L) observed in streamflow 1 and 2 months after application. Drought during study may have decreased amount lost to stream.

Table 2.3. Process and matrix distribution studies reviewed—*Continued*

Reference(s)	Matrix	Sampling dates	Location(s)	Compounds	Comments
Pieper, 1979	w	1975	**Montana:** Four streams in Beaverhead National Forest	Carbaryl	Evaluation of carbaryl concentrations in streams after aerial application to nearby forests. Concentrations ranged from 2 to 260 µg/L during the first 3 h after application.
West and others, 1979	w	1976–78	**Michigan, New York, Florida:** Ponds at Ithaca, New York; Orlando, Florida; Lake City, Michigan; Gatun Lake, **Panama**	Fluridone	Study of the environmental fate of fluridone (aquatic herbicide) applied to four lakes/ponds. A 5-day half-life was observed, with losses attributed to uptake by plants, deposition in sediments, and photolysis. A low potential for bioaccumulation was observed.
Johnsen and Warskow, 1980	w	1970's	**Arizona:** Small stream (Tangle Creek)	Picloram	Study of dissipation of picloram directly injected into a stream.
Miller and Bace, 1980	w	1970's	**Mississippi:** Piedmont stream	Hexazinone	Surface water monitoring after aerial application of hexazinone pellets (0.8 kg/ha). Concentrations observed after application: 2,400 µg/L (30 min), 1,100 µg/L (1 h), 490 µg/L (2 h), <20 µg/L (5 days). Contamination is due to direct deposition to stream.
Peterman and others, 1980	w	1976–77	**Wisconsin:** Fox River	DDT Dieldrin HCH Chlordane	Semiquantitative evaluation of organochlorine contamination of water, sediments, and biota. The study analyzed 250 samples and identified 105 compounds.
Rohde and others, 1980	w	1974–75	**Georgia:** Southeastern Coastal Plain	Trifluralin	Runoff monitored from small agricultural watershed. Total losses were 0.17 percent and 0.03 percent of the applied amount in 1974 and 1975, respectively. Highest concentrations occurred after first rainfall after application. Results imply that use of buffer strips will reduce trifluralin movement to surface waters.
Schroeder and Sturges, 1980	w	5/76	**Wyoming:** Loco and Sane Creeks, near Saratoga	2,4-D	Monitoring study of concentrations of 2,4-D after aerial spraying for control of big sagebrush. A buffer strip of 30 meters helped to minimize contamination from spray drift. Maximum concentration of 2,4-D was 5 µg/L, which decreased to <1 µg/L within a few days.

Table 2.3. Process and matrix distribution studies reviewed—Continued

Reference(s)	Matrix	Sampling dates	Location(s)	Compounds	Comments
Stanley and Trial, 1980	w	1979	**Maine:** 18 streams	Carbaryl	Carbaryl concentrations were monitored in the water of 18 streams after aerial application of nearby forests. Peak concentrations ranged from 0.4 to 16 µg/L. Concentrations rapidly decreased with rate constants of 0.005 to 0.068 per h. The disappearance rate represents the rate at which carbaryl was flushed from the system, as well as disappearance caused by degradation or volatilization.
Carroll and others, 1981	w	1976–77 1977–78	**Louisiana:** Agricultural plots	Permethrin (cis- and trans-)	Study of runoff losses and persistence of permethrin under field conditions (cotton). In 1976 (low rainfall year), runoff concentrations were all <0.2 µg/L, and total runoff losses were <0.01 percent. In 1977 (high rainfall year), concentrations exceeded 0.39 µg/L in 3 of 15 events, and total runoff losses were <1 percent. Authors conclude that even under severe runoff conditions, runoff concentrations were not high enough to be harmful to aquatic species.
McDowell and others, 1981	w f	1973–75	**Mississippi**	Toxaphene	Concentrations of toxaphene in runoff from cotton fields monitored over three growing seasons. Ninety-three percent of total lost in runoff associated with suspended sediment in 1975. Total losses were estimated as 1 percent (1974) and 0.5 percent (1975) of the amount applied in the watershed.
Morris and Jarman, 1981	w	1977–78	**Oklahoma:** Kerr Lake	2,4-D (butoxyethyl ester) 2,4-Dichlorophenol	Monitoring study after direct application of 2,4-D ester for control of Eurasian watermilfoil. 2,4-D ester detected in only 2 of 240 samples, 2,4-dichlorophenol (degradation product) was not detected in 158 samples. Detection limit for 2,4-D ester is <3 µg/L.
Rohde and others, 1981	w	1974–75	**Georgia:** Agricultural plots	Atrazine	Study of persistence of atrazine in soil and movement in surface runoff and subsurface drains for different application rates. Very little atrazine in surface runoff if no runoff occurs within 26 days of application. No atrazine was detected in water flowing in subsurface drains (no detection limit given). Surface runoff losses accounted for 0.2 to 0.3 percent of the amount of applied atrazine in 1974 and 1.1 to 1.6 percent of the amount of applied atrazine in 1975.
Gangstad, 1982	w	1970's	Various large reservoirs	2,4-D	Analysis of data from four reservoirs after application of 2,4-D for control of Eurasian watermilfoil. Author suggests that use of 2,4-D in reservoirs at normal rates will not result in concentrations above the established criteria of 100 µg/L.

Table 2.3. Process and matrix distribution studies reviewed—*Continued*

Reference(s)	Matrix	Sampling dates	Location(s)	Compounds	Comments
Gold and Loudon, 1982	w	3/81–11/82	**Michigan:** Agricultural plots	Atrazine	Comparison of losses of nutrients and atrazine from tile-drained fields under conservation and conventional tillage. Overall atrazine losses for the two systems were 1.0 percent (conservation) and 2.2 percent (conventional) of the amount applied.
Jaffe and others, 1982	f s	9/71–1/72	**Tennessee:** Wolf River near Memphis	HCB Heptachlor Aldrin Chlordane Dieldrin Endrin	Study of distribution of organochlorines in river. Vertical and horizontal distribution in the water column determined. Distribution between water and suspended sediment also determined.
Mayack and others, 1982	w	4/79–1/80	**Georgia:** Stream draining forested watershed	Hexazinone (plus two metabolites)	Study of impact of hexazinone on stream and forest invertebrates after application to a forested watershed. Less than 5 percent of the watershed represented by the sampling point was treated with hexazinone. Hexazinone residues in streamflow ranged from <1 to 44 µg/L. Peaks in residues were infrequent and of short duration, partly attributed to dilution with water from untreated portions of the watershed. No accumulation of hexazinone was observed in stream organisms, nor were there observable changes in species composition or diversity.
Norris and others, 1982	w	9/71–1/72	**Oregon:** Boyer Ranch stream, near Roseburg	Silvex Picloram 2,4-D	Study of the environmental fate and distribution of herbicides used for brush control on hilly pastures. Picloram and 2,4-D discharge in streamflow represented 0.35 and 0.014 percent of the amount applied, respectively.
Stoltz and Pollock, 1982	w	7/80–9/80	**Idaho:** Irrigation canals	Methoxychlor	Methoxychlor added directly to canal for control of black flies. Treatment resulted in initial concentration of 300 µg/L. Samples taken 75 mi downstream; maximum concentration of 1.4 µg/L. Authors conclude that treatment poses no problems for fish in receiving river (Snake River).

Table 2.3. Process and matrix distribution studies reviewed—Continued

Reference(s)	Matrix	Sampling dates	Location(s)	Compounds	Comments
Hickman and others, 1983	w s	1980–81	**Oregon:** Runoff from agricultural watersheds in western Oregon	Diclofop-methyl Trifluralin	Study of runoff from agricultural fields in a high-winter-rainfall zone. Losses of diclofop-methyl were much higher in 1980 (3.9–7.1 percent of amount applied) than in 1981 (<0.1–0.7 percent). Greater loss in 1980 attributed to timing of first runoff event with respect to application, higher soil moisture in 1980, differences in magnitude of storm events, and soil crust formation in 1980. Loss of trifluralin in runoff represented 0.9 percent at site with no subsurface drainage and <0.1 percent at site with subsurface drainage. Relatively high trifluralin loss (0.9 percent) attributed to rainfall within 7 h of application. Subsurface drainage reduced runoff losses for both compounds.
Hoeppel and Westerdahl, 1983	w	1981	**Georgia:** Lake Seminole near Reynoldsville	2,4-D (DMA) 2,4-D (BEE) 2,4-dichlorophenol Dimethylnitros-amine	Study of dissipation of two forms of 2,4-D directly applied to lake for control of Eurasian watermilfoil. Concentrations in water, sediment, and fish were measured at four sites in the lake. Water concentrations at all four sites were below the detection limit (10 µg/L) 13 days after application. Concentrations of degradation products were below the detection limit (10 µg/L) at all sites throughout the study.
Lietman and others, 1983	w f	5/79–12/80	**Pennsylvania:** Six sites in Pequea Creek Basin	Ametryn Atrazine Atratone Cyanazine Cyprazine Prometryn Prometone Propazine Simazine Simetone Simetryn Alachlor	USGS study of effects of differing land uses (residential, forest, corn cultivation, pasture) on water quality of streams receiving runoff.
Neary and others, 1983	w	4/79–5/80	**Georgia:** Four forested watersheds	Hexazinone	Residues in runoff water monitored for 26 storm events throughout year. Concentrations in runoff highest in first rainfall after application (average concentration was 442 µg/L). Concentrations declined rapidly after 1 month (4 storms). Total losses averaged 0.53 percent of the amount applied.

Table 2.3. Process and matrix distribution studies reviewed—*Continued*

Reference(s)	Matrix	Sampling dates	Location(s)	Compounds	Comments
O'Connor and others, 1983	w	1966–78	**Virginia**: James River estuary	Kepone	Study of distribution of kepone in estuarine system. Sediment also analyzed. A model of the estuarine system is developed, and a comparison is made between the observed distribution and model predictions. Discharge of kepone to Chesapeake Bay is shown to have ceased with termination of kepone production.
Smith and others, 1983	w	1980–81	**Louisiana**: Two sugar cane plots	Azinphos-m. Fenvalerate	Runoff from plots monitored after rain events. Concentrations of fenvalerate exceeded acutely toxic levels for aquatic organisms in some cases.
Wu and others, 1983		1976–78	**Maryland**: Rhode River watershed	Atrazine Alachlor	Study of runoff from eight watersheds (16–253 ha). Atrazine loading represented from 0.05 to 2 percent of amount applied in basins. Alachlor loading was <0.1 percent of amount applied. Concentrations in runoff were much lower than concentrations reported in previous studies of smaller field plots.
Glotfelty and others, 1984	w	5/80–10/80 4/81–10/81 5/82–10/82	**Chesapeake Bay**: Wye River estuary	Atrazine Simazine	Study of herbicide movement from fields to estuary. Studies of runoff from typical field and adsorption to sediment included. Simazine and atrazine detected in all Wye River samples. Highest concentrations occurred after critical rain events. Concentration ranges: atrazine, 0.9 to 14.6 µg/L; simazine, 0.1 to 1.5 µg/L. Runoff effectively ceased 4 to 6 weeks after application.
Mayeux and others, 1984	w	4/79	**Texas**: Rangeland watersheds near Riesel	Picloram	Study of runoff losses of picloram after application to Bermudagrass covered watershed. About 6 percent of the amount applied was detected in the stream receiving runoff.
Nutter and others, 1984	w	4/79–9/80	**Georgia**: Forested watersheds	Hexazinone	Evaluation of CREAMS model predictions of hexazinone movement in forested watersheds. Model underpredicted stream residues after ~75 days, possibly because the model did not account for movement within the soil. Measured concentrations of hexazinone in the stream remained at 10 to 20 µg/L for ~275 days.

Table 2.3. Process and matrix distribution studies reviewed—*Continued*

Reference(s)	Matrix	Sampling dates	Location(s)	Compounds	Comments
Oliver and Charlton, 1984	s w	5/81–11/81 5/82–11/82	**Great Lakes:** Lake Ontario Niagara River vicinity	HCH (α) Lindane DDE DDT Mirex Chlordane (α) Chlordane (γ)	Study of organochlorine concentrations in settling particulates in Lake Ontario, and relations with Lake Ontario bed sediment concentrations and Niagara River water concentrations. For HCH and chlordane, only a small percentage (<10 percent) of the amount entering from the Niagara River appears in the settling particulates. Higher amounts of mirex (>35 percent of input from Niagara River), DDT (>32 percent), and DDE (19 percent) appear in the settling particulates. A general lack of agreement was observed between measured field partition coefficients and coefficients derived from semiempirical equations on the basis of physical and chemical properties of the compounds.
Baker and others, 1985	w s	5/83–10/83	**Great Lakes:** Lake Superior	DDE HCB	Study of benthic nepheloid layer in western Lake Superior and its effects on concentrations and residence time of organochlorines in the lake. DDE found to be an effective tracer of sediment resuspension. Variations in relative amounts of PCB congeners present in the nepheloid layer indicate that loss of organochlorines to bottom sediments is not necessarily a permanent sink, but rather a selective removal process.

Table 2.3. Process and matrix distribution studies reviewed—*Continued*

Reference(s)	Matrix	Sampling dates	Location(s)	Compounds	Comments
El-Shaarawi and others, 1985	w	1978–84	**New York:** Great Lakes, Niagara River	Aldrin DDD (p,p') DDE (p,p') DDT (o,p') DDT (p,p') Dieldrin Endosulfan (α) Endosulfan (β) Endosulfan sulfate Endrin HCB HCH (α) HCH (β) Lindane Heptachlor Hept. epox. Mirex Oxychlordane	Statistical evaluation and loading estimates for organics and trace elements from existing data. Significantly higher concentrations were observed in the lower river (Lake Ontario end) compared to the upper river (Lake Erie end) for a number of compounds.
Kaiser and others, 1985; Platford and others, 1985	f s m	5/83–6/83	**Great Lakes:** Detroit River	HCB DDE	Studies investigated distribution of hydrophobic organochlorines among the filtered water, dissolved organic carbon, and surface microlayer phases. Also investigated distribution between bottom sediments and sediment porewater.
Neary and others, 1985	w	5/78–9/79	**North Carolina:** Forest stream	Picloram	Study of movement of picloram manually broadcast as pellets in a mixed-oak forest. Stream concentrations of picloram monitored for ~18 months after application. Picloram detected sporadically at low levels, although maximum concentration was detected in small stream ~14 months after application (10 µg/L). Authors conclude that the observed levels would cause no adverse affects on water quality.
Cunningham and Myers, 1986	w	1981	**Florida:** Saltmarsh near Cocoa Beach	Diflubenzuron 4-Chlorophenyl-urea (degradation product)	Field study examining dissipation of compound and degradation product in a mosquito breeding supratidal lagoon. Measurable concentrations still existed in water and bed sediments after 14 days.

Table 2.3. Process and matrix distribution studies reviewed—Continued

Reference(s)	Matrix	Sampling dates	Location(s)	Compounds	Comments
Lewis and others, 1986	w	6/84–11/84 1/85–2/85 2/86	**Georgia:** Chalet Stream	2,4-D (BEE)	Study of seasonal effects on microbial transformation rates.
Neary and others, 1986	w	4/79–5/80	**Georgia:** Four 1-ha watersheds in northern Georgia piedmont	Hexazinone	Study of runoff (stormflow) concentrations after treatment with hexazinone. Concentrations peaked in first storm event after application (442 µg/L), then declined to undetectable levels within 7 months.
Yoo and others, 1986	w	4/85–10/85	**Alabama:** Agricultural plots (cotton)	Aldicarb Pendimethalin Paraquat	Study of runoff losses with three different tillage practices. Pendimethalin concentration in runoff highest (14 µg/L) in first rain event, decreased rapidly in subsequent events. Aldicarb detected in first event only (10–51 µg/L). No paraquat detected in runoff.
Hardy and others, 1987	m	1985	**Washington:** Puget Sound	HCH (α) Lindane HCH (δ) DDT Aldrin (and other OCs)	Study of the occurrence of organic compounds and trace elements in the surface microlayer of Puget Sound and toxicity to fish eggs. About 28 percent of the samples contained quantifiable organochlorine pesticides. Compounds listed were those detected.
LeBel and others, 1987	w d	1982–83	**Great Lakes:** Four sites in Great Lakes used for drinking water supplies	HCH (α)	Raw and treated water analyzed as part of a validation study of large XAD columns capable of analyzing 1,500 liter samples. Other industrial organic compounds analyzed as well. HCH detected at nanograms per liter levels in raw and treated water at all sites.
Lum and others, 1987	s	1981–86	**Great Lakes:** Niagara River, St. Lawrence River	Mirex	Study of the relative importance of migrating eels and suspended sediments in the transport of mirex from Lake Ontario to the St. Lawrence River estuary. Transport of mirex out of the lake by migrating eels is estimated to be almost twice that because of suspended particulate flux.
Wang and others, 1987b	w	9/84–6/85	**Florida:** Indian River Lagoon	Fenthion	Study of deposition and persistence of fenthion aerially applied for mosquito control near the estuary. Concentrations in estuary declined to <0.01 µg/L 24 h after application in four separate tests. Tidal flushing may have transported some of the residue away from the test site.

Table 2.3. Process and matrix distribution studies reviewed—*Continued*

Reference(s)	Matrix	Sampling dates	Location(s)	Compounds	Comments
Wang and others, 1987a	w	1984–85	**Florida:** Salt marsh, Indian River County	Fenthion Malathion Dibrom	Study of concentrations and dissipation rates following aerial application for mosquito control. Concentrations decreased to less than the detection limit (0.01–0.05 µg/L) within 48 h. Only 0.1 to 0.4 percent of applied fenthion deposited on water surface. Approximately 19 percent of applied malathion and 93 percent of applied dibrom reached water surface.
Deubert and Kaczmarek, 1989	f	1987–88 weekly	**Massachusetts:** Cranberry bogs and drainage	Parathion	Study of residues of parathion released from cranberry bogs. More than 90 percent of the initial amounts disappear from ditch water within 3 to 4 days after application. Residues in water at the outlets of watersheds downgradient from the bogs (0.2–0.9 µg/L, mean 0.39) were lower than in the bogs themselves (0.1–5.8 µg/L, mean 1.1).
Lau and others, 1989	f s	1/86	**Great Lakes:** St. Clair and Detroit Rivers	HCB Organochlorines	Study of relative amounts of chlorinated contaminants transported by water, suspended sediments, and bed sediments. Results indicate that suspended sediments can transport a significant portion, especially near sources of industrial discharges.
Lavy and others, 1989	w	5/84–5/86	**West Virginia:** Forested watershed	Hexazinone	Study of movement of hexazinone in a forested watershed and resulting stream concentrations. Automatic samplers enabled daily water sampling from the stream. Stream water concentrations ranged from ~0.1 to 16 µg/L during the 24-month period. Approximately 1.8 percent of applied hexazinone was detected in stream water during the first year and 1.2 percent in the second year. Mass entering the stream and stream concentrations 20 to 24 months after application were comparable to mass and concentrations measured 6 to 12 months after application. Concentrations remained well below levels known to affect fish throughout study.
Michael and others, 1989	w	4/81–1/82	**Alabama:** Forested coastal plain area, east-central Alabama	Picloram	Study of movement of picloram after aerial application to forested watershed. Soils, forest runoff, and stream water were analyzed. Picloram observed in stream water after 200 days; concentrations ranged from <2 to 241 µg/L.

Table 2.3. Process and matrix distribution studies reviewed—*Continued*

Reference(s)	Matrix	Sampling dates	Location(s)	Compounds	Comments
Miltner and others, 1989	w	1984	**Ohio:** Influent water to treatment plants on the Maumee and Sandusky Rivers	Atrazine Alachlor Simazine Metribuzin Metolachlor Carbofuran Linuron	Study of efficacy of water treatment for removal of hydrophilic pesticides from surface waters. Samples of pretreatment and posttreatment river water and lab spiked river water were analyzed. Conventional treatment demonstrated poor removal of most of these compounds. Tests with granular or powdered activated carbon showed much better removal.
Neary and Michael, 1989	w	1984	**Florida:** Forested watershed near Gainesville	Sulfometuron-methyl	Study of movement of the herbicide from a forested watershed to surface water and ground water. Concentrations were monitored for 203 days after application. Low concentrations (<7 µg/L) were observed in streamflow after the first rainfall only (3 days after application).
Spalding and Snow, 1989	w	5/88	**Nebraska:** Shell Creek	Alachlor Atrazine Butylate Cyanazine Disulfoton EPTC Metolachlor Metribuzin Propachlor Trifluralin	Study of pesticide levels in Platte River tributary during spring runoff event. Twelve samples taken over 40-hr period. Concentrations up to 89 µg/L (atrazine) were detected. Maximum concentrations of atrazine, alachlor, and cyanazine occurred in early stages, preceding peak discharge. Concentration peaks coincide with suspended solids, but not with nitrate. Compounds shown were those detected. Compounds analyzed for, but not detected, included carbaryl, carbofuran, chlorpyrifos, diazinon, endrin, fonofos, isofenphos, lindane, methoxychlor, parathion, terbufos, and toxaphene.
Troelstrup and Perry, 1989	w	5/92 10/92	**Minnesota:** Southeast area streams, Root River and tributaries	Atrazine	Study examined a wide range of variables and perturbations in small-stream ecosystems. Results indicate that monitoring of water quality must consider a finer spatial scale than that suggested by the aquatic ecoregion approach.
Watson and others, 1989	w	6/85–9/85 6/86–9/86	**Montana:** Two mountain stream watersheds	Picloram	Fate and transport study to determine persistence and surface water concentrations resulting from roadside applications. Mass balance included. Picloram in streams less than detection limit (0.5 µg/L) throughout study.

Table 2.3. Process and matrix distribution studies reviewed—*Continued*

Reference(s)	Matrix	Sampling dates	Location(s)	Compounds	Comments
Yin and Hassett, 1989	w s	10/83–10/84 (bi-monthly)	**Great Lakes:** Lake Ontario, Oswego River	Mirex	Study of speciation of mirex in surface waters. Particulate fraction small. In the water phase, 8 to 22 percent was free dissolved, with remainder bound to dissolved organic carbon. Total mirex concentration ranges: Oswego River (8–21 pg/L) and Lake Ontario (10–21 pg/L).
Bengtson and others, 1990	w	1987	**Louisiana:** Agricultural plots	Atrazine Metolachlor	Examination of runoff as a function of subsurface (tile) drainage. Runoff losses of atrazine and metolachlor were reduced by 55 and 51 percent, respectively, with subsurface drainage.
Bero and Gibbs, 1990	—	1945–80	**New York:** Hudson River	DDT Chlordane	Discussion of mechanisms of chemical transport in the Hudson River. Historical data and estimates of fluxes of DDT and chlordane in the river are presented.
Blevins and others, 1990	w	1984–85	**Kentucky:** Agricultural plots	Atrazine	Study of runoff losses of sediment, nutrients, and atrazine from plots using three different tillage systems. Total atrazine losses were 0.01 to 0.04 and 0.19 to 0.64 percent of the amount applied in 1984 and 1985, respectively. Atrazine losses were lowest with chisel-plow tillage, highest with conventional tillage.
Buttle, 1990	f s	1987	**Ontario, Canada:** Agricultural plots	Metolachlor	Study of metolachlor losses from field plots using different agricultural practices. Incorporation and contour plowing reduced runoff losses. Total losses from four plots were all <1 percent of the amount applied. Sediment transported 20 to 46 percent of the metolachlor lost in runoff.
Felsot and others, 1990	f s	1983–85	**Illinois:** Agricultural plots	Alachlor Carbofuran Terbufos Terbufos sulfoxide Terbufos sulfone	Study of the influence of tillage system and contouring practice on runoff losses (silt loam soil). Conservation tillage systems reduced losses of all three herbicides. Contouring alone significantly reduced losses, regardless of tillage. Terbufos and metabolites were transported mainly in the sediment phase; alachlor and carbofuran were transported in solution phase.

Table 2.3. Process and matrix distribution studies reviewed—*Continued*

Reference(s)	Matrix	Sampling dates	Location(s)	Compounds	Comments
Nicosia and others, 1990	w	1988 (spring/summer)	**California:** Agricultural drainage canals from rice fields, Sacramento Valley	Carbofuran	Study to determine and compare discharge of carbofuran from rice fields and sugar beet fields and to examine disappearance of soil-incorporated carbofuran from rice paddy soil and water. Rice fields contributed approximately 11 times more carbofuran to drains than beet fields. A total of 2 to 11 percent of applied carbofuran was discharged in runoff water in the 3-month period after fields were flooded. Soil half-lives of carbofuran were 1.5 months or longer, based on data from the 70-day period after flooding. Half-lives in paddy water ranged from 18 to 26 days in the three fields studied.
Southwick and others, 1990	w	4/87–8/87	**Louisiana:** Agricultural plots	Atrazine Metolachlor	Study of concentrations of herbicides in runoff from field plots with and without subsurface drainage. Runoff concentrations of both compounds were significantly higher from plots without subsurface drainage in the first runoff event. Runoff losses for the season were more than twice as high for both compounds from the plots without subsurface drainage. This was due to both reduced runoff from drained fields and to the reduced concentrations of the herbicides in the first runoff event. Average runoff losses of atrazine for the season were 1.4 percent and 3.2 percent of the amount applied for drained and nondrained fields, respectively. Metolachlor losses were 1.1 percent and 2.4 percent for drained and nondrained fields, respectively.
Trim and Marcus, 1990	—	1978–88	**South Carolina:** Coastal zone	Pesticides in general	Study of fish kills from 1978 to 1988. Those attributable to agricultural use of pesticides reduced dramatically because of changes in land-use and agricultural practices. Pesticides were responsible for ~19 percent of fish kills in tidal saltwaters during this period. Pesticides were rarely detected in the ambient monitoring program in effect. The results of the study indicate that regulations and ongoing monitoring have reduced point source impacts and that ambient trend data are not useful for early detection of potential problems or for identification of sources.

Table 2.3. Process and matrix distribution studies reviewed—*Continued*

Reference(s)	Matrix	Sampling dates	Location(s)	Compounds	Comments
Wauchope and others, 1990	w s	1986	Small agricultural plots	Sulfometuron-methyl Cyanazine	Effects of herbicide formulation and grass cover on amounts lost in runoff studied. Total losses of both compounds were 1 to 3 percent of amount applied, regardless of formulation or presence of grass cover. Grass cover retarded initiation of runoff, reducing losses when runoff water volumes held constant. Results compared to GLEAMS (Ground Water Loading Effects of Agricultural Management Systems) model simulations, with excellent agreement.
Wess and others, 1990	w	1987–88	**Florida:** Two ponds	Fluridone NMF (degradation product)	Study of persistence of fluridone (aquatic herbicide sonar) in pond water and potential formation of N-methylformamide (NMF). No NMF detected at any time after application (detection limit is 2 µg/L). Fluridone concentrations gradually declined from ~100 to <2 µg/L after 324 days. Pellet formulation maintained concentration of ~20 µg/L throughout most of the study period before dissipating by 324 days.
Evans and others, 1991	—	1982	**Great Lakes:** Lake Michigan	Toxaphene DDT DDE	Study of biomagnification of organochlorines in food web. Benthic organisms, plankton, fish (sculpin), suspended particulates, and bed sediments were analyzed. DDE was the most strongly biomagnified compound. The role of benthic and epibenthic organisms in recycling of OCs is discussed.
Fleck and others, 1991	w	11/88–12/88	**California:** Runoff from artichoke fields, Monterey County	Endosulfan	Study of off-target movement of aerially applied endosulfan, via spray drift and rain runoff. Runoff concentrations ranged from 2.2 to 13 µg/L. Authors conclude that these concentrations could result in adverse impacts on water quality, unless diluted sufficiently with uncontaminated water.
Kelly and others, 1991	w	1980's	**Great Lakes:** Lake Erie	Dieldrin DDT	Assessment of atmospheric and tributary inputs of a number of toxic substances to Lake Erie. Results indicate that ~30 percent of the total input of DDT and dieldrin is via the atmosphere. The Detroit River is the source of ~50 percent of the DDT and dieldrin inputs to the lake. Other tributaries (7 Canadian, 14 U.S.) contribute ~25 percent of the total DDT and ~13 percent of the total dieldrin inputs. Lake Erie provides a net source of DDT to the atmosphere throughout the year via volatilization.

Table 2.3. Process and matrix distribution studies reviewed—*Continued*

Reference(s)	Matrix	Sampling dates	Location(s)	Compounds	Comments
Nicosia and others, 1991a	w	4/88–7/88	**California:** Rice fields	Carbofuran	Study of carbofuran discharges from flooded rice fields and dissipation and fate of incorporated carbofuran. Total discharge in runoff water ranged from 1.7 to 11 percent of the amount applied in three fields within 60 days of application. Soil incorporation may reduce the amount of carbofuran released to paddy water.
Nicosia and others, 1991b	w	4/89–7/89	**California:** Rice fields in Sacramento Valley, agricultural drains, and Sacramento River	Bensulfuron-methyl (BSM)	Study of dissipation in rice paddy water and levels in drains and river water receiving paddy water. Exponential decline in concentration of BSM in paddy water observed. Dissipation half-life for three fields averaged 2 days. Low level residues (0.5–2 µg/L) were detected in agricultural drains for as long as a month after application. No BSM was detected in the Sacramento River, presumably because of dilution.
Segawa and others, 1991	w	2/90–6/90	**California:** Rivers, ponds, reservoir in Los Angeles County	Malathion Maloxon (degradation product)	Monitoring of malathion residues after 1989–90 aerial spraying for Mediterranean fruit flies. Concentrations of malathion and maloxon in ponds within the sprayed area averaged 49.4 µg/L and 0.8 µg/L, respectively. Concentrations of malathion and maloxon in swimming pools within the sprayed area averaged 9.4 µg/L and 16.5 µg/L, respectively. Malathion concentrations in surface waters within the sprayed area, but not directly sprayed, ranged from <0.1 to 6.2 µg/L. Concentrations of malathion in rain runoff samples collected within and outside the sprayed area ranged from <0.1 to 44 µg/L. Authors conclude that runoff resulting from rainfall may be a problem, as water quality criteria were exceeded in several areas. No monitoring of biota was conducted to assess the effects of these concentrations.

Table 2.3. Process and matrix distribution studies reviewed—*Continued*

Reference(s)	Matrix	Sampling dates	Location(s)	Compounds	Comments
Spencer and Cliath, 1991	w	1980's(?)	**California:** Runoff from farm plots	DCPA Dinitramine Prometryn Trifluralin Chlorpyrifos Methidathion Fenvalerate Permethrin Parathion M. parathion Diazinon Malathion Methomyl Sulprofos Endosulfan Ethylan EPTC	Multiyear, systematic study of pesticide runoff from large fields of cotton, sugar beets, lettuce, alfalfa, onions, and melons. Runoff concentrations dependent on chemical properties, agricultural practices, and rate of application. Generally, 1 to 2 percent of applied herbicides and <1 percent of applied insecticides lost in runoff.
Wang, 1991	w	9/86–10/86	**Florida:** Indian River	Malathion	Malathion concentration in water monitored after aerial and ground-level spraying of an impoundment adjacent to the waterway for mosquito control. Samples collected for 48 h after application. Maximum concentrations were 5 µg/L after aerial spray and 1.3 µg/L after ground-level spray. Concentrations gradually declined to <0.01 µg/L within 48 h (aerial) and 12 h (ground).
Heinis and Knuth, 1992	w	1990's(?)	**Minnesota:** Littoral enclosures in pond	Esfenvalerate	Esfenvalerate applied to surface of littoral enclosures in natural pond water. Samples of water, bed sediments, macrophytes, and fish were taken. Ninety percent of active ingredient was lost within 24 h.
Mudambi and others, 1992	w	1982 1984	**Great Lakes:** Lake Ontario, Oswego River, New York	Mirex Photomirex	Study of mirex-photomirex relationships in Lake Ontario. Photomirex/mirex ratios in sediments and water column of Lake Ontario and the St. Lawrence River much higher than in mirex source sediments, indicating that photomirex is formed in the water column. Measured ratios in water, biota, and sediment indicate that mirex and photomirex enter the food chain from the upper layer of the water column.

Table 2.3. Process and matrix distribution studies reviewed—*Continued*

Reference(s)	Matrix	Sampling dates	Location(s)	Compounds	Comments
Pantone and others, 1992	w s	1988–89	**Minnesota:** Agricultural plot	Atrazine	Comparison of runoff losses with preemergence and postemergence application of atrazine. Significant differences in runoff losses were observed; losses from postemergence plots were 2 to 20 times lower, depending on time after application. Differences were primarily due to decreased volume and flow rate of runoff water from postemergence plots.
Paterson and Schnoor, 1992	w	7/89–10/89	**Iowa:** Field plots near Amana	Alachlor Atrazine	Mass-balance study of herbicides on field plots. Losses in runoff, leaching, plant uptake, and transformations were measured. Pesticides were applied to a barren plot, a corn-planted plot, and a plot with poplar trees as a riparian buffer. Runoff losses of alachlor and atrazine accounted for 12 and 10 percent of the applied amounts, respectively. Maximum concentrations of alachlor and atrazine in runoff were 375 and 200 µg/L, respectively.
Squillace and Thurman, 1992	w f	5/84–11/85	**Iowa:** Cedar River watershed	Alachlor Atrazine Cyanazine Metolachlor Metribuzin	Concentrations of herbicides measured at six sites in the basin. Atrazine transport from the basin estimated as 1.5 to 5 percent of the amount applied. Overland flow responsible for ~94 percent of river load of atrazine and ground water inputs responsible for ~6 percent. Atrazine concentrations correlated with river discharge per unit drainage area.
Moody and Goolsby, 1993	w	5/90	Mississippi River	Alachlor Ametryn Atrazine Cyanazine Deethylatr. Deisoatr. Metolachlor Metribuzin Prometon Prometryn Propazine Simazine Terbutryn	Spatial variability of concentrations of triazine herbicides measured in lower Mississippi River after intense rainstorm passed through the basin. Samples taken every 16 km from Baton Rouge to Ohio River confluence from May 26–29, 1990. Slugs of water with increased atrazine concentration observed because of inputs from tributaries. Cross-channel gradients in atrazine concentration existed below major tributaries.
Kolpin and Kalkhoff, 1993	f	1990	**Iowa:** Roberts Creek, northeastern Iowa	Atrazine Deethylatr. Deisoatr.	Atrazine concentrations measured at two points in the stream seven times from May through November. Results indicate that decrease in atrazine concentrations occurs over the 11-mi stretch, and authors attribute loss to degradation via photolysis.

CHAPTER 3

Overview of Occurrence and Distribution
of Pesticides in Relation to Use

3.1 OCCURRENCE

The studies reviewed in this work show that pesticides have been detected in every region of the United States where surface waters have been analyzed (Tables 2.1, 2.2, and 2.3). The 98 pesticides and 20 pesticide transformation products targeted in the studies in Tables 2.1 and 2.2 are listed in Table 2.5. Of these 118 compounds, 76 have been detected in one or more surface water bodies in at least one study. In terms of pesticide classes, 31 of 52 targeted insecticides, 28 of 41 herbicides, 2 of 5 fungicides, and 15 of 20 pesticide transformation products were detected in surface waters.

In order to examine general patterns of occurrence of pesticides in surface waters, results from the state and local studies are combined with data from national and multistate studies. Table 2.5 shows the frequency of detection for pesticides targeted in these studies. It should be noted that the data shown in Table 2.5 pertain only to those compounds that have been selected for analysis and reported in the scientific literature, and do not imply that other pesticides and other transformation products are not present in surface waters. It is likely that if other pesticides were targeted for analysis in areas where they are used, many of them would be observed in surface waters. Even for the pesticides listed in Table 2.5 with zero or few detections, these results do not necessarily imply absence from surface waters throughout the United States.

Although no individual study analyzed for every pesticide, or even representative pesticides from every class, the studies, taken together, show that a wide variety of pesticides are present, at least for certain times of the year, in surface waters throughout the United States. The process and matrix distribution studies (Table 2.3) do not provide much information on the occurrence and distribution of pesticides in surface waters, as they generally do not address ambient conditions. The state and local studies (Table 2.2) provide considerable information on the occurrence and distribution of pesticides, but the results of these studies are difficult to compare because of the large variability in sampling sites, dates of sampling, analytical methods, target analytes, and detection limits. The best existing data to assess the occurrence and distribution of pesticides in surface waters, and the relation to pesticide use on a broad scale, are the national and multistate studies (Table 2.1). The studies that will be used for this purpose are briefly described below.

From 1957 to 1968, the Federal Water Quality Administration, or FWQA (U.S. Department of the Interior), collected samples from about 100 rivers in the United States (Figure 2.1) for analysis of pesticides and other organic compounds (Weaver and others, 1965; Breidenbach and others, 1967; Green and others, 1967; Lichtenberg and others, 1970). All rivers were sampled during September each year, except in 1968 when samples were collected in June. The study primarily targeted the organochlorine insecticides (OCs), with selected organophosphorus insecticides (OPs) added in 1967 and 1968. With this extensive data set, the temporal, seasonal, and geographic variations in pesticide concentrations were evaluated.

The U.S. Geological Survey (USGS) conducted a long-term study of nine OCs and selected transformation products and three phenoxy acid herbicides from 1965 to 1971 at about 20 sites (Figure 2.1) on major rivers and streams of the western United States (Brown and Nishioka, 1967; Manigold and Schulze, 1969; Schulze and others, 1973). Samples were collected monthly throughout the year. Near the end of the study, three OPs—methyl parathion, parathion, and diazinon—were added as analytes. DDT was the insecticide most commonly detected, and 2,4-D was the herbicide most commonly detected. All stations had at least one detection of one of the target analytes, and some had very frequent detections.

A cooperative study conducted by the USGS and the U.S. Environmental Protection Agency (USEPA) examined pesticides in water and bed sediments of rivers throughout the United States between 1975 and 1980 (Gilliom, 1985; Gilliom and others, 1985). The study examined 21 pesticides and transformation products (11 OCs, 6 OPs, 1 triazine herbicide, and 3 phenoxy acid herbicides) at more than 150 sites (Figure 2.2). They observed pesticides in water in less than 10 percent of the samples, but detection limits in this study were relatively high (see Section 2.7). Almost all of these detections were for the OCs.

In the four Great Lakes shared by Canada and the United States, Canadian scientists conducted large surveys of pesticides in 1974 (Glooschenko and others, 1976) and in 1986 (Stevens and Neilson, 1989). The 1974 study targeted 15 OCs and 17 OPs at 34 sites in Lakes Superior and Huron (Figure 2.2). Very few of the targeted analytes were observed in the water column. The 1986 study targeted 17 organochlorine pesticides and transformation products at 95 sites in the lakes (Figure 2.3), with much lower detection limits. Dieldrin, α-HCH, γ-HCH, and heptachlor epoxide were detected at every site. Chlordane (*cis* and *trans*) and *p,p'*-DDE also were detected frequently.

During the early 1980's, the USEPA's National Urban Runoff Program (NURP) (Cole and others, 1984) extended their range of analytes to include selected pesticides and the priority pollutants for a limited period in 19 cities (Figure 2.3). Eighty-six samples of urban runoff water were analyzed for OCs, acrolein, and pentachlorophenol. The most frequently detected compound was α-HCH (20 percent of the samples), and the only compound that exceeded a freshwater acute toxicity criteria was pentachlorophenol (in one sample). These results are difficult to relate to pesticide use, but are the only national data on the occurrence of pesticides in urban runoff water.

Starting in 1975, and continuing through the 1980's, Ciba-Geigy Corporation (Ciba-Geigy 1992a,c,d, 1994a) monitored atrazine concentrations at a number of sites on streams and rivers throughout the Mississippi River Basin. Many of the sites were sampled during 1975–1976, and again during the mid-1980's. One site on the Mississippi River (at Vicksburg, Mississippi) was continuously sampled from 1975 to 1989, with a sampling frequency of one to five samples per month. Data from this site, which serves to integrate the atrazine inputs throughout most of the Mississippi River Basin, represent the longest continuous record of pesticide concentrations found in any of the studies reviewed. The Ciba-Geigy reports also include data from Monsanto Company, which conducted a monitoring program beginning in 1985 in several streams and lakes in the Mississippi River Basin. The focus of the Monsanto program was on concentrations of alachlor and other herbicides in raw and finished water at 24 midwestern water utilities. The atrazine data from both companies' monitoring programs, along with results of monitoring by federal and state agencies, water utilities, and universities, are summarized in several Ciba-Geigy technical reports (Ciba-Geigy, 1992a,c,d, 1994a). The emphasis of these reports is on relating the detected concentrations to the USEPA maximum contaminant level (MCL) for atrazine of 3 mg/L (annual mean concentration). Atrazine was detected frequently at nearly all the sites sampled, with a detection frequency of 60 to 100 percent of samples, depending on the site. Annual mean concentrations were less than the MCL at 94 percent of the sites over the entire sampling period.

In the late 1980's and continuing into the 1990's, the USGS has conducted a number of large-scale studies of current-use pesticides in the Mississippi River Basin. All of these studies targeted triazine and acetanilide herbicides, and some included other high-use pesticides. In 1989 and 1990, rivers and streams with relatively small watersheds (100 to 60,000 mi^2) were sampled in spring (preplanting), summer (postplanting), and autumn (postharvest, low river discharge) at 147 sites throughout the Midwest (Goolsby and Battaglin, 1993), as shown in Figure 2.3. Water was analyzed for 11 triazine and acetanilide herbicides and 2 atrazine transformation products. Atrazine was detected at every site, and 10 of the 13 compounds were detected at one or more sites. In 1991 and 1992, three sites on the mainstem of the Mississippi River and sites on six major tributaries (Platte, Missouri, Minnesota, Illinois, Ohio, and White Rivers) were sampled one to three times per week for 18 months (Goolsby and Battaglin, 1993; Larson and others, 1995), as shown in Figure 2.4. The water was analyzed for 27 high-use pesticides (15 herbicides and 12 insecticides). Triazine and acetanilide herbicides were observed most frequently. The OPs and most of the other compounds were rarely observed. In 1992, a survey of 76 reservoirs in the midwestern United States was conducted (Goolsby and others, 1993). See Figure 2.4. The reservoirs were sampled in late April to mid-May, late June to early July, late August to early September, and late October to early November. Eleven triazine and acetanilide herbicides and selected transformation products were targeted and observed in the reservoirs. Atrazine was detected in 92 percent of the reservoirs during the summer months.

The studies previously summarized will be used to relate agricultural use of specific compounds to their occurrence and geographical distribution in surface waters. For a few compounds, the relation is strong and well defined. For most, however, the relation is weak or nonexistent because of the lack of adequate data (pesticide use or surface water observations) or the fact that environmental processes prevent significant amounts of certain pesticides from reaching or persisting in surface waters.

3.2 NATIONAL PESTICIDE USE

Pesticides are introduced purposely into the environment for many reasons. They are commonly used in agriculture, forestry, transportation (weed control along roadsides and railways), urban and suburban areas (control of pests in homes, buildings, gardens, and lawns), lakes and streams (control of aquatic flora and fauna), and various commercial and industrial settings. Total annual pesticide use in the United States increased steadily during the 1960's and early 1970's, but has been quite stable for the past 20 years at approximately 1.1 billion pounds of active ingredient (lb a.i.). The majority of pesticide use is in agriculture, which accounts for 70 to 80 percent of total pesticide use each year since 1978 (Aspelin, 1994). From a national perspective, agricultural pesticide use provides the greatest potential for contamination of surface waters. It is the primary focus in this book for most of the comparisons between pesticide use and pesticide occurrence in surface waters. Non-agricultural uses of pesticides are also substantial, however, and may be the dominant source to surface waters in some areas. For certain pesticides, non-agricultural use accounts for a large proportion of the total use, and exceeds agricultural use in some cases. In general, however, much less data are available on non-agricultural use of pesticides than on agricultural use. A national survey of non-agricultural pesticide use—the 1993 Certified/Commercial Pesticide Applicator Survey (C/CPAS)—has been conducted by USEPA, and is soon to be published. This survey will provide crucial information on the types and amounts of pesticides being applied in non-agricultural settings. Included in the survey are data on aquatic pest control, ornamental and turf pest control, rights-of-way pest control, public health pest control, and pest control in commercial, industrial, and institutional settings (Aspelin, 1994). In the following sections, the available data on pesticide use in five major areas—agriculture, urban use, forestry, roadways and rights-of-way, and aquatic use—are discussed.

AGRICULTURAL USE

Pesticides used agriculturally in quantities of 8,000 lb a.i. or more in 1988–1991 (Gianessi and Puffer, 1991, 1992a,b) are listed in Table 3.1. Also included in this table are use estimates reported for 1966 (Eichers and others, 1970) and 1971 (Andrilenas, 1974). One hundred pesticides (57 herbicides, 31 insecticides, 12 fungicides) were used in quantities greater than 500,000 lb a.i. during 1988–1991 for agricultural purposes. Agricultural pesticide-use data for the major pesticide groups are summarized in Table 3.2. Several major trends in agricultural pesticide use are evident from Table 3.2. Herbicide use has increased substantially since the 1960's and now accounts for approximately 75 percent of the total agricultural use of pesticides. Total insecticide use has declined slightly, and a major shift in the types of compounds used has taken place, as organophosphorus and other insecticides have largely replaced the organochlorine compounds. Fungicide use has increased slowly in the last 2 decades, but still represents only a small fraction—approximately 6 percent—of total agricultural pesticide use. Trends in agricultural use of the different classes of compounds, and of specific compounds, are discussed in Section 3.3.

From a national perspective, agricultural use of pesticides is heaviest in fairly well defined areas, corresponding to regions of the most intense agricultural activity and to regions where specific crops are grown. This is illustrated in Figure 3.1, in which expenses for agricultural chemicals (excluding fertilizer) in 1987 are plotted for the entire United States. Pesticide use on a county basis (in lb a.i. applied per square mile) is shown in Figures 3.2 through 3.36 for the conterminous United States. These maps were generated from data compiled by Gianessi and Puffer (1991, 1992a,b) and reflect estimated agricultural use in 1988–1991. Total

Table 3.1. Estimates of agricultural pesticide use in the United States

[Compounds are listed in order of 1988-91 use, except for organochlorine insecticides, which are listed in order of 1971 use. Estimates are in pounds active ingredient x 1000. 1966 data from Eichers and others (1990); 1971 data from Andrilenas (1974). 1988–1991 data from Gianessi and Puffer (1991, 1992a,b). γ, gamma; λ, lambda. —, no use reported]

Compound	1966	1971	1988-91	Compound	1966	1971	1988-91
INSECTICIDES				Heptachlor	1,536	1,211	—
Organochlorine:				Endosulfan	791	882	1,992
Toxaphene	34,605	37,464	—	HCH, γ	704	650	66
DDT	27,004	14,324	—	Dieldrin	724	332	—
Aldrin	14,761	7,928	—	DDD	2,896	244	—
Methoxychlor	2,578	3,012	109	Strobane	2,016	216	—
Chlordane	526	1,890	—	Dicofol	—	—	1,718
Endrin	571	1,427	—	Others	347	293	—

Table 3.1. Estimates of agricultural pesticide use in the United States—*Continued*

Compound	1966	1971	1988-91
Organophosphorus:			
Chlorpyrifos			16,725
Methyl parathion	8,002	27,563	8,131
Terbufos	—	—	7,218
Phorate	326	4,178	4,782
Fonofos	—	—	4,039
Malathion	5,218	3,602	3,188
Disulfoton	1,952	4,079	3,058
Acephate	—	—	2,965
Dimethoate	—	—	2,960
Parathion	8,452	9,481	2,848
Azinphos-methyl	1,474	2,654	2,477
Diazinon	5,605	3,167	1,710
Ethoprop	—	—	1,636
Ethion	2,007	2,326	1,249
Profenfos	—	—	1,224
Methamidophos	—	—	1,135
Phosmet	—	—	1,055
Dicrotophos	—	807	963
Sulprofos	—	—	874
Fenamiphos	—	—	763
Mevinphos	—	—	463
Methidathion	—	—	402
Oxydemeton-methyl	—	—	370
Naled	—	—	224
Trichlorfon	1,060	617	16
Dichlorvos	912	2,434	—
Ronnel	391	479	—
Others	2,710	9,319	—
Other Insecticides[1]:			
Carbaryl	12,392	17,838	7,622
Carbofuran	—	2,854	7,057
Propargite	—	—	3,786
Aldicarb	—	—	3,573
Cryolite	—	—	2,970
Methomyl	—	1,077	2,952
Thiodicarb	—	—	1,714
Permethrin	—	—	1,122
Oxamyl	—	—	726
Fenbutatin oxide	—	—	560
Formetanate HCl	—	—	414
Esfenvalerate	—	—	287

Compound	1966	1971	1988-91
Tefluthrin	—	—	197
Cypermethrin	—	—	188
Trimethacarb	—	—	131
λ-Cyhalothrin	—	—	110
Cyfluthrin	—	—	105
Oxythioquinox	—	—	102
Amitraz	—	—	75
Fenvalerate	—	—	73
Metaldehyde	—	—	44
Tralomethrin	—	—	43
Diflubenzuron	—	—	42
Bifenthrin	—	—	28
Abamectin	—	—	12
Cyromazine	—	—	9
Others	502	37	—
HERBICIDES			
Triazine and Acetanilide:			
Atrazine	23,521	57,445	64,236
Alachlor	—	14,754	55,187
Metolachlor	—	—	49,713
Cyanazine	—	—	22,894
Metribuzin	—	—	7,516
Propazine	580	3,171	4,015
Propachlor	2,269	23,732	3,989
Simazine	193	1,738	3,964
Prometryn	—	—	1,807
Terbutryn	—	—	1,113
Ametryn	—	—	186
Others	—	450	1,022
Phenoxy:			
2,4-D	40,144	34,612	33,096
MCPA	1,669	3,299	4,338
2,4-DB	—	—	1,368
2,4,5-T	760	1,530	—
Other Herbicides:			
EPTC	3,138	4,409	37,191
Trifluralin	5,233	11,427	27,119
Butylate	—	5,915	19,107
Pendimethalin	—	—	12,521
Glyphosate	—	—	11,595
Dicamba	222	430	11,240
Bentazon	—	—	8,211
Propanil	2,589	6,656	7,516

Table 3.1. Estimates of agricultural pesticide use in the United States—*Continued*

Compound	1966	1971	1988-91
Other	21,479	27,686	5,635
MSMA	—	—	5,065
Molinate	—	—	4,408
Ethalfluralin	—	—	3,518
Triallate	—	—	3,509
Paraquat	—	—	3,025
Chloramben	3,765	9,555	3,019
Picloram	—	—	2,932
Clomazone	—	—	2,715
Bromoxynil	—	—	2,627
Linuron	1,425	1,803	2,623
Fluometuron	—	3,334	2,442
Dacthal	—	—	2,219
Diuron	1,624	1,234	1,986
Norflurazon	—	—	1,768
DSMA	—	—	1,705
Acifluorfen	—	—	1,475
Diclofop	—	—	1,452
Oryzalin	—	—	1,426
Thiobencarb	—	—	1,359
Cycloate	—	—	1,175
Benefin	—	—	1,167
Bromacil	—	—	1,155
Asulam	—	—	1,088
Imazaquin	—	—	1,073
Diphenamid	—	—	929
Vernolate	—	3,739	855
Sethoxydim	—	—	792
Fluazifop-butyl	—	—	731
Napropamide	—	—	699
Naptalam	999	3,332	655
Pebulate	150	1,062	653
Bensulide	—	—	633
Profluralin	—	—	621
Tebuthiuron	—	—	608
Oxyfluorfen	—	—	599
Dieathatyl ethyl	—	—	502
Dalapon	38	1,043	453
Amiben	3,765	9,555	—

Compound	1966	1971	1988-91
Alanap	999	3,332	—
Nitralin	14	2,706	—
Fluorodifen	—	1,330	—
Norea	239	1,323	—
FUNGICIDES			
Chlorothalonil	—	—	9,932
Mancozeb	—	—	8,66
Captan	6,869	6,490	3,710
Maneb	4,443	3,878	3,592
Ziram	—	—	1,889
Benomyl	—	—	1,344
PCNB	—	—	800
Iprodione	—	—	741
Fosetyl-Al	—	—	689
Metiram	—	—	641
Metalaxyl	—	—	635
Thiophanate-methyl	—	—	527
Triphenyltin-hydroxide	—	—	415
Ferbam	2,945	1,398	337
DCNA			286
Dodine[2]	1,143	1,191	275
Propiconazole	—	—	274
Thiram	—	—	238
Triadimefon	—	—	149
Anilazine	—	—	144
Thiabendazole	—	—	139
Myclobutanil	—	—	124
Etridiazole	—	—	104
Vinclozolin	—	—	103
Streptomycin	—	—	88
Triforine	—	—	81
Fenarimol	—	—	58
Oxytetracycline	—	—	37
Carboxin	—	—	15
Dinocap[2]	1,143	1,191	14
Zineb	6,903	1,969	—
Others	3,334	10,814	—

[1]Including acaricides, miticides, and nematocides.
[2]Use for these two compounds was reported as combined totals for 1966 and 1971.

agricultural pesticide use (Figure 3.2) is divided into total herbicide, insecticide, and fungicide use in Figures 3.3, 3.4, and 3.5, respectively. These figures do not include the use of oil (as an insecticide) or sulfur and copper (as fungicides). Maps showing agricultural use of selected pesticides, by county, in the 48 conterminous United States are shown in Figures 3.6 through 3.36. Individual pesticides were selected to show the geographic use patterns of the major, high-use compounds, as well as compounds whose use is specific to certain regions or crops. To represent the geographic patterns with the most detail, the use levels shown by the different gradations of shading are different in each of the maps of total use of pesticides, herbicides, insecticides, and fungicides (Figures 3.2 through 3.5). The different scales are necessary because of large differences in the amounts of the three types of pesticides used agriculturally. For the maps of individual pesticides (Figures 3.6 through 3.36), the scale is the same on all the maps, so that agricultural use of all the compounds can be compared, regardless of pesticide type. Thus, in the maps for individual pesticides, many of the maps of insecticide and fungicide use are relatively light compared to the maps for herbicides, reflecting the generally lower amounts of these compounds applied agriculturally. Because the data used for these maps are tabulated by county, a certain amount of distortion may occur in some of the large counties that have distinct land-use areas. For instance, certain counties in southern California and Arizona contain irrigated areas of intense agriculture, as well as large expanses of desert with virtually no farming. On the pesticide use maps, one shading is used for the entire county, even though the actual use may be restricted to a relatively small area. This problem occurs more in the western states, where counties are generally much larger than in the eastern part of the United States. Despite this, the maps provide a good overview of the general-use patterns of agricultural pesticides across the United States, and can be used to help identify areas that have the greatest potential for contamination of surface waters. They will be referred to throughout this book to relate the occurrence of specific pesticides in surface waters to agricultural use (see page 176 for a list of general observations made from the maps).

Table 3.2. Summary of estimated agricultural pesticide use in the United States

[Estimates are in pounds active ingredient x 1000. 1966 data from Eichers and others (1970); 1971 data from Andrilenas (1974); 1988–1991 data from Gianessi and Puffer (1991, 1992a,b)]

	1966		1971		1988–91	
	Use	Percent of Total Use	Use	Percent of Total Use	Use	Percent of Total Use
HERBICIDES						
Triazine and Acetanilide	27,000	23	100,000	42	220,000	48
Phenoxy	43,000	37	39,000	16	39,000	8
Other	46,000	40	100,000	42	200,000	44
Total Herbicide Use	116,000	100	239,000	100	459,000	100
INSECTICIDES						
Organochlorine	89,000	64	70,000	46	3,900	4
Organophosphorus	37,000	27	61,000	40	70,000	65
Other[1]	12,000	9	22,000	14	34,000	31
Total Insecticide Use	138,000	100	153,000	100	107,900	100
FUNGICIDES						
Total Fungicide Use[2]	27,000	100	27,000	100	38,000	100

[1]Use data for other insecticides does not include use of oil.
[2]Excluding fungicidal use of sulfur and copper.

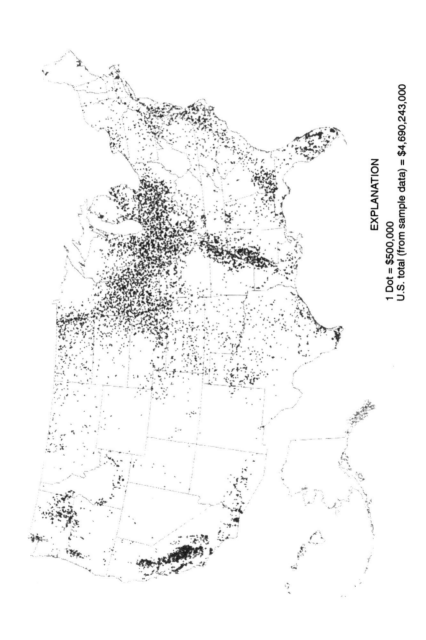

EXPLANATION

1 Dot = $500,000
U.S. total (from sample data) = $4,690,243,000

Figure 3.1. Geographic distribution of expenditures for agricultural chemicals, excluding fertilizer, in 1987. Reprinted from U.S. Department of Commerce (1990).

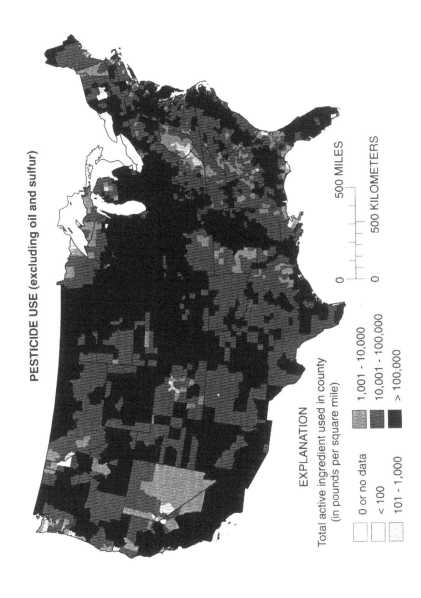

Figure 3.2. Annual estimated agricultural pesticide (herbicides, insecticides, and fungicides) use in the conterminous United States, by county, 1988–1991. Noncrop and postharvest uses of pesticides, insecticidal use of oil, and fungicidal use of copper and sulfur are not included. Data are from Gianessi and Puffer (1991, 1992a,b).

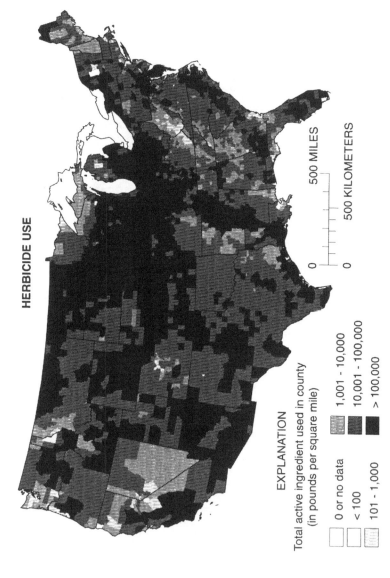

HERBICIDE USE

EXPLANATION

Total active ingredient used in county
(in pounds per square mile)

0 or no data

< 100

101 - 1,000

1,001 - 10,000

10,001 - 100,000

> 100,000

0   500 MILES

0   500 KILOMETERS

Figure 3.3. Annual estimated agricultural herbicide use in the conterminous United States, by county, 1988–1991. Noncrop use of herbicides is not included. Data are from Gianessi and Puffer (1991).

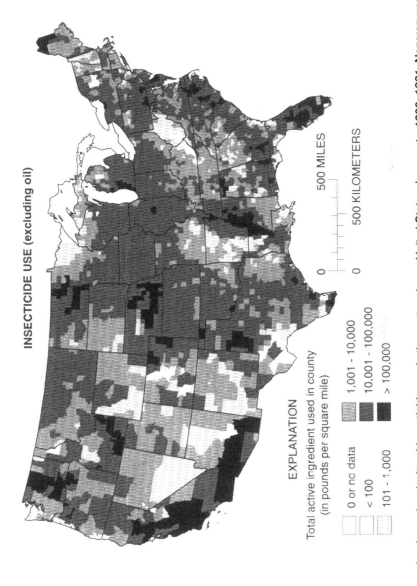

Figure 3.4. Annual estimated agricultural insecticide use in the conterminous United States, by county, 1988–1991. Noncrop use of insecticides and insecticidal use of oil are not included. Data are from Gianessi and Puffer (1991).

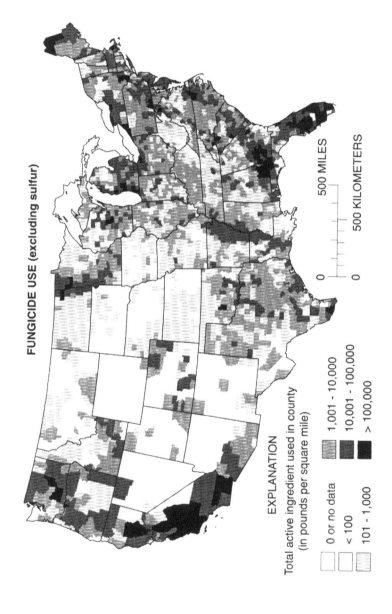

Figure 3.5. Annual estimated agricultural fungicide use in the conterminous United States, by county, 1988–1991. Noncrop use, seed treatments, postharvest use, and fungicidal use of copper and sulfur are not included. Data are from Gianessi and Puffer (1992b).

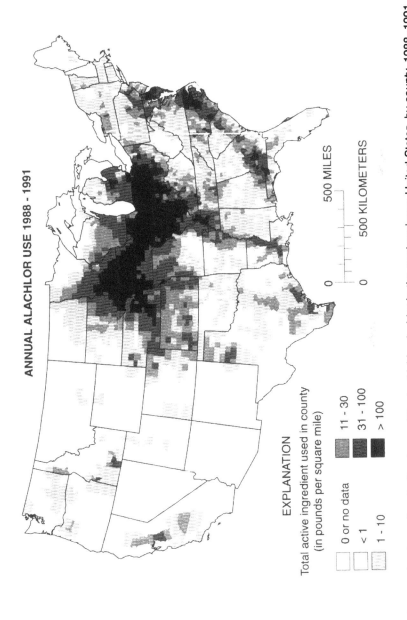

Figure 3.6. Annual estimated agricultural use of the herbicide alachlor in the conterminous United States, by county, 1988–1991. Data are from Gianessi and Puffer (1991).

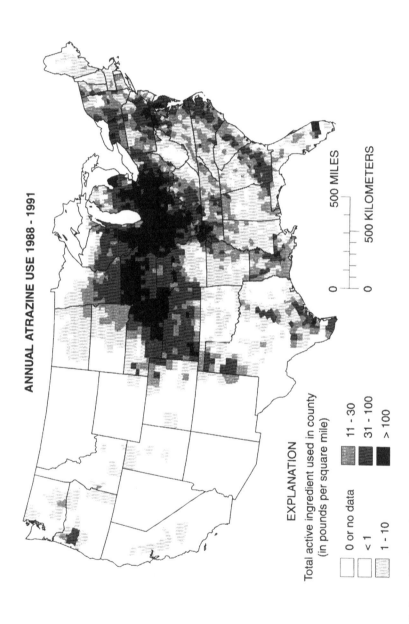

Figure 3.7. Annual estimated agricultural use of the herbicide atrazine in the conterminous United States, by county, 1988–1991. Data are from Gianessi and Puffer (1991).

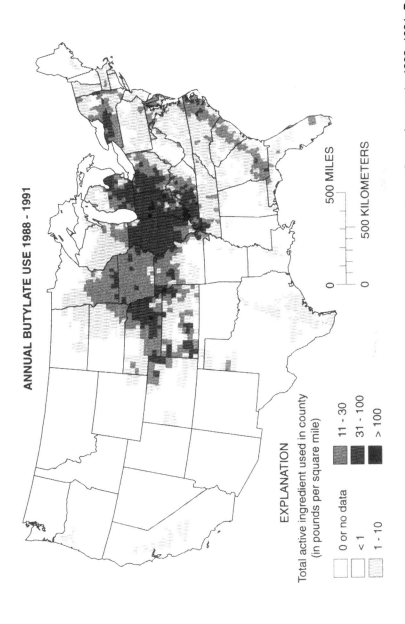

Figure 3.8. Annual estimated agricultural use of the herbicide butylate in the conterminous United States, by county, 1988–1991. Data are from Gianessi and Puffer (1991).

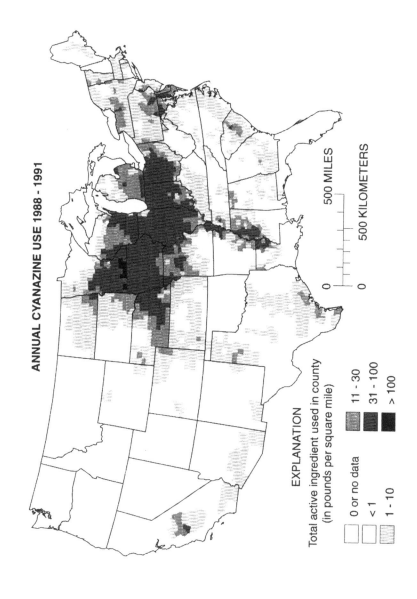

ANNUAL CYANAZINE USE 1988 - 1991

EXPLANATION

Total active ingredient used in county
(in pounds per square mile)

0 or no data

< 1

1 - 10

11 - 30

31 - 100

> 100

0   500 MILES

0   500 KILOMETERS

Figure 3.9. Annual estimated agricultural use of the herbicide cyanazine in the conterminous United States, by county, 1988–1991. Data are from Gianessi and Puffer (1991).

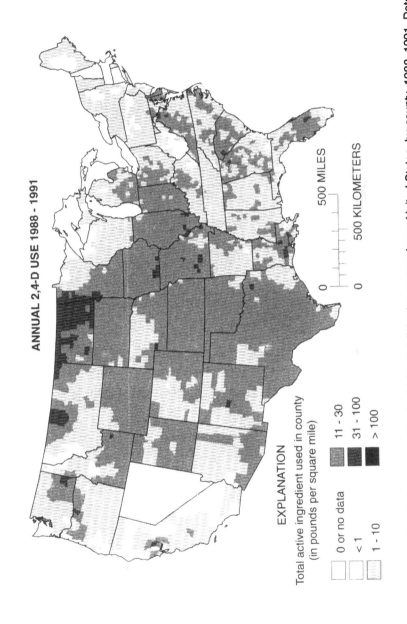

Figure 3.10. Annual estimated agricultural use of the herbicide 2,4-D in the conterminous United States, by county, 1988–1991. Data are from Gianessi and Puffer (1991).

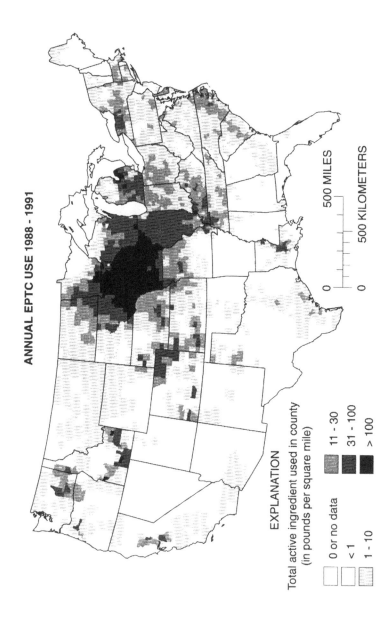

Figure 3.11. Annual estimated agricultural use of the herbicide EPTC in the conterminous United States, by county, 1988–1991. Data are from Gianessi and Puffer (1991).

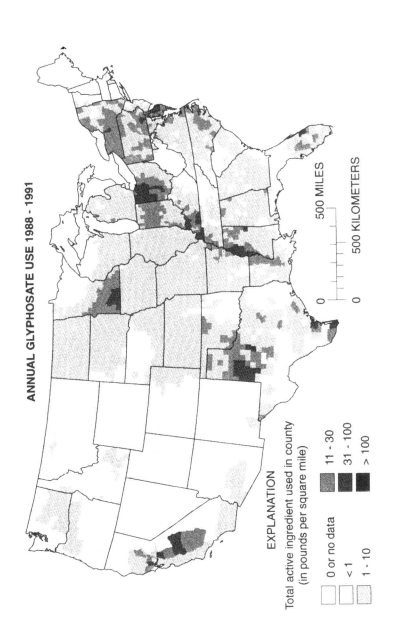

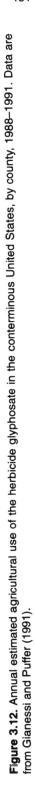

Figure 3.12. Annual estimated agricultural use of the herbicide glyphosate in the conterminous United States, by county, 1988–1991. Data are from Gianessi and Puffer (1991).

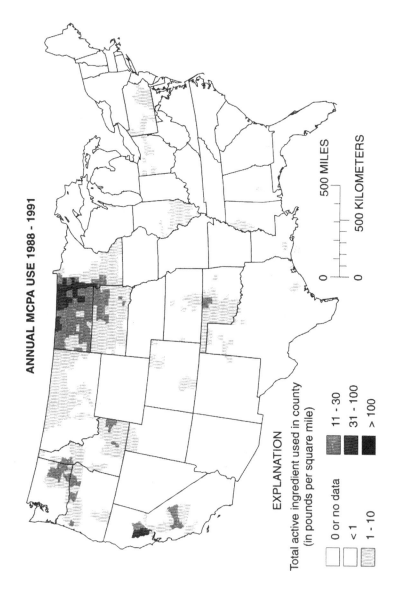

ANNUAL MCPA USE 1988 - 1991

EXPLANATION

Total active ingredient used in county
(in pounds per square mile)

0 or no data

11 - 30

<1

31 - 100

1 - 10

> 100

0   500 MILES

0   500 KILOMETERS

Figure 3.13. Annual estimated agricultural use of the herbicide MCPA in the conterminous United States, by county, 1988–1991. Data are from Gianessi and Puffer (1991).

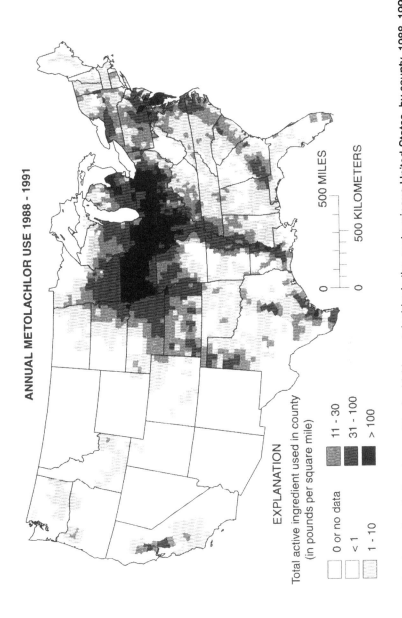

ANNUAL METOLACHLOR USE 1988 - 1991

EXPLANATION

Total active ingredient used in county
(in pounds per square mile)

0 or no data

< 1

1 - 10

11 - 30

31 - 100

> 100

0   500 MILES

0   500 KILOMETERS

Figure 3.14. Annual estimated agricultural use of the herbicide metolachlor in the conterminous United States, by county, 1988–1991. Data are from Gianessi and Puffer (1991).

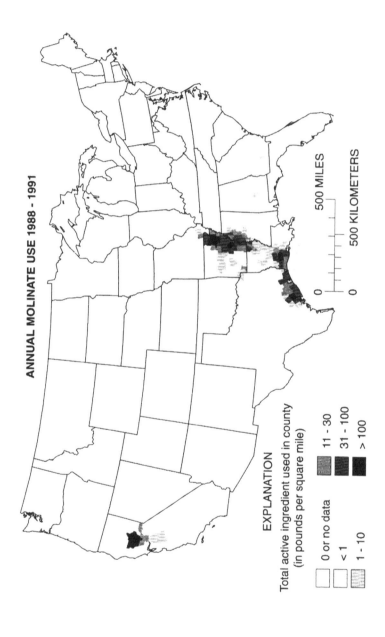

Figure 3.15. Annual estimated agricultural use of the herbicide molinate in the conterminous United States, by county, 1988–1991. Data are from Gianessi and Puffer (1991).

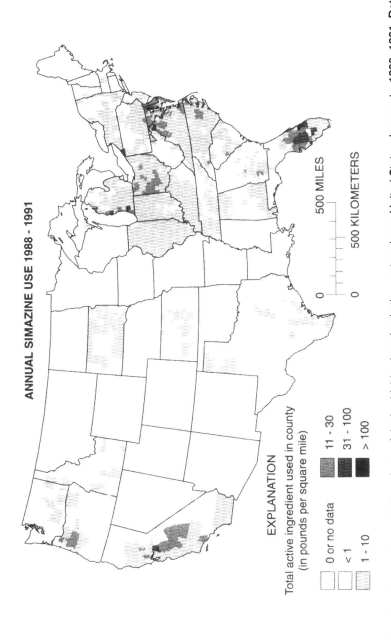

Figure 3.16. Annual estimated agricultural use of the herbicide simazine in the conterminous United States, by county, 1988–1991. Data are from Gianessi and Puffer (1991).

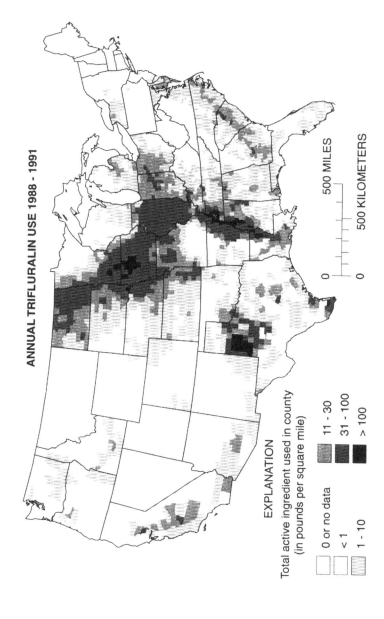

ANNUAL TRIFLURALIN USE 1988 - 1991

EXPLANATION

Total active ingredient used in county
(in pounds per square mile)

☐ 0 or no data
☐ < 1
▨ 1 - 10
▨ 11 - 30
■ 31 - 100
■ > 100

0 ___ 500 MILES

0 ___ 500 KILOMETERS

Figure 3.17. Annual estimated agricultural use of the herbicide trifluralin in the conterminous United States, by county, 1988–1991. Data are from Gianessi and Puffer (1991).

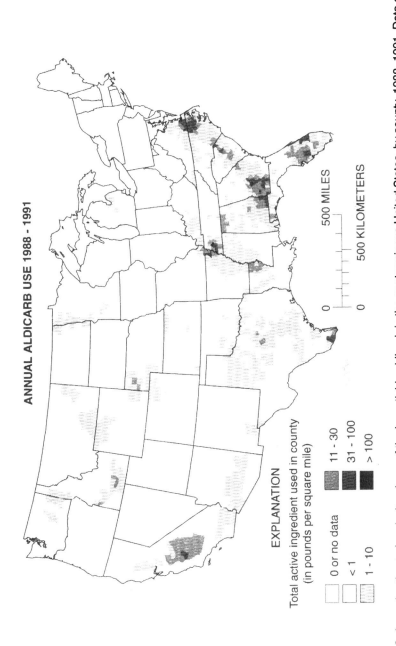

Figure 3.18. Annual estimated agricultural use of the insecticide aldicarb in the conterminous United States, by county, 1988–1991. Data are from Gianessi and Puffer (1992a).

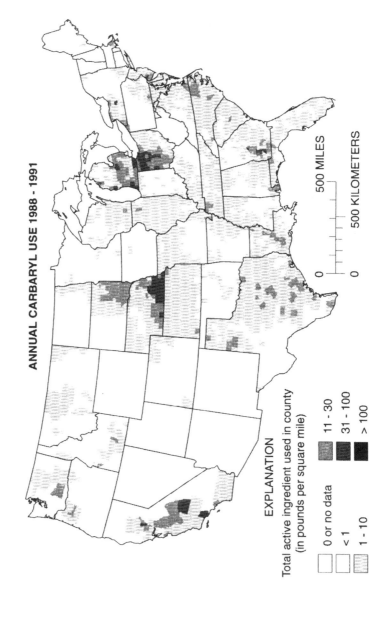

ANNUAL CARBARYL USE 1988 - 1991

EXPLANATION

Total active ingredient used in county
(in pounds per square mile)

0 or no data

< 1

1 - 10

11 - 30

31 - 100

> 100

0   500 MILES

0   500 KILOMETERS

Figure 3.19. Annual estimated agricultural use of the insecticide carbaryl in the conterminous United States, by county, 1988–1991. Data are from Gianessi and Puffer (1992a).

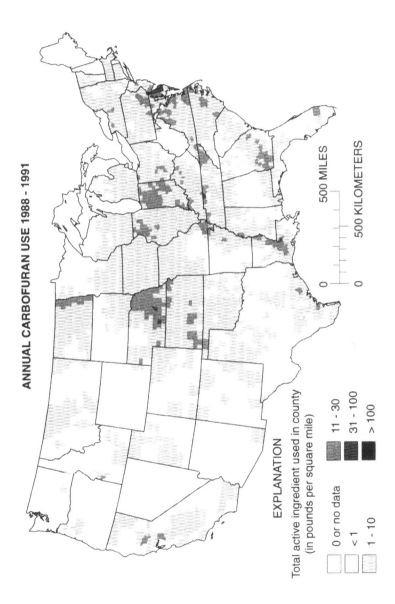

Figure 3.20. Annual estimated agricultural use of the insecticide carbofuran in the conterminous United States, by county, 1988–1991. Data are from Gianessi and Puffer (1992a).

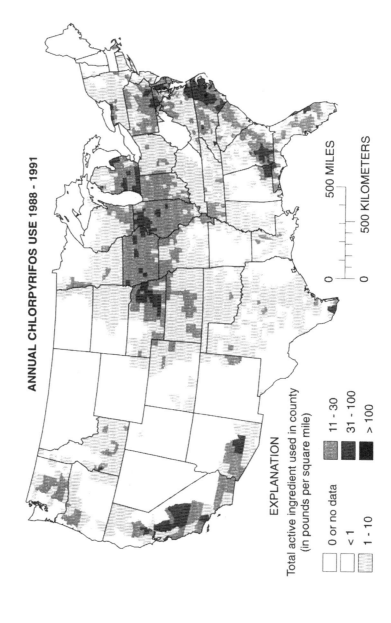

Figure 3.21. Annual estimated agricultural use of the insecticide chlorpyrifos in the conterminous United States, by county, 1988–1991. Data are from Gianessi and Puffer (1992a).

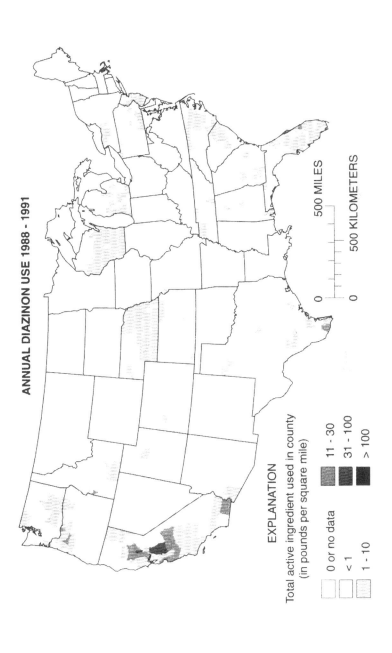

Figure 3.22. Annual estimated agricultural use of the insecticide diazinon in the conterminous United States, by county, 1988–1991. Data are from Gianessi and Puffer (1992a).

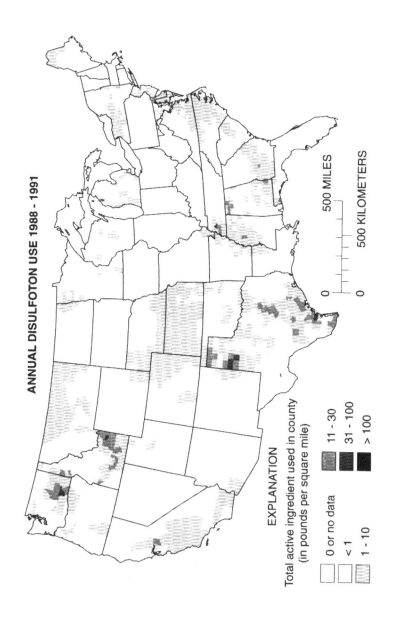

ANNUAL DISULFOTON USE 1988 - 1991

EXPLANATION

Total active ingredient used in county
(in pounds per square mile)

0 or no data	11 - 30
< 1	31 - 100
1 - 10	> 100

Figure 3.23. Annual estimated agricultural use of the insecticide disulfoton in the conterminous United States, by county, 1988–1991. Data are from Gianessi and Puffer (1992a).

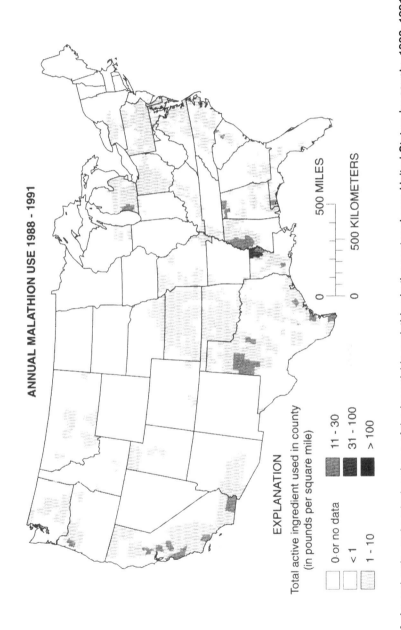

ANNUAL MALATHION USE 1988 - 1991

EXPLANATION

Total active ingredient used in county
(in pounds per square mile)

0. or no data	11 - 30
< 1	31 - 100
1 - 10	> 100

500 MILES

500 KILOMETERS

0

0

Figure 3.24. Annual estimated agricultural use of the insecticide malathion in the conterminous United States, by county, 1988–1991. Data are from Gianessi and Puffer (1992a)

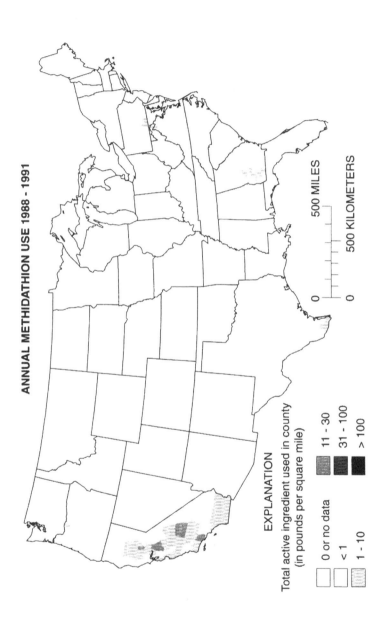

Figure 3.25. Annual estimated agricultural use of the insecticide methidathion in the conterminous United States, by county, 1988–1991. Data are from Gianessi and Puffer (1992a).

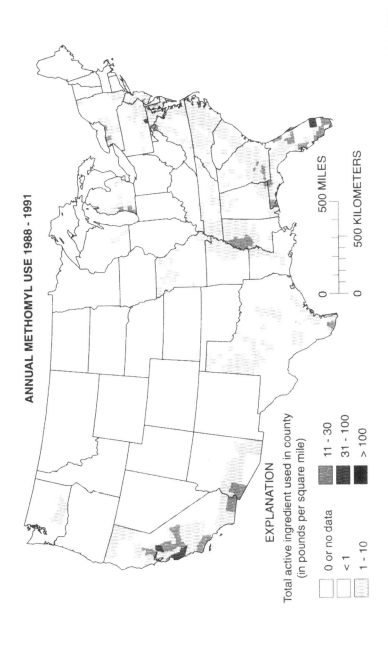

Figure 3.26. Annual estimated agricultural use of the insecticide methomyl in the conterminous United States, by county, 1988–1991. Data are from Gianessi and Puffer (1992a).

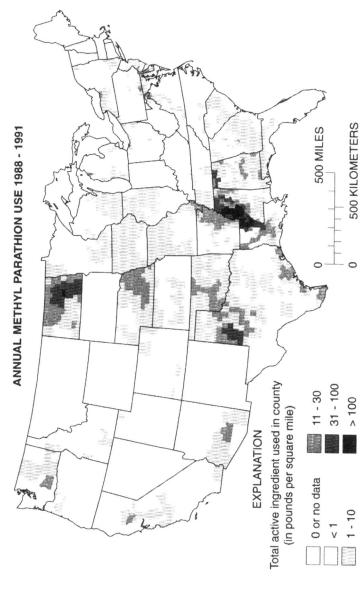

Figure 3.27. Annual estimated agricultural use of the insecticide methyl parathion in the conterminous United States, by county, 1988–1991. Data are from Gianessi and Puffer (1992a).

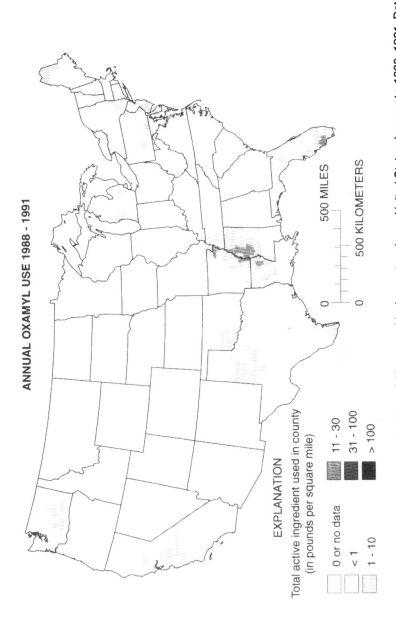

Figure 3.28. Annual estimated agricultural use of the insecticide oxamyl in the conterminous United States, by county, 1988–1991. Data are from Gianessi and Puffer (1992a).

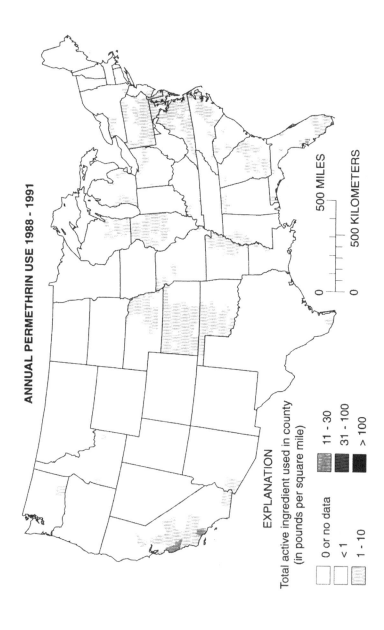

Figure 3.29. Annual estimated agricultural use of the insecticide permethrin in the conterminous United States, by county, 1988–1991. Data are from Gianessi and Puffer (1992a).

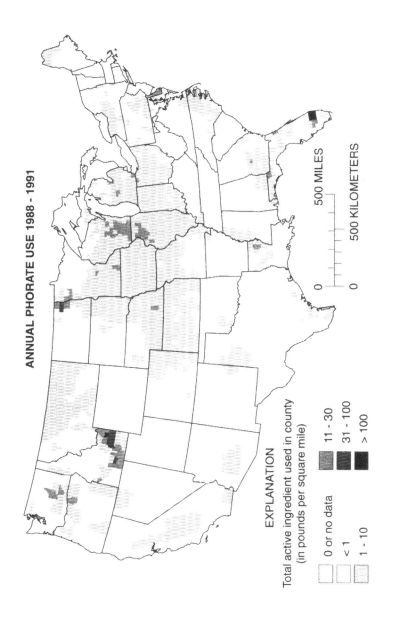

Figure 3.30. Annual estimated agricultural use of the insecticide phorate in the conterminous United States, by county, 1988–1991. Data are from Gianessi and Puffer (1992a).

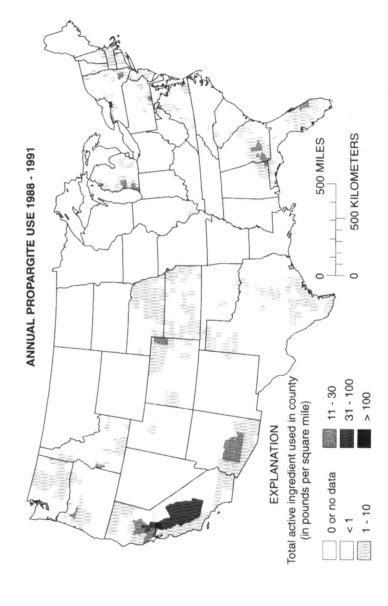

ANNUAL PROPARGITE USE 1988 - 1991

EXPLANATION

Total active ingredient used in county
(in pounds per square mile)

☐ 0 or no data

☐ < 1

☐ 1 - 10

▨ 11 - 30

▨ 31 - 100

■ > 100

0   500 MILES

0   500 KILOMETERS

Figure 3.31. Annual estimated agricultural use of the insecticide propargite in the conterminous United States, by county, 1988–1991. Data are from Gianessi and Puffer (1992a).

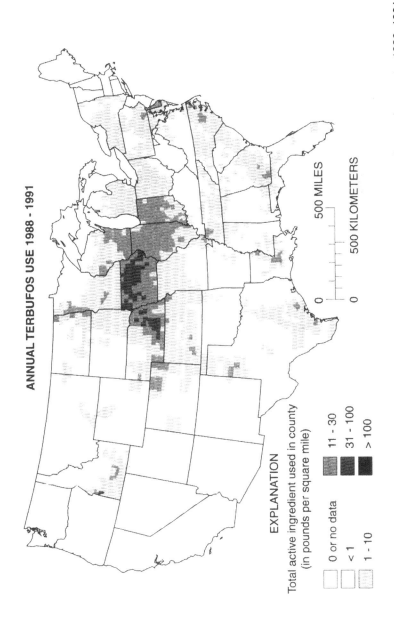

ANNUAL TERBUFOS USE 1988 - 1991

EXPLANATION

Total active ingredient used in county
(in pounds per square mile)

0 or no data	11 - 30
< 1	31 - 100
1 - 10	> 100

500 MILES

500 KILOMETERS

Figure 3.32. Annual estimated agricultural use of the insecticide terbufos in the conterminous United States, by county, 1988–1991. Data are from Gianessi and Puffer (1992a).

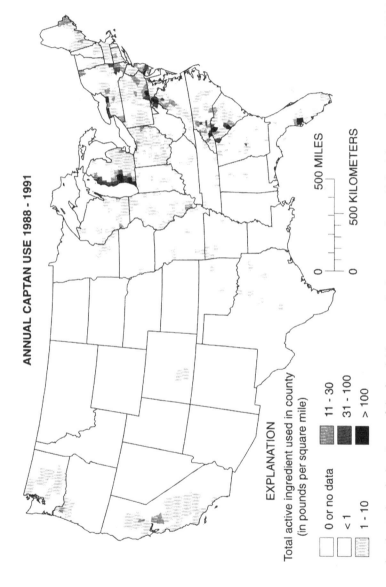

Figure 3.33. Annual estimated agricultural use of the fungicide captan in the conterminous United States, by county, 1988–1991. Data are from Gianessi and Puffer (1992b).

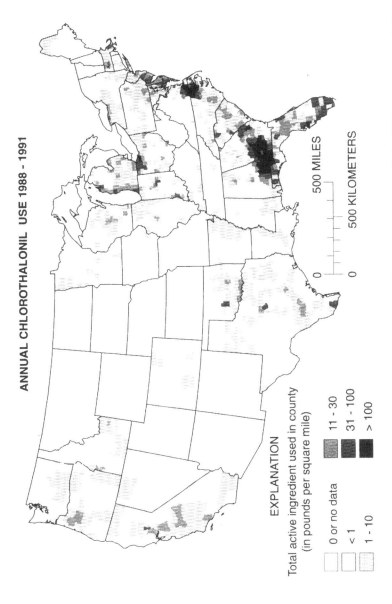

ANNUAL CHLOROTHALONIL   USE 1988 - 1991

EXPLANATION

Total active ingredient used in county
(in pounds per square mile)

0 or no data

< 1

1 - 10

11 - 30

31 - 100

> 100

500 MILES

500 KILOMETERS

0

0

Figure 3.34. Annual estimated agricultural use of the fungicide chlorothalonil in the conterminous United States, by county, 1988–1991. Data are from Gianessi and Puffer (1992b)

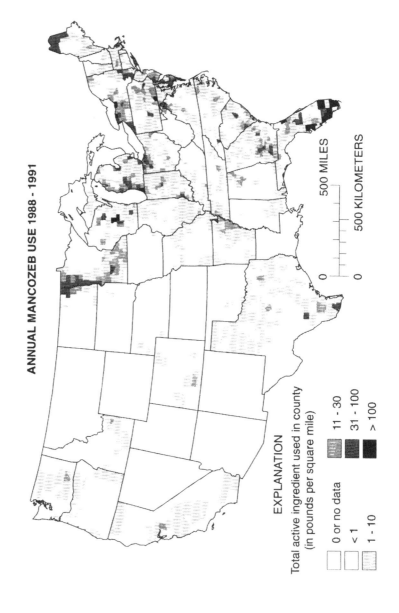

ANNUAL MANCOZEB USE 1988 - 1991

EXPLANATION

Total active ingredient used in county
(in pounds per square mile)

- 0 or no data
- < 1
- 1 - 10
- 11 - 30
- 31 - 100
- > 100

0   500 MILES

0   500 KILOMETERS

Figure 3.35. Annual estimated agricultural use of the fungicide mancozeb in the conterminous United States, by county, 1988–1991. Data are from Gianessi and Puffer (1992b).

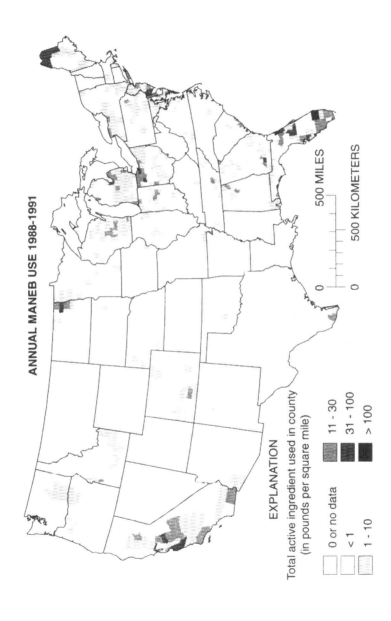

Figure 3.36. Annual estimated agricultural use of the fungicide maneb in the conterminous United States, by county, 1988–1991. Data are from Ganessi and Puffer (1992b).

Several general observations concerning patterns of agricultural pesticide use can be made from the maps:

1. Pesticides are used for agricultural purposes in virtually every county in the conterminous United States (Figure 3.2).
2. The largest areas of heavy use occur in the Midwest, California, Florida, the lower Mississippi River Valley, and coastal areas of the Southeast (Figure 3.2).
3. In terms of the mass applied, herbicide use is much higher than insecticide or fungicide use in most areas of the country (Figures 3.3, 3.4, and 3.5).
4. Heaviest use of a number of the high-use herbicides, including alachlor, atrazine, butylate, cyanazine, EPTC, and metolachlor, occurs in the corn belt of the midwestern United States (Figure 3.6 through 3.9, 3.11, and 3.14).
5. Fungicide use is much more fragmented than herbicide or insecticide use, with heavy use on specific crops in relatively small areas. Heaviest fungicide use occurs primarily in parts of Alabama, California, Florida, Georgia, Michigan, and Washington, (Figure 3.5).
6. The geographic range of use varies widely between pesticides. Some compounds have widespread use in most areas of the United States, such as the herbicides 2,4-D, EPTC, and glyphosate, and the insecticides carbaryl, carbofuran, chlorpyrifos, and methyl parathion (Figures 3.10, 3.11, 3.12, 3.19, 3.20, 3.21, and 3.27). Others, such as the herbicide molinate and the insecticides methidathion and oxamyl, are used almost entirely in relatively small regions, corresponding to areas where specific crops are grown (Figures 3.15, 3.25, and 3.28).

The limitations in the agricultural-use data must be considered when attempting to relate occurrence of a specific pesticide in surface waters to its use in a particular area. The data from 1964 to 1976 were reported geographically in terms of regional use only (Eichers and others, 1970; Andrilenas, 1974). Use totals were estimated for each of the 10 U.S. Department of Agriculture (USDA) farm production regions, which contain 2 to 11 states each. Total use was obtained by extrapolating data for specific pesticide and crop combinations from a statistically chosen sample of farms within each region, based on the proportion of total regional acreage of that crop represented by the sampled farms. The validity of the estimates depends on how representative the sampled farms were in terms of pesticides used and application rates, but the regional totals were judged to be reasonably accurate by the authors of the reports. An important point to realize, however, is that use of particular pesticides within these large regions was highly variable geographically. Thus, use within a particular river basin in a region may not be well accounted for by these data. Only broad, general patterns can be distinguished when detections and concentrations of pesticides in surface waters are compared with regional agricultural use.

The use data from 1988 to 1991, on the other hand, are reported for each county in the conterminous 48 states (Gianessi and Puffer, 1991, 1992a,b). Since many of the surface water bodies considered in this book have drainage areas much larger than most counties, variation in use within basins can be accounted for with this more detailed use data. Thus, recent observations of pesticides in surface waters can be compared with much more detailed—and probably more accurate—pesticide-use data than observations from the 1960's and 1970's. However, the use of this data also has limitations. Pesticide use totals for each county were obtained by using estimates for the percentage of each crop treated with a particular pesticide and the average application rate on that crop. Appropriate revisions were made by Gianessi and Puffer (1991, 1992a,b) so that the data reflect practices for 1988–1991. Crop acreage in each county, however, was estimated from the 1987 Census of Agriculture. Any change in crop acreage, pesticides used on particular crops, or application rates since the time the data were collected will affect the accuracy of the use totals. For example, the recommended application rates for atrazine decreased considerably between 1990 and 1992, and the use of cyanazine has reportedly increased by as much as 25 percent in the midwestern United States during the same period (Goolsby and others, 1994). The 1988–1991 use estimates for these compounds may be some-what inaccurate for comparisons with recent observations of pesticides in surface waters, but they are the most recent estimates available.

PESTICIDE USE IN URBAN AREAS

Pesticide use in urban areas of the United States has undergone major changes over the last several decades. The growth of suburban areas, the rise of the lawn care industry, the development of new herbicides and insecticides, and the virtual replacement of OCs with alternative compounds, have influenced both the amounts and types of pesticides applied in urban areas. The amounts applied are large—the professional applicator and consumer markets for pesticides were each estimated at $1.1 billion in sales, at the manufacturers level, in 1991, compared to $4.9 billion in sales in the agricultural market (Hodge, 1993). A 1981 survey of professional pesticide applicators identified 1,073 pesticide products containing 338 different active ingredients. Total use by professional applicators in 1981 (applications to lawns, trees, and structures) was estimated at 47 million lb a.i. (Immerman and Drummond, 1984). Data from a 1990 USEPA survey of households across the United States indicate that approximately 73 percent of households (69 million out of 94 million) used some type of pesticide during 1990 (Aspelin and others, 1992). This survey includes both indoor and outdoor use. Estimates of the volume of home and garden pesticide use indicate that it has been relatively stable in recent years, with 65 to 88 million lb a.i. applied each year from 1979 to 1993 (Aspelin, 1994). Again, these estimates include both indoor and outdoor use.

Major pesticides used in urban areas are shown in Table 3.3, based on data from the *National Home and Garden Pesticide Use Survey*, which estimated household use (outdoor only) in 1989–1990 (Whitmore and others, 1992). This survey was not designed to collect quantitative information on the actual amounts of individual pesticides used, but rather on the number of pesticide products each surveyed household had on hand and the number of times they were used in the previous year. Use by professional pest control firms in 1981 was estimated in the National Urban Pesticide Applicator Survey (NUPAS) conducted by the USEPA (Immerman and Drummond, 1984). The compounds listed in the NUPAS reflect use in three sectors of the professional applicator industry—lawn care, tree care, and treatment of structures. Treatment of structures with insecticides, much of which may have been indoors, accounted for more than 50 percent of the total amount of pesticides used by professional applicators in 1981. The data from the NUPAS (not shown) is somewhat out of date, but it is the most recent compilation available on use by professional applicators in urban areas. Both the home and garden survey and the NUPAS indicate that insecticide use accounted for the largest portion of urban pesticide use. More recent estimates (1991) indicate that herbicides and insecticides now account for approximately 50 percent and 30 percent, respectively, of total pesticide sales in the professional market, whereas insecticide sales account for approximately 75 percent of total pesticide sales in the consumer market (Hodge, 1993). A comparison of the most widely used agricultural pesticides (Table 3.1) with the most widely used home and garden pesticides (Table 3.3) shows that the types of pesticides used in agriculture differ considerably from those used in urban areas. The overlap in the top 50 agricultural pesticides and the top 50 urban-use pesticides is only 20 percent.

Table 3.3. Rankings of urban pesticides by estimated outdoor use during 1989–1990 and detection frequency in reviewed studies

[Ranking of pesticide use from the *National Home and Garden Pesticide Use Survey* (Whitmore and others, 1992). Only compounds with reported outdoor home and garden use are included. Percent detections are based on data from studies included in Tables 2.1 and 2.2. Blank cells indicate compound was not targeted in any of the reviewed studies]

Pesticide	Rank within Pesticide Group (Outdoor Applications)	Percent of Sites with Detections	Percent of Samples with Detections
INSECTICIDES			
Organochlorine:			
Dicofol	21		
Methoxychlor	22	12	4
Lindane	29	31	29
Endosulfan	32	2	3
Paradichlorobenzene	34		
Chlordane	35	18	28
Heptachlor	40	11	7
Pentachlorophenol	45	73	73
Organophosphorus:			
Diazinon	1	18	14
Chlorpyrifos	4	6	1
Acephate	11		
Malathion	13	4	4
Dichlorvos	16	18	5
Disulfoton	18	0	0
Oxydemeton-methyl	30		
Azinphos-methyl	38	0	0
Phosmet	42		
Dimethoate	43	0	0
Phosalone	44		
Isofenphos	46		
Other insecticides[1]:			
Propoxur	2		
Allethrin (total)	3		
Pyrethrins	5		
Resmethrin	6		
Sumithrin	7		
Carbaryl	8	25	10
Tetramethrin	9		
Metaldehyde	10		
Permethrin	12		
Diethyltoluamide	14		
MGK 264	15		
Hydramethylon	17		
Cyfluthrin	19		
Rotenone	20		
Fenvalerate	23	100	
Fenbutatin-oxide	24		
Allethrin (isomer unspecified)	25		
Methoprene	26		

Table 3.3. Rankings of urban pesticides by estimated outdoor use during 1989–1990 and detection frequency in reviewed studies—Continued

Pesticide	Rank within Pesticide Group (Outdoor Applications)	Percent of Sites with Detections	Percent of Samples with Detections
Bendiocarb	27		
Dienochlor	28		
Methomyl	31	0	0
Warfarin	33		
Tricosene	36		
Methiocarb	37		
Brodifacoum	39		
Bromadiolone	40		
HERBICIDES			
Triazine and Acetanilide:			
Prometon	14	27	17
Atrazine	22	87	78
Phenoxy:			
2,4-D	1	51	21
MCPP	2		
Mecoprop	11		
Other herbicides:			
Glyphosate	3		
Acifluorfen	4		
Dicamba	5	25	9
Oryzalin	13		
Chlorflurenol, methyl ester	15		
Triclopyr	17		
Oxyfluorfen	19		
Diquat dibromide	20		
Trifluralin	21	23	10
Fluazifop-butyl	23		
Dacthal	24		
Pendimethalin	25	93	8
MSMA	26		
Dichlobenil	29		
Benefin	30		
Paraquat	31	0	0
Metam-sodium	32		
Bensulide	33		
Endothall, di-Na salt	36		
Amitrole	37		
DSMA	38		
FUNGICIDES			
Captan	1	0	0
Triforine	2		
Folpet	2		
Benomyl	4		
Zineb	4		
Chlorothalonil	6	0	0
Maneb	7		
Dinocap	8		

Table 3.3. Rankings of urban pesticides by estimated outdoor use during 1989–1990 and detection frequency in reviewed studies—Continued

Pesticide	Rank within Pesticide Group (Outdoor Applications)	Percent of Sites with Detections	Percent of Samples with Detections
Thiram	9		
Thymol	10		
Anilazine	11		
PCNB	12	0	0

[1]Other Insecticides: Includes miticides, acaricides, and nematocides.

PESTICIDE USE IN FORESTRY

Pesticides have been used in the forests of the United States for decades. During the 1950's and 1960's, pesticides used were primarily chlorinated insecticides, such as DDT and endrin, which have since been banned in the United States. During the 1970's and 1980's, new insecticide compounds and biological agents replaced the organochlorines for control of insects, and the use of herbicides for vegetation control became more common in forestry. Since the early 1990's, use of pesticides, particularly herbicides, has apparently declined in some sectors of the forestry industry. The amount of pesticides applied in forestry, however, has always been a small fraction of the amount used in agriculture. Similarly, the area treated with pesticides each year is much smaller in forestry than in agriculture.

Data on the actual amounts of pesticides applied and the areas involved are difficult to obtain because of the lack of a national database of non-agricultural pesticide use and the varied ownership of forested land. Forested land in the United States is owned or administered by the U.S. Forest Service (USFS), the Bureau of Land Management, states, counties, municipalities, farmers, individual land owners, and private companies. In Minnesota, for example, ownership of forested land involved in silviculture (i.e., trees destined for harvesting) during 1990–1991 was distributed as follows: state–33 percent, county–28 percent, national forest–22 percent, forest industry–10.6 percent, and Native American–6.4 percent (Minnesota Environmental Quality Board, 1992). Comprehensive data are available on pesticide use by the National Forest Service, which administers 191 million acres of the approximately 800 million acres of forested land in the United States (U.S. Forest Service, 1993). Pesticides were used on less than 0.2 percent of national forest land in 1993 and on less than 1 percent each year since the mid-1970's when detailed reporting was begun (U.S. Forest Service, 1978, 1985, 1989, 1990, 1991, 1992, 1993). Whether this level of use is representative of use on the remainder of forested land in the United States is not clear. Only two states—California and Virginia—have collected statistics on pesticide use on all forested land within their borders, and these statistics are compared with USFS data in Table 3.4. In California, approximately 75,000 lb a.i. of pesticides were applied to 0.2 percent of its forested land in 1991 (Johnson, 1988; California Department of Food and Agriculture, 1991). These figures apparently do not include treatment with bacterial or viral insecticides. In Virginia, approximately 79,000 lb a.i. of herbicides were applied to 0.4 percent of its forested land in 1993 (Artman, 1994). In addition, approximately 0.5 to 1.5 percent of forested land in Virginia has been treated for gypsy moth suppression each year from 1990 to 1993 (U.S. Forest Service, 1994b). Data from these two states also indicate that the pesticide chemicals used on these forested areas are essentially the same compounds used on USFS land (Table 3.4). On the basis of the data from these two states, it appears that pesticide use on national forest land is representative of use on the remainder of forested land in the United States, and that a small percentage of the total forested area in the United States is treated with any type of pesticide in a given year. Pesticide use on 0.2 to 1.5 percent of forested land would imply that between 1.6 and 12 million acres (of about 800 million acres total) are treated with some type of pesticide in a given year (even this may be a high estimate, as land receiving applications of more than one pesticide is counted more than once in the USFS data). For comparison, agricultural applications of atrazine and alachlor covered approximately 50 million and 27 million acres, respectively, during 1988–1991 (Gianessi and Puffer, 1991).

Pesticide chemicals used most commonly in forestry in recent years are shown in Table 3.4. The total mass of herbicides, insecticides, fungicides, and fumigants applied on national forest land is shown in Figure 3.37 for 1977–1993. While these values probably represent only a fraction of the total amount applied on United States forests, they can be used to indicate the general trends in pesticide use in forestry.

The herbicides with highest use in the late 1970's and early 1980's were 2,4-D, picloram, and hexazinone. Use of these compounds has declined in recent years, however, and triclopyr is now the herbicide with highest use. Overall, there has been a significant decline in herbicide use over the last decade in the national forests. This is partially due to a 1984 ban on aerial application

Table 3.4. Estimated pesticide use on forested land

[Estimates are in pounds active ingredient x 1000, unless otherwise indicated. National forest data from U.S. Forest Service (1993); California data from California Department of Food and Agriculture (1991); Virginia data from Artman (1994). Insecticides: Bt, *Bacillus thuringiensis* var. *kurstaki.* ~. number is approximate; nd, no data availabale; —, no use reported]

Pesticide	Pesticide use in national forests	Pesticide use on all forested land in selected state	
	1992	California (1991)	Virginia (1993)
HERBICIDES:			
Triclopyr	47	19	2
2,4-D	12	9	—
Hexazinone	18	20	8
Glyphosate	17	6	50
Picloram	8	—	—
Imazapyr	2	—	18
Fosamine	2	—	1
Dacthal	2	—	—
Dicamba	1	—	—
Diuron	1	—	—
INSECTICIDES:			
Bt	~300,000 acres	nd	~96,000 acres
Dimilin	60 acres	nd	3 (~113,000 acres)
Carbaryl	97	nd	nd
Malathion	5	nd	nd
FUNGICIDES/FUMIGANTS:			
Dazomet	35	—	nd
Methyl bromide	40	16	nd
Borax	27	—	nd
Chloropicrin	27	3	nd
Chlorothalonil	5	—	nd

of herbicides in national forests and the increased costs of preparing the required environmental impact statements (Wehr and others, 1992). A number of national forests, particularly in the upper Midwest, have suspended all herbicide use for the past several years. The decline in herbicide use in the national forests has probably not occurred in other sectors of the forestry industry, although data on this are scarce.

In contrast to herbicide use, insecticide use in forestry is focused much more on controlling outbreaks of specific pests in localized areas and is not a routine part of normal silvicultural practice. As in agriculture, there has been a dramatic change in the types of insecticides used in forestry over the last 30 years. DDT and other OCs were used extensively in the 1950's and 1960's. During the 1970's and 1980's several OP compounds (malathion, azinphos-methyl, trichlorfon, and acephate) and carbamate compounds (carbaryl and carbofuran) were used. More recently, a bacterial agent (*Bacillus thuringiensis* var. *kurstaki* [Bt]) has become the main insecticide used to control outbreaks of several major insect pests, including the gypsy moth, spruce budworm, and various cone and seed insects. Use of Bt accounted for 25 to 90 percent of the total acreage treated with insecticides on USFS land each year from 1984 to 1993. The apparent decline in insecticide use shown in Figure 3.37 is largely due to the replacement of many of the more traditional insecticides by Bt. (Amounts of Bt used are not included in Figure 3.37.) Carbaryl still is used in relatively large quantities by the USFS, primarily for control of grasshoppers and crickets on rangeland. Diflubenzuron (dimilin) is used for control of gypsy moths in the eastern United States (Artman, 1994), although use data from the USFS do not reflect this.

Fungicides and fumigants used most commonly in forestry include methyl bromide, dazomet, chloropicrin, and borax. Use of these compounds, primarily on nursery stock and in seed orchards, has remained relatively stable in the national forests over the last 15 years (Figure 3.37). Whether this is true for other forested land is not known.

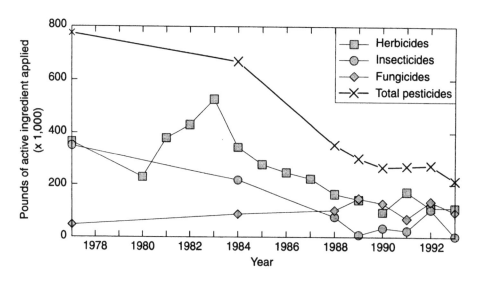

Figure 3.37. Pesticide use on national forest land, 1977–1993. Insecticide amounts do not include use of *Bacillus thuringiensis* var. *kurstaki* (Bt). Data are from U.S. Forest Service (1978, 1985, 1989, 1990, 1991, 1992, 1993, 1994a).

PESTICIDE USE ON ROADWAYS AND RIGHTS-OF-WAY

A variety of herbicides is used in roadway applications to control weeds and grasses for safety and aesthetic purposes and as firebreaks in some areas. The particular compound applied in a given location is often a local choice based on the type of weeds and the weather. Some of the most frequently used herbicides include 2,4-D and other phenoxy herbicides, picloram, and triclopyr. Occasionally, insecticides, such as fonofos, are applied to roadsides to help control the movement of pests, such as grasshoppers, during infestations. Some states have programs in which the majority of roadsides are sprayed with herbicides each year. Other states have a more conservative approach to pesticide use, in which only problem areas—perhaps 5 percent of the total roadside area—are sprayed. In some locations, particularly in some national forests, the use of herbicides along roadsides is prohibited. Data on the amounts of pesticides applied to roadways and rights-of-way generally are not available at this time. Pesticide use on rights-of-way is included in the USEPA's C/CPAS.

AQUATIC PESTICIDE USE

In some instances, pesticides are applied directly to surface waters for control of algae, macrophytes, insects, fish parasites, and sometimes even fish themselves. The most commonly used aquatic pesticides are herbicides to control algae and macrophytes (such as water hyacinth, hydrilla, and Eurasian watermilfoil). Many states have extensive ongoing programs for control of aquatic plants in reservoirs and canals. A summary of registered aquatic herbicides and their target plants has been compiled (Westerdahl and Getsinger, 1988). Commonly used compounds include acrolein, diquat, endothall, glyphosate, dalapon, 2,4-D, MCPA, fluridone, diuron, simazine, hexazinone, chlorthiamid, dichlobenil, and copper sulfate—probably the most common one (Bowmer, 1987). Quantitative data on national aquatic pesticide use are generally not available at this time. Most aquatic applications of pesticides are carried out by federal, state, and local government agencies, or through permits issued by these agencies. Documentation of aquatic pesticide use on a national scale is poor, since management programs through the United States operate independently, with no organized system of information exchange. Aquatic pesticide use will be included in the USEPA's C/CPAS.

In the past, measures used to control mosquitoes in parts of the United States included spraying surface water breeding areas with DDT, and later with OPs such as fenthion and malathion (Wang and others, 1987b; Metropolitan Mosquito Control District, 1993). More recently, bacterial agents, such as *Bacillus thuringiensis* var. *israelensis* (Bti), and growth regulators, such as methoprene, have been used increasingly to control mosquito larvae in surface water bodies without the potential negative impacts of more traditional chemical insecticides. These materials are applied to known breeding areas, such as marshes and areas of standing water in a variety of forms, including briquets, pellets, and granules.

Fish parasites, such as the lamprey, are somewhat controlled in the Great Lakes by the use of the insecticide 3-trifluoromethyl-4-nitrophenol (TFM), which is applied during spawning times in the lakes' tributaries (Carey and Fox, 1981). Rough fish are sometimes controlled with piscicides, such as synthetic pyrethroids (permethrin, cypermethrin, and others), antimycin, and rotenone. Also, it should be noted that physical controls (such as water drawdown) and biological controls (such as grass carp and specific viruses and bacteria) are commonly used to control pests in surface waters.

Pesticides are often used in aquaculture, although water bodies used for this purpose are not normally considered part of the natural surface water system. Aquaculture is the farming of aquatic organisms (such as catfish, shrimp and salmon) for human consumption. The herbicides 2,4-D, diquat, endothall, fluridone, simazine, and xylene, and the piscicides antimycin and rotenone are approved for use in aquaculture (Fong and Brooks, 1989), although little data are available on the amounts used in this application.

3.3 OCCURRENCE AND DISTRIBUTION IN RELATION TO USE

INTRODUCTION

The geographic distribution of pesticides in surface waters can be best evaluated—with varying degrees of success—by comparing results from individual national and multistate occurrence studies to national and regional use patterns for specific compounds. The majority of the occurrence data for these comparisons are for the OCs and the triazine and acetanilide herbicides, although there are limited data for comparisons with other classes of pesticides. Studies listed in Table 2.1 are used as the basis for most of the comparisons. For high-use pesticides not addressed in large-scale studies, applicable smaller-scale studies (Table 2.2) will be used when possible. The regional and national use patterns are dominated by agricultural applications, which are the primary focus at this scale of analysis. The significance of other sources of pesticides (such as urban use and forestry), and sources not directly related to use (such as ground water, bed sediments, and the atmosphere), is discussed in Chapter 4.

ORGANOCHLORINE INSECTICIDES

The use of OCs in the United States began in the 1940's and continued into the 1970's, with peak use occurring in the late 1950's and early 1960's. The use of most OCs in the United States was banned or severely restricted in the early to mid-1970's, as potential human health concerns and the adverse ecological effects of some of these compounds became apparent. The resulting dramatic shift in agricultural insecticide use is apparent in Table 3.2. The only organochlorine compound still widely used in United States agriculture is endosulfan, which is applied to cotton, fruits, nuts, berries, and vegetables throughout the United States (Gianessi and Puffer, 1991).

In general, the organochlorines are hydrophobic compounds with extremely low water solubility and strong sorption tendencies. Most of these are resistant to degradation in both soil and water, and most have very long environmental lifetimes (Howard, 1991; Howard and others, 1991). In water, OCs tend to be associated with suspended sediments and bed sediments. Exceptions to these generalizations are lindane (γ-HCH), which is relatively water soluble (6 mg/L), and endosulfan, which can undergo biodegradation and hydrolysis in water much more quickly than most of the other OCs. The estimated half-life of endosulfan in surface water is 0.2 to 9 days (Howard and others, 1991). The OCs remain the focus of considerable attention long after most uses were curtailed in the United States because of their continued presence in bed sediments of surface waters, in soil contaminated from past applications, and in the atmosphere.

The continued focus on these compounds also is due to their toxicity. Most of the OCs are classified as probable human carcinogens by the USEPA, based on results of animal studies (Nowell and Resek, 1994). In addition, a number of these compounds accumulate in tissues of aquatic organisms and have been shown to biomagnify in the food chain. Chronic criteria concentrations established by the USEPA for the protection of aquatic organisms are very low for most OCs, well below detection limits for these compounds in many of the reviewed studies (see Section 5.2).

Several OCs have been common contaminants in surface waters since at least the 1950's, when the first systematic monitoring took place (Breidenbach and others, 1967). The most commonly detected compounds during the 1960's and 1970's were dieldrin, endrin, heptachlor epoxide, DDT and its transformation products DDD and DDE, and lindane and other isomers of hexachlorocyclohexane (HCH). Detection frequencies for DDT, DDD, and DDE are based on detections of both the p,p'- and o,p'-isomers, as many studies did not report detections of each isomer separately. In the studies conducted by the FWQA from 1958 to 1968, dieldrin was detected in 38 to 90 percent of the samples each year. Detection frequencies for other compounds commonly detected in these studies were as follows: endrin–13 to 40 percent, DDT–15 to 30 percent, DDD–13 to 60 percent, and DDE–5 to 40 percent. Heptachlor epoxide and HCH were monitored from 1964 to 1968, and detected in 5 and 7 percent of samples, respectively. Detection limits for the OCs in these studies were 0.001 to 0.002 μg/L. In the USGS studies of streams in the western United States, conducted from 1965 to 1971, the same group of compounds was observed, but detection frequencies were somewhat lower, and declined markedly during the 6-year period (Table 2.1). Results from these two studies cannot be compared directly, however, because of differences in sampling frequency, detection limits, and sampling sites. Many of the same compounds were detected in the Mississippi and Missouri Rivers in another study conducted during the same period (see Schafer and others, 1969 in Table 2.1).

Results from the USGS western streams studies (Brown and Nishioka, 1967; Manigold and Schulze, 1969; Schulze and others, 1973) are compared with data on regional agricultural use of selected OCs in Figures 3.38, 3.39, and 3.40. Figure 3.38 compares use of DDT with the detection frequency of total DDT (DDT, DDE, and DDD). Figure 3.39 compares use of aldrin plus dieldrin with the detection frequency of dieldrin (dieldrin was used as an insecticide, and is also the major transformation product of aldrin). Figure 3.40 compares the use of lindane with its detection frequency. These three figures indicate that the regional-use data cannot adequately relate occurrence of the OCs in surface waters with agricultural use. For every compound in these figures, examples can be found of sites in supposedly high-use areas that had very low detection frequencies and of sites in supposedly low-use areas with high detection frequencies. Little correlation is evident whether the use data are expressed as total amount applied in the region, amount applied per acre of cropland, or amount applied per acre of all land in the region. The most likely reason for the apparent discrepancies is variability of use within the large regions delineated in the use data. For example, sites on two rivers (the Yakima River in Washington and the Gila River in Arizona) had consistently high detection frequencies for most of the targeted OCs, regardless of the use-level reported. Both of these rivers drain areas in which irrigation is used to support intensive agricultural activity, and it is likely that pesticide use within these river basins was considerably higher than the regional average. More recent pesticide use data show heavy use of insecticides in both areas (Figure 3.4), although different compounds are now used.

Another reason for the apparent lack of correlation between agricultural use of OCs and occurrence of these compounds in rivers may be their tendency to associate with particles in the water column. The studies from the 1960's and 1970's analyzed whole-water samples. That is, samples were not filtered before analysis, and any pesticides sorbed to suspended sediments were included in the analysis. For most of the OCs, the sorbed phase represents a significant portion of the total amount in the water column. Thus, much of the variation in the concentrations of

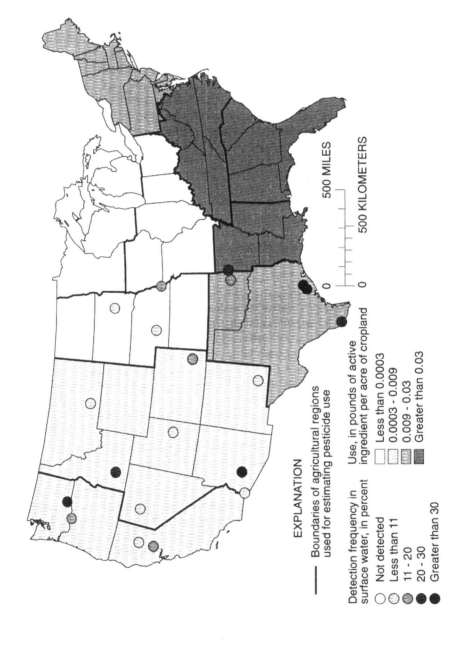

Figure 3.38. Regional agricultural use of DDT in 1971, and detection frequency of DDT, DDD, and DDE in rivers and streams of the western United States from 1967 to 1971. Use data are from Manigold and Schulze (1969) and Schulze and others (1973). Detection frequencies are from Andrilenas (1974).

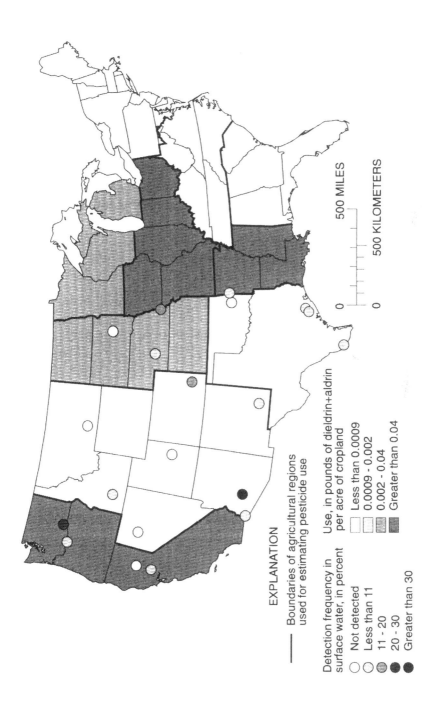

Figure 3.39. Combined regional agricultural use of aldrin and dieldrin in 1971, and detection frequency of dieldrin in rivers and streams of the western United States from 1967 to 1971. Use data are from Andrilenas (1974). Detection frequencies are from Manigold and Schulze (1969) and Schulze and others (1973).

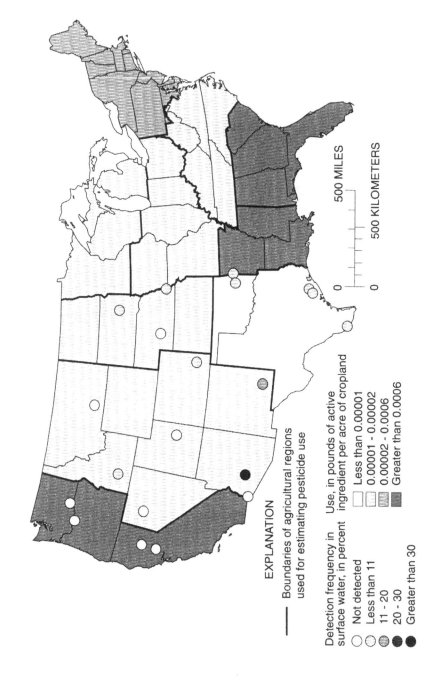

Figure 3.40. Regional agricultural use of lindane in 1971, and detection frequency of lindane in rivers and streams of the western United States from 1967 to 1971. Use data are from Andrilenas (1974). Detection frequencies are from Manigold and Schulze (1969) and Schulze and others (1973).

organochlorine compounds in these studies may have been due to variations in suspended-sediment concentrations at the time of sampling. Analysis of suspended-sediment and bed-sediment concentrations is a more appropriate method of monitoring the occurrence of OCs in surface waters. None of the national-scale studies described above measured concentrations of OCs in suspended sediments. In the USGS/USEPA national study (Gilliom and others, 1985) conducted from 1975 to 1980, both whole-water and bed-sediment samples were analyzed. The detection frequencies for the OCs were much higher in bed-sediment samples than in whole-water samples. For several compounds (chlordane, DDT, endrin, heptachlor, and toxaphene), a moderately strong positive correlation was evident between detection frequency in bed sediments and regional agricultural use. The occurrence of pesticides in bed sediments of surface waters of the United States will be discussed in a companion text (Nowell, 1996).

As mentioned earlier, agricultural use of most OCs in the United States was banned or severely restricted in the early 1970's. Because of the persistence of many of these compounds, they continue to be detected in surface waters of the United States. However, few large-scale studies have been conducted in recent years in which OCs were included as analytes. In a 1986 survey of organochlorine compounds in the Great Lakes (see Stevens and Nielson [1989] in Table 2.1), α-HCH, lindane, heptachlor epoxide, and dieldrin were detected in all 95 samples, although concentrations were very low. Data retrieved from USEPA's STOrage and RETrieval water quality database (STORET) show that a number of organochlorine compounds were detected routinely in ambient waters between 1975 and 1982 (Staples and others, 1985), as shown in Table 2.1.

The OCs were targeted in more than 70 of the state and local studies listed in Table 2.2. Differences in detection limits, sampling frequency, and sampling sites among these studies prevent direct comparison of their results in most cases. Taken together, however, these studies give a general picture of which compounds have been detected most often in surface waters since their use was restricted. Studies published since 1975, in which site- and sample-detection frequencies were given (Table 2.2), show that a number of OCs are still present in many surface water bodies of the United States. In these studies, α-HCH was detected most often—in 47 percent of samples and at 42 percent of sites. Chlordane was detected in 42 percent of samples and at 23 percent of sampling sites. Other compounds commonly detected were dieldrin (22 percent of sites), lindane (27 percent of sites), heptachlor epoxide (13 percent of sites), DDT (13 percent of sites), methoxychlor (11 percent of sites), and DDE (11 percent of sites). Studies listed in Table 2.2 that were used to calculate these detection frequencies had combined totals of 200 to 400 sampling sites and 1,000 to 3,000 samples for most of these compounds. For state and local studies conducted from 1982 to the present, detection frequencies for these compounds are similar. The continued detection of these compounds is not due to any significant lowering of detection limits in the more recent studies. Detection limits in most of the studies since 1975 were similar to, or higher than, detection limits in the large-scale studies of the 1960's and early 1970's (Table 2.2).

Concentrations of all the OCs were nearly always low in the reviewed studies. Maximum concentrations, when reported, rarely exceeded 1 µg/L and were usually below 0.2 µg/L in natural surface waters. The significance of these levels is discussed further in Chapter 5. High concentrations of these hydrophobic compounds in water would not be expected, as they tend to associate with particles in the water column and eventually settle and accumulate in bottom sediments. Occasional detections of higher concentrations in some of the reviewed studies may have been due to especially high concentrations of suspended sediment, if whole-water samples were analyzed.

ORGANOPHOSPHORUS INSECTICIDES

The OPs came into wide-scale use in the United States in the late 1960's and 1970's. The total amount of OPs used in agriculture in the United States has remained relatively stable over the last two decades, as use of some compounds declined and new compounds came into wide use (Table 3.2). The proportion of total insecticide use accounted for by OPs has increased steadily, however, as use of OCs declined. By the late 1980's, OPs accounted for approximately 65 percent of total insecticide use, and 7 of the top 11 insecticides were OPs, in terms of mass applied (Gianessi and Puffer, 1992a). The most commonly used OPs of recent years (chlorpyrifos, methyl parathion, terbufos, phorate, and fonofos) are used primarily on corn, cotton, alfalfa, sorghum, citrus crops, apples, potatoes, and peanuts. They are used on a myriad of other crops in lesser amounts, in nearly every state in the nation. Several OPs also have significant non-agricultural use, including diazinon, chlorpyrifos, malathion, and acephate (Table 3.3). Maps showing agricultural use of a number of the high-use OPs are included in Section 3.2.

The OPs as a group vary considerably in chemical and environmental properties (Goss, 1992). A number of the high-use OPs (diazinon, ethion, disulfoton, fonofos, and phorate) have a relatively high potential to move from agricultural fields to runoff water in either the dissolved or sorbed phase. Others (such as chlorpyrifos and malathion) are unlikely to be transported in runoff. Aquatic persistence is variable as well. Half-lives in surface water may be as short as 1 to 3 days for azinphos-methyl, dichlorvos, phorate, and disulfoton, but as long as weeks for diazinon, methyl parathion, parathion, and terbufos (Howard, 1991; Howard and others, 1991). Concern about the presence of OPs in surface waters stems from the relatively high toxicity of some of the most commonly used compounds. Phorate, disulfoton, parathion, fonofos, azinphos-methyl, and methyl parathion are all regarded as toxic to mammals, having an oral LD_{50} (the dosage of a chemical needed to produce death in 50 percent of the treated test animals) of less than 100 mg/kg for the rat (Fukuto, 1987). USEPA criteria for safety of freshwater aquatic biota, for both acute and chronic exposures, are less than 0.1 µg/L for those OPs (azinphos-methyl, chlorpyrifos, malathion, and parathion) for which criteria have been established (Nowell and Resek, 1994).

In general, OPs have not been frequently detected in surface waters. National-scale studies from the late 1960's and the 1970's included several OPs as analytes (Table 2.1). In the USGS study of western streams (Schulze and others, 1973), four OPs were targeted in samples collected monthly from June 1970 to September 1971. Diazinon, parathion, and methyl parathion were detected at 3 to 5 of the 20 sites, but in only 1 to 2 percent of samples, with detection limits of 0.01 µg/L. Samples collected from over 100 sites on rivers throughout the United States in September 1967 (low flow) and in June 1968 (high flow, after pesticide application) contained no detectable residues of any of six targeted OPs with detection limits of 0.01 to 0.025 µg/L (Lichtenberg and others, 1970). In the USGS/USEPA national study conducted from 1975 to 1980 (Gilliom and others, 1985), quarterly samples were collected at 160 to 180 sites on rivers throughout the United States. Seven OPs were included as analytes, but reporting limits were relatively high—0.1 to 0.5 µg/L. Detection frequencies were low for all OPs, ranging from 0 to 1.2 percent of the nearly 3,000 samples. Diazinon, however, was detected at nearly 10 percent of the sites (1.2 percent of samples). Diazinon had the lowest reporting limit (0.1 µg/L) of all the OPs in this study, and it is difficult to say whether the apparently higher incidence of occurrence was real or a result of this difference in reporting limits. The small number of detections of OPs in this study generally prevented analysis of the relation between geographical occurrence and agricultural use. A 1974 study analyzing water from 33 sites in the upper Great Lakes (Lakes

Superior and Huron) included 17 OPs as analytes (Glooschenko and others, 1976). No OPs were detected in any of the samples, with detection limits ranging from 0.003 to 0.05 µg/L.

Several more recent studies have included OPs as analytes, primarily in the Midwest (Goolsby and Battaglin, 1993; Richards and Baker, 1993; Larson and others, 1995). The data from the Mississippi River Basin study (Goolsby and Battaglin, 1993; Larson and others, 1995) can be used to illustrate the general patterns observed in this region, where a number of OPs are heavily used in agriculture. Eleven OPs were targeted in this study, in which two sites on the main stem of the Mississippi River and six sites near the mouths of major tributaries were sampled two to three times per week from May 1991 to March 1992 (Figure 2.4). Detection limits ranged from 0.002 to 0.02 µg/L for the OPs. Detection frequencies again were very low for most of the OPs (Table 3.5), with four of the compounds (phorate, parathion, disulfoton, and azinphos-methyl) not detected in any of the 316 samples. Four more compounds (methyl parathion, terbufos, ethoprop, and malathion) were detected in less than 2 percent of samples, at only one or two of the eight sites. Diazinon, fonofos, and chlorpyrifos were detected at most or all of the sites, but concentrations were always low, with none higher than 0.1 µg/L. Total riverine flux for the 11-month period, expressed as a percentage of agricultural use in each of the subbasins, also was very low (most less than 0.1 percent of the amount applied) for nearly all detected OPs. This was substantially lower than the percentages for many of the herbicides analyzed in the same study. The exception to this pattern was diazinon, which had a consistently higher percentage in several of the basins. This occurred in the basins with the highest population densities and major urban centers, and the high flux percentages were attributed to urban use of diazinon, which was not accounted for in the agricultural-use data used in the calculation of the percentages (see Section 5.3).

The OPs were included as analytes in 39 of the state and local studies in Table 2.2. In only one of these studies has an OP concentration exceeded a human health-based water quality criteria—maximum concentrations of terbufos in several Lake Erie tributaries were slightly above the Health Advisory Level (HAL) of 1 µg/L for several years between 1982 and 1991 (Baker, 1988b; Richards and Baker, 1993). Taken together, the studies listed in Table 2.2 give a general picture of the occurrence of OPs in surface waters from the 1960's to the early 1990's. Using only those studies for which site-specific data were given, eight OPs were detected at more than 5 percent of sampling sites. Seven of these compounds (chlorpyrifos, phorate, parathion, terbufos, malathion, methyl parathion, and fonofos) were among the 10 highest agricultural-use OPs in the United States in the late 1980's (Gianessi and Puffer, 1992a). Chlorpyrifos, the highest agricultural-use OP, was detected at approximately 7 percent of sites, but was observed consistently in less than 1 percent of samples. Diazinon, the twelfth highest agricultural-use OP, was detected at the highest percentage of sites (22 percent), but also has significant use in urban areas. Several of the studies in which diazinon was detected included urban storm drains as sampling sites.

Recent studies have reported the occurrence of several OPs in rivers draining the intensely farmed Central Valley of California (Kuivila and Foe, 1995). Distinct pulses of elevated concentrations of diazinon and methidathion were observed in the San Joaquin and Sacramento Rivers in January and February of 1993, following periods of rainfall in the area. These compounds are two of several pesticides sprayed on orchards in the area during the winter dormant period. Concentrations of diazinon in the pulses were in the 0.1 to 0.4 µg/L range in the Sacramento River and up to 1.1 µg/L in the San Joaquin River, and elevated concentrations continued downstream into San Francisco Bay. These levels are well above the National Academy of Sciences' (NAS) diazinon guideline (0.009 µg/L) for the protection of aquatic life (Nowell and Resek, 1994). Seven-day bioassay tests conducted in this study demonstrated that

Table 3.5. Agricultural use and riverine flux as a percentage of use for 26 pesticides in the Mississippi River Basin, 1991

[Data from Larson and others, 1995. Total reported agricultural use in the basin, in metric tons (1,000 kilograms) (Gianessi and Puffer, 1991, 1992a). Flux calculated by substituting zero for concentrations below the detection limit. no use, no agricultural use reported in basin; nd, no data; no det., no samples with concentrations above the detection limit; h, herbicide; i, insecticide. <, less than]

| Pesticide | Rivers Sampled | | | | | | | |
| | Minnesota | | White | | Illinois | | Platte | |
	Use in basin	Flux as percent of use	Use in basin	Flux as percent of use	Use in basin	Flux as percent of use	Use in basin	Flux as percent of use
PESTICIDES WITH LARGE RUNOFF POTENTIAL								
Atrazine (h)	290	0.62	710	0.95	2,000	1.9	1,600	0.84
Butylate (h)	160	no det.	270	0.01	750	0.02	560	<0.01
Carbofuran (i)	11	no det.	42	0.05	95	0.38	190	0.09
Diazinon (i)	0.0	no use	0.08	20.0	2.4	4.0	39	0.02
Disulfoton (i)	0.39	no det.	0.0	no det.	0.24	no det.	39	no det.
Fonofos (i)	31	0.45	20	0.04	120	0.09	100	0.03
Linuron (h)	0.54	no det.	44	0.01	86	no det.	3.7	no det.
Metolachlor (h)	510	0.65	430	0.48	1,800	0.93	440	0.74
Metribuzin (h)	32	0.23	54	0.21	140	0.23	38	0.33
Pendimethalin (h)	160	<0.01	41	no det.	290	<0.01	92	<0.01
Phorate (i)	0.0	no det.	0.52	no det.	0.67	no det.	47	no det.
Propargite (i)	47	<0.01	11	0.06	73	no det.	66	no det.
Simazine (h)	0.96	5.2	9.5	5.0	67	0.97	4.6	2.4
Trifluralin (h)	510	<0.01	79	<0.01	630	0.01	240	0.01
PESTICIDES WITH MEDIUM RUNOFF POTENTIAL								
Alachlor (h)	1,400	0.20	850	0.15	2,000	0.46	1,000	0.31
Azinphos-methyl (i)	0.06	no det.	0.72	no det.	1.2	no det.	0.18	no det.
Carbaryl (i)	8.6	nd	3.2	<0.01	30	0.01	140	<0.01
Chlorpyrifos (i)	53	0.15	60	no det.	280	0.07	300	0.01
Cyanazine (h)	450	1.3	210	0.76	750	3.1	460	2.6
EPTC (h)	1,500	<0.01	42	0.02	650	0.05	460	<0.01
Ethoprop (i)	0.99	.08	0.0	no det.	0.01	no det.	37	no det.
Methyl parathion (i)	8.5	.10	0.08	no det.	9.4	no det.	140	no det.
Parathion (i)	3.3	no det.	0.0	no det.	0.0	no det.	110	no det.
Propachlor (h)	22	2.0	0.0	no use	35	0.09	70	0.20
Terbufos (i)	56	0.05	42	no det.	150	<0.01	260	no det.
PESTICIDES WITH SMALL RUNOFF POTENTIAL								
Malathion (i)	0.0	no use	0.98	0.12	0.90	no det.	9.4	no det.

water sampled during the pulses was indeed acutely toxic to daphnia. The estimated total flux of diazinon in the Sacramento River represented 0.5 to 1.7 percent of the amount of diazinon applied in the Sacramento Valley during January and February, when the dormant season spraying occurs. This is similar to the losses observed for preemergent herbicides in the Midwest and agrees well with predictions of runoff losses from fields when rainfall occurs shortly after application (Wauchope, 1978). The fact that similar losses occurred with these two very different

Table 3.5. Agricultural use and riverine flux as a percentage of use for 26 pesticides in the Mississippi River Basin, 1991—*Continued*

Pesticide	Rivers Sampled							
	Missouri		Ohio		Mississippi			
					at Thebes, Illinois		at Baton Rouge, Louisiana	
	Use in basin	Flux as percent of use	Use in basin	Flux as percent of use	Use in basin	Flux as percent of use	Use in basin	Flux as percent of use
PESTICIDES WITH LARGE RUNOFF POTENTIAL								
Atrazine (h)	6,300	1.2	4,800	1.2	1,500	0.96	13,000	1.5
Butylate (h)	1,900	<0.01	1,900	0.03	470	no det.	4,200	<0.01
Carbofuran (i)	650	0.09	370	0.08	53	0.28	930	0.14
Diazinon (i)	84	0.07	6.6	1.10	18	no det.	110	0.13
Disulfoton (i)	200	no det.	3.3	no det.	11	no det.	220	no det.
Fonofos (i)	270	0.02	160	0.01	65	0.11	660	0.02
Linuron (h)	55	no det.	270	no det.	15	no det.	240	no det.
Metolachlor (h)	3,500	0.68	3,400	0.45	1,700	0.53	11,000	0.80
Metribuzin (h)	210	0.41	400	0.09	72	0.10	610	0.36
Pendimethalin (h)	810	0.01	480	<0.01	330	<0.01	2,100	<0.01
Phorate (i)	100	no det.	6.6	no det.	0.32	no det.	120	no det.
Propargite (i)	300	0.05	130	no det.	64	0.08	490	no det.
Simazine (h)	52	1.4	220	3.9	3.9	10.5	170	1.6
Trifluralin (h)	2,400	0.01	750	<0.01	810	no det.	5,400	<0.01
PESTICIDES WITH MEDIUM RUNOFF POTENTIAL								
Alachlor (h)	4,700	0.16	4,700	0.12	2,100	0.30	12,000	0.27
Azinphos-methyl (i)	1.5	no det.	14	no det.	4.3	no det.	11	no det.
Carbaryl (i)	810	<0.01	200	0.08	36	no det.	920	0.01
Chlorpyrifos (i)	1,100	no det.	490	<0.01	250	0.02	2,300	<0.01
Cyanazine (h)	2,000	2.0	1,400	0.82	1,600	0.57	6,200	1.6
EPTC (h)	4,200	<0.01	810	0.02	3,900	0.01	12,000	0.01
Ethoprop (i)	85	no det.	3.9	no det.	1.4	no det.	98	no det.
Methyl parathion (i)	120	no det.	19	no det.	15	no det.	150	no det.
Parathion (i)	360	no det.	0.08	no det.	4.3	no det.	580	no det.
Propachlor (h)	1,100	0.01	19	no det.	066	0.05	1,400	<0.01
Terbufos (i)	640	no det.	280	no det.	130	no det.	1,300	no det.
PESTICIDES WITH SMALL RUNOFF POTENTIAL								
Malathion (i)	85	<0.01	45	no det.	0.23	no det.	100	no det.

pesticide applications (herbicides on corn and soybeans, and an insecticide sprayed on orchards) demonstrates that agricultural practices (such as application technique) and weather can be major factors in determining the amounts of pesticides lost in runoff. Two other OPs used as dormant sprays in this area—malathion and chlorpyrifos—were rarely detected in the rivers during this period. Malathion is known to degrade quickly in soil, and chlorpyrifos has a much lower water solubility (2 mg/L) than diazinon (40 mg/L) and methidathion (250 mg/L), resulting in a low runoff potential for both compounds (Becker and others, 1989; Goss, 1992).

From the studies reviewed, it appears that the combination of low runoff potential of some OPs, and the short aquatic lifetime of others, has precluded any significant occurrence of most OPs in surface waters in most areas of the United States. It should be noted, however, that USEPA criteria levels and NAS guidelines for protection of aquatic life are very low for some of these compounds (see Section 6.1). Detection limits in most of the reviewed studies are close to, or in some cases above, these levels. Because of the widespread use of these compounds, their use in non-agricultural settings, and the low detection frequency in most of the reviewed studies, the relation between regional use patterns of OPs and their occurrence in surface waters is unclear. An important exception is the dormant spray use of diazinon and methidathion in California described previously, where the relation is relatively clear. It also appears that urban uses of diazinon are resulting in detectable levels in surface waters in some areas, although this is difficult to quantify due to the lack of data on non-agricultural use.

TRIAZINE AND ACETANILIDE HERBICIDES

Triazine and acetanilide herbicides have been used in the United States since the 1940's. The amounts used in agriculture have risen dramatically in the last 30 years, increasing by more than a factor of eight between 1966 and 1991 (Table 3.2). More than 215 million lb a.i. of these compounds were used in agriculture during 1988–1991, accounting for approximately 47 percent of total herbicide use and 36 percent of total agricultural pesticide use. In the late 1980's, 8 of the top 20 herbicides, in terms of mass applied, were triazines or acetanilides (Gianessi and Puffer, 1991). Estimates of individual compound use are given in Table 3.1. Of the commonly used triazines, atrazine, cyanazine, and simazine are used primarily on corn; metribuzin is used primarily on soybeans; and propazine was used almost exclusively on sorghum. The use of propazine was discontinued in the late 1980's (Gianessi and Puffer, 1991). Atrazine also is used on sorghum. The most commonly used acetanilides (alachlor, metolachlor, and propachlor) are primarily used on corn, soybeans, and sorghum. The heaviest agricultural use of triazine and acetanilide herbicides is in Illinois, Indiana, Iowa, Minnesota, Wisconsin, Nebraska, Missouri, and Ohio (the corn belt). Several of these compounds also are used extensively in Kansas, Texas, and several states in the southeastern United States. Maps showing agricultural use for a number of the high-use triazine and acetanilide herbicides are included in Section 3.2.

As a group, the commonly used triazine and acetanilide herbicides have moderate to high water solubility and relatively low soil-sorption coefficients, and several are relatively persistent in soil (Wauchope and others, 1992). As a result, they have a moderate to high potential for loss from fields through surface runoff, primarily in the dissolved phase (Goss, 1992). In addition, most are chemically stable in water (not hydrolyzed) and are unlikely to volatilize from water. In general, triazines are somewhat more resistant to biodegradation than acetanilides (Muir, 1991). Aquatic half-lives, as measured in laboratory studies, range from 10 days for propachlor to more than 90 days for atrazine (Muir, 1991), although considerable variation can be found in the experimental results for individual chemicals, depending on the experimental conditions.

As use of these compounds has increased, concern about their presence in surface waters also has grown. Although triazines and acetanilides, like most herbicides, have low acute toxicity to most animals, the potential effects on human health remain an area of concern. Alachlor is classified as a probable human carcinogen, and several others (atrazine, cyanazine, metolachlor, propazine, and simazine) are classified as possible human carcinogens by the USEPA (Nowell and Resek, 1994). The USEPA has established MCLs in drinking water for alachlor, atrazine, and simazine of 2, 3, and 4 µg/L, respectively (Nowell and Resek, 1994). Several of these herbicides have been detected in surface waters used for drinking water supplies (Wnuk and others, 1987; Goolsby and Battaglin, 1993; Larson and others, 1995). In addition to concerns about human health effects, the effects of long-term, low-level concentrations of these compounds, or combinations of these compounds, on aquatic ecosystems is largely unknown.

Several of the recent regional studies mentioned in Section 3.1 have provided important information on the occurrence and geographic distribution of some of the most commonly used triazine and acetanilide herbicides. The availability of detailed agricultural-use data from approximately the same period allows a number of conclusions to be drawn about the relation between agricultural use of these compounds and their occurrence in surface waters. Selected findings from these studies will be used to illustrate these conclusions.

In the 1989–1990 USGS study (Figure 2.3) of numerous small river basins throughout the Midwest (Goolsby and Battaglin, 1993), herbicides used in the basins were detected at 98 to 100 percent of the sites in the postplanting samples. This region has the highest use of triazine and acetanilide herbicides in the United States (Figures 3.6, 3.7, 3.9, 3.14, and 3.16). Detection limits in this study were 0.2 µg/L for cyanazine and 0.05 µg/L for all other analytes. Atrazine, alachlor, and metolachlor were the most frequently detected herbicides in both years, with detections at 81 to 100 percent of the sites in the postplanting samples. Of the herbicides targeted in the study, these three compounds were the highest-use herbicides in this region, although use totals for each basin were not available. Concentrations in most 1989 postplanting samples ranged from 1 to 10 µg/L for atrazine and from 0.2 to 4 µg/L for alachlor, metolachlor, and cyanazine. Maximum concentrations in the 1989 postplanting samples were 108 µg/L for atrazine; 40 to 60 µg/L for alachlor, metolachlor, and cyanazine; and 1 to 8 µg/L for simazine, propazine, and metribuzin. Concentrations were much lower in the preplanting and postharvest samples, as were detection frequencies for most of the compounds. The results from this study show the widespread nature of the occurrence (during 1989–1990) of these herbicides in this region of intensive row-crop agriculture. The seasonal aspects of pesticide occurrence in streams are discussed in Section 5.1.

Similar results were obtained in the 1992 study of the midwestern reservoirs shown in Figure 2.4 (Goolsby and others, 1993). At least 1 of the 14 targeted triazine and acetanilide herbicides (and their transformation products) was detected in 82 to 92 percent of the 76 sampled reservoirs during the four sampling periods from April through November. These compounds occurred most frequently in reservoirs in the areas of the highest herbicide use. In addition, the locations of reservoirs in which MCLs for atrazine and alachlor were exceeded correspond to areas of highest use of these compounds (Figure 3.41).

Additional inferences can be made about relations between pesticide use and occurrence of pesticides in rivers when basin-wide pesticide use totals are available. In the Mississippi River Basin study (Figure 2.4) of large rivers (Goolsby and Battaglin, 1993; Larson and others, 1995), pesticide use totals for each basin were combined with data on pesticide concentrations and river discharge to calculate the total flux of each targeted pesticide at each sampling site (Table 3.5). For most of the triazine and acetanilide herbicides, the amounts detected in the rivers generally represented 0.2 to 2 percent of the amounts applied to fields in each basin (Figure 3.42). Atrazine, metolachlor, metribuzin, and alachlor percentages were quite similar in each of the basins. Approximate ranges of the percentages were 1 to 1.5 for atrazine, 0.5 to 1.0 for metolachlor, 0.2 to 0.4 for metribuzin, and 0.1 to 0.5 for alachlor. The results for these compounds imply that a relatively constant amount of each is lost from agricultural fields to surface waters, across a wide range of spatial scales. Percentages for cyanazine, propachlor, and simazine varied substantially among basins. The variability in the percentages of these three compounds can be partially accounted for in each case. The anomalously high percentages for simazine in some basins may

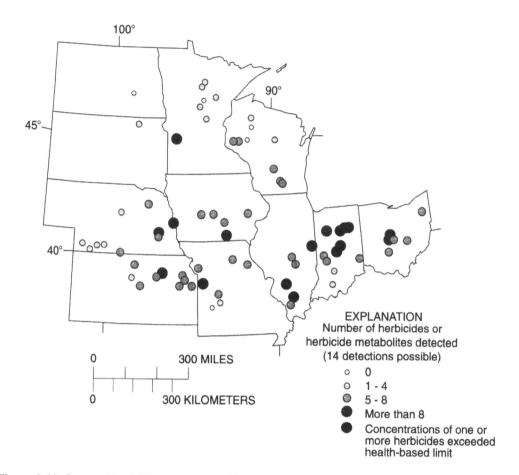

Figure 3.41. Geographic distribution of herbicide and metabolite detections in midwestern reservoirs, and locations of reservoirs in which concentrations of one or more herbicides exceeded a U.S. Environmental Protection Agency maximum contaminant level or health advisory level for drinking water. Samples collected from late June to early July 1992 (see Table 2.1 for the analytes targeted in this study). Redrawn from Goolsby and others (1993).

have been due to use of simazine not accounted for in the agricultural-use data used to calculate the percentages in this study. The USEPA estimated that non-agricultural use of simazine accounted for 30 to 55 percent of total simazine use in the late 1980's (Gianessi and Puffer, 1991). Cyanazine has a somewhat different use pattern than the other triazines and acetanilides, which are applied primarily as preemergent herbicides. Cyanazine is often applied after the spring runoff (Schottler and others, 1994), which could add variability to the amount lost to surface waters. Propachlor had low use in most of the basins, and a lower detection frequency than the other triazine and acetanilide herbicides, resulting in a higher level of uncertainty in the calculated flux value for some of the basins.

A much larger fraction of the applied triazine and acetanilide herbicides was detected in the rivers in this study than in the other herbicides monitored (butylate, linuron, EPTC, pendimethalin, and trifluralin). This is shown graphically in Figure 3.43, in which the fluxes of all the herbicides monitored in this study are plotted in relation to the amount of each compound

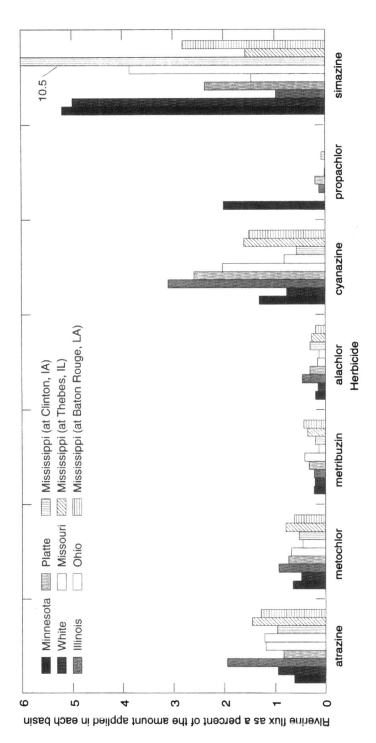

Figure 3.42. Riverine flux of herbicides at three sites on the Mississippi River and at sites on six major tributaries in 1991, expressed as a percentage of the amount applied agriculturally in each basin. Data are from Larson and others (1995).

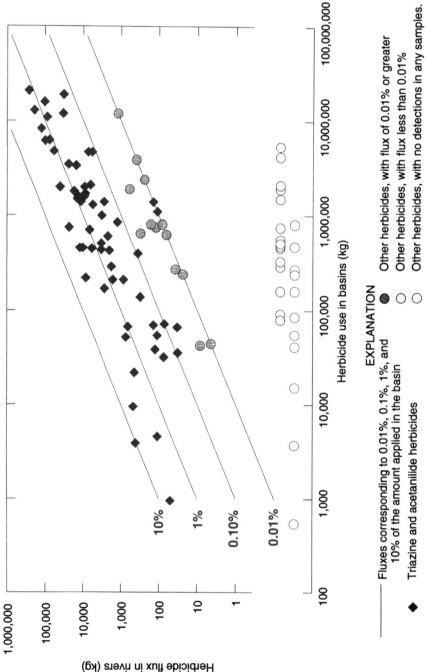

Figure 3.43. Riverine flux of herbicides in relation to the amount applied agriculturally in the drainage basins at three sites on the Mississippi River and at sites on six tributaries in 1991. Lines represent fluxes corresponding to 0.01, 0.1, 1, and 10 percent of the amount of herbicides applied in the basins. ● – Triazine and acetanilide herbicides. ◐ – Other herbicides, with flux of 0.01 percent or greater. ○ – Other herbicides, with flux less than 0.01 percent. ○ – Other herbicides, with no detections in any samples. Data from Larson and others (1995).

applied in each of the basins. Fluxes of the triazines and acetanilides, expressed as a percentage of the amount applied, were consistently higher than fluxes of the other herbicides in the study (the three anomalously low data points for the triazine and acetanilide herbicides are for propachlor, in basins where use was low and detections less frequent, resulting in large uncertainty in the calculated flux). Fluxes of the triazines and acetanilides were in the 0.1 to 10 percent range, whereas fluxes of the other herbicides were all less than 0.1 percent. Application technique may be the most important factor in the differences observed. The triazines and acetanilides are normally applied preemergent to bare soil or residual vegetation from the prior year's crop. Most of the other herbicides in the study are incorporated into soil when applied (Wauchope and others, 1992). Incorporation greatly decreases the potential for loss from agricultural fields in runoff (Leonard, 1990).

More is known about the presence of atrazine in surface waters of the Mississippi River Basin than is known about any other pesticide in any part of the nation. Ciba-Geigy Corporation (Ciba-Geigy, 1992a,c,d,f, 1994a) has reviewed the results of numerous atrazine monitoring programs conducted on small and large rivers, lakes, and reservoirs of the central United States during 1974–1993. These programs were conducted by state and federal agencies, water utilities, universities, Monsanto Company, and Ciba-Geigy. Selected data from these reviews are discussed in the sections on seasonal patterns and health implications in Chapters 5 and 6. Results pertaining to occurrence and distribution of atrazine in surface waters are summarized briefly here.

Ciba-Geigy has summarized the results of monitoring for atrazine in Iowa from 1974 to 1992 (Ciba-Geigy, 1994a) and in Illinois from 1975 to 1988 (Ciba-Geigy, 1992f). These two states are in the heart of the corn belt, and are probably representative of conditions in Indiana, Ohio, and parts of Nebraska, Wisconsin, Minnesota, and Michigan. Illinois and Iowa ranked first and fourth, respectively, in atrazine use during 1988–1991 (Gianessi and Puffer, 1991). National atrazine use (1988–1991) is shown in Figure 3.7.

In Iowa, 69 surface water bodies were monitored at 86 sites between 1974 and 1992, with a total of 2,780 samples analyzed (Ciba-Geigy, 1994a). The duration of monitoring at each site varied between 1 and 13 years. Sampling frequency was variable, with many sites sampled quarterly or monthly, whereas others were sampled as many as 10 times per month during the spring and summer. The detection limit for many of the sites was 0.1 μg/L, but ranged from 0.02 to 3.0 μg/L. Overall, atrazine was detected in 81 percent of the samples. The maximum concentrations of atrazine (up to 71 μg/L) occurred primarily in May, June, and July. The overall distribution of concentrations was skewed toward the lower values, with 1.6 percent greater than 10 μg/L, 8 percent greater than 3 μg/L, and 21 percent greater than 1 μg/L. At sites where the sampling frequency was sufficient to calculate an annual mean concentration, 2.5 percent had annual means higher than the MCL of 3 μg/L.

In Illinois, the data reviewed by Ciba-Geigy (Ciba-Geigy, 1992g) is less extensive, but the results were very similar. Ciba-Geigy, Monsanto, and the Illinois Environmental Protection Agency (IEPA) monitored atrazine concentrations in 39 rivers and streams, and 3 lakes and reservoirs. Sampling was done over varying lengths of time during 1975–1976 and 1985–1988. Again, atrazine was frequently detected—in 98, 91, and 80 percent of samples analyzed by Ciba-Geigy, Monsanto, and IEPA, respectively. Detection limits were 0.1 μg/L (Ciba-Geigy), 0.2 μg/L (Monsanto), and 0.05 μg/L (IEPA). Maximum concentrations at most sites occurred in April, May, June, or July and ranged from less than 0.05 to 30 μg/L. Concentration distributions were skewed toward lower values, with 3 percent of annual means and 7 percent of individual sample concentrations greater than 3 μg/L. In the IEPA program, four sites in non-agricultural or low atrazine-use watersheds served as controls. Atrazine was detected in 68 percent of samples at these control sites, however, with concentrations up to 5 µg/L, and annual mean concentrations of 40 to 80 percent of the average annual mean from the other sites. Reasons for the relatively high level of detections at these sites were not discussed in the review.

The data from Illinois and Iowa agree well with Ciba-Geigy and Monsanto data from other states in the central United States and with atrazine data from long-term studies of Lake Erie tributaries in Ohio (Richards and Baker, 1993). The picture that emerges is of widespread, low-level contamination of surface waters by atrazine, often lasting throughout the year. This appears to be true for the regions of the United States where atrazine is used commonly. During much of the year, concentrations in nearly all monitored surface waters were well below 3 µg/L, and usually less than 1 µg/L. In virtually all the rivers and streams monitored, however, concentrations rose in the spring and early summer after atrazine application, often exceeding 3 µg/L for periods of days to weeks. Peak concentrations were usually higher in smaller streams draining agricultural areas than in larger rivers that integrate the inputs from larger, more heterogeneous areas. Annual mean concentrations seldom exceeded the MCL of 3 µg/L. The topics of seasonal patterns, concentrations exceeding established water quality criteria values, and the potential effects of the observed concentrations are discussed in more detail in Sections 5.1, 6.1, and 6.2.

In summary, the relation between agricultural use of the triazine and acetanilide herbicides and their occurrence and distribution in surface waters is relatively clear. Their use as preemergent herbicides in the central United States results in a fairly predictable level of contamination of surface waters. Several recent developments in the use of these compounds, however, raise new questions. First, recommended atrazine application rates have been lowered substantially in the last few years, from a maximum of 4 lb a.i. per acre before 1990, to 1.6 to 2.5 lb a.i. per acre (depending on tillage practices) in late 1992 (Goolsby and others, 1994). Second, use of cyanazine on corn and sorghum increased by more than 25 percent between 1989 and 1992, and use of metolachlor has also reportedly increased (Goolsby and others, 1994). Finally, a new acetanilide herbicide—acetochlor—was registered conditionally for use on corn in 1994 (U.S. Environmental Protection Agency, 1994a). An explicit condition of this registration states that the introduction of acetochlor must result in a 33 percent reduction (equivalent to 66.3 million lb a.i.), within 5 years, in the use of certain herbicides (atrazine, alachlor, metolachlor, EPTC, butylate, and 2,4-D) on corn (based on 1992 use levels). The effects of these changes on the occurrence of herbicides in surface waters are unknown at this time.

PHENOXY ACID HERBICIDES

The only phenoxy acid herbicides with significant current agricultural use in the United States are 2,4-D, esters of 2,4-D, and MCPA (Table 3.1). During the 1960's and 1970's, 2,4,5-T also was used agriculturally. Silvex (2,4,5-TP) was used in smaller amounts before 1984, when its use was restricted in the United States. During 1989–1991, 2,4-D ranked fifth among agricultural herbicides and MCPA ranked seventeenth in terms of mass applied. In addition, the USEPA estimates non-agricultural use of 2,4-D to be 38 to 87 percent of agricultural use (Gianessi and Puffer, 1991). If the high estimate for non-agricultural use is correct, total 2,4-D use ranked second among all herbicides during 1988–1991. The total amount of phenoxy compounds used in United States agriculture has remained nearly stable over the last two decades, although their proportion of the total herbicide use has dropped as use of triazine and acetanilide herbicides has increased (Table 3.2). The phenoxy compound 2,4-D is used primarily on pasture land, hay, wheat, corn, and barley, with heaviest use in the plains states, the Midwest, and the Southwest (Figure 3.10). MCPA is used primarily on wheat, barley, oats, and rice, with heaviest use in the northern plains states and California (Figure 3.13). Phenoxy compounds, primarily 2,4-D and MCPP, are also used heavily for lawn care in urban areas by both homeowners and professional applicators. These two compounds were the herbicides used most often by homeowners, according to a 1990 survey (Whitmore and others, 1992). Occurrence of these compounds in surface waters as a result of application in urban areas is discussed in Sections 4.1 and 5.3.

Physical and chemical properties of phenoxy herbicides vary considerably, depending on the structure of the particular chemical applied. 2,4-D is applied in several forms, including esters, salts, and the acid form. Water solubility of the esters generally is very low, whereas most of the salts are highly soluble. The acid form of 2,4-D has a relatively high water solubility of 890 mg/L (Wauchope, 1992). In general, the phenoxys do not sorb strongly to soil, and their dissipation rate in soil is relatively fast (half-lives in soil of approximately 7 to 10 days). They generally have a medium potential for movement in surface runoff (Goss, 1992). In water, the esters of 2,4-D may be hydrolyzed rapidly to the acid form, depending on the pH. In neutral water, half-lives of 2,4-D esters have been reported as ranging from 4.5 to 23 days. Hydrolysis of the esters is faster in alkaline water and slower in acidic water. The acid form of 2,4-D is relatively stable in water, with reported half-lives ranging from 6 to 170 days (Muir, 1991). Under conditions in which photolysis can occur, aquatic persistence may be lower, with half-lives of 2 to 4 days reported (Howard and others, 1991). DeMarco and others (1967) investigated the behavior of 2,4-D in a simulated stratified impoundment. They showed that 2,4-D is much more persistent in cold, deoxygenated water than in warm, oxygenated water, and that 2,4-D concentrations remained constant in a biologically inactive control for more than 100 days. The reported aquatic persistence of MCPA also is variable, with half-lives of 4 to more than 25 days reported. The reported aquatic half-life of MCPP is 7 to 10 days, apparently because of biodegradation (Howard and others, 1991). For the phenoxy compounds as a group, aquatic persistence appears to vary considerably, depending on pH, temperature, season, and the concentrations of suspended sediments and dissolved organic carbon.

Acute toxicity of phenoxy compounds to humans and aquatic organisms is relatively low (Que Hee and Sutherland, 1981). The USEPA MCL for 2,4-D in drinking water is 70 µg/L, and the NAS recommended maximum concentration in water for protection of aquatic life is 3 µg/L (Nowell and Resek, 1994). No corresponding values have been established for MCPA or MCPP.

Compared to the other high-use pesticide classes, relatively few recent studies have examined the occurrence of phenoxy compounds in surface waters. Only four large-scale studies conducted during the 1960's and 1970's included phenoxy compounds as analytes. Three USGS studies conducted on streams of the western United States during 1965–1966, 1966–1967, and 1968–1971 (Brown and Nishioka, 1967; Manigold and Schulze, 1969; Schulze and others, 1973) and one national-scale USGS study during 1975–1980 (Gilliom and others, 1985) quantified 2,4-D, 2,4,5-T, and silvex (2,4,5-TP). The earliest USGS western rivers study, in 1965–1966 (Brown and Nishioka, 1967), did not detect any phenoxy acid herbicides in any samples from 11 sites. Monthly samples were collected from these sites and nine additional sites in the 1966–1967 and 1968–1971 studies. Detection rates were 12, 8, and 4 percent (1966–1967) and 17, 18, and 8 percent (1968–1971) for 2,4-D, 2,4,5-T, and silvex, respectively. This apparent increase in detections of 2,4-D from the 1965–1966 study is somewhat misleading, however. The analytical detection limit decreased from 0.1 µg/L in the 1965–1966 study to 0.02 µg/L in the later studies. If only the concentrations above the 1965–1966 detection limit are considered, and only at the sites common to all the studies, the detection frequencies for 2,4-D would have been 5.7 percent and 3.3 percent of samples in the 1966–1967 and 1968–1971 studies, respectively. Half of the increase in both of the later studies is because of detections at one site (the Yakima River at

Kiona, Washington). In addition, sample handling techniques improved during and after the earliest study, when there was a delay of 2 to 3 weeks (apparently unrefrigerated) between sample collection and analysis (Brown and Nishioka, 1967). There may have been degradation of 2,4-D between collection and analysis of some samples in the earliest study. Thus, the findings in this series of studies indicate that 2,4-D levels in western rivers did not increase significantly during this period, except in the Yakima River. Concentrations of 2,4-D in the Yakima River show a distinct seasonal pattern each year from 1967 to 1971, with elevated concentrations (0.1 to 0.4 µg/L) detected from April through September (see Section 5.1).

The combined detection frequencies for 2,4-D at each site from the 1966–1967 and 1968–1971 studies may be compared with regional use of 2,4-D during this period. Figure 3.44 shows the frequency of occurrence at the 20 sites, and regional agricultural use of 2,4-D in 1971, expressed as pounds active ingredient applied per year per acre of cropland. National use of 2,4-D did not change appreciably between 1966 and 1971 (Table 3.1), and for Figure 3.44, the assumption is made that 2,4-D use also did not change significantly in the different regions shown. The frequency of 2,4-D detections in the rivers shows little correlation with the regional use totals. Sites with low and high detection frequencies were in both the high and low use regions. No correlation is evident whether the use data are expressed as total pounds active ingredient applied in the region, applied per acre of cropland, or applied per acre of all land in the region. Use data on this scale are obviously not sufficient to show a relation between use and occurrence in these rivers. For example, using the same 2,4-D use level for the Humboldt River Basin in northern Nevada and the Snake River Basin in southern Idaho is an oversimplification. Data on recent use of 2,4-D, on a statewide basis, show that over 14 times more 2,4-D was applied in Idaho than in Nevada during 1989–1991 (Gianessi and Puffer, 1991), most of it in the area drained by the Snake River (Figure 3.10). The relatively high use value for the mountain states may be a good estimate for the Snake River Basin, but is probably much too high for the Humboldt River Basin. In addition, the drainage basins of some of the rivers monitored in these studies include land in more than one of the regions. For example, the Arkansas River sampling site at Van Buren, Arkansas, receives inputs from areas in the mountain, northern plains, and southern plains regions, which had quite different 2,4-D use levels, according to the 1971 data. It is very likely that better correlation between use of 2,4-D and occurrence in these rivers would be seen if use data for this period were available at the county or river-basin scale.

The USGS/USEPA national study from 1975 to 1980 (Gilliom and others, 1985) quantified 2,4-D, 2,4,5-T, and silvex in approximately 1,760 samples from throughout the United States, with detection frequencies of 0.2, 0.1, and 0.1 percent, respectively. Detection limits in this study were 25 (2,4-D) and 100 (2,4,5-T, silvex) times higher than detection limits in the studies depicted in Figure 3.44. These higher detection limits were undoubtedly the cause of the very low frequency of detections in the later study. The effect of differences in the detection limits in these studies is discussed in Section 2.7. Thirty of the state and local studies in Table 2.2 included 2,4-D as an analyte. Most of these also included 2,4,5-T and silvex. In these studies, mostly from the 1980's, these compounds were detected relatively often—2,4-D was detected at nearly 50 percent of sites (30 percent of samples), 2,4,5-T was detected at approximately 30 percent of sites (20 percent of samples), and silvex was detected at approximately 11 percent of sites (3 percent of samples). Detection limits in most of these studies ranged from 0.01 to 0.1 µg/L for all three compounds, but were higher in a few cases. Not all studies provided data on individual sites or samples, and the percentages given should be used only for an overview of the relative frequency of occurrence during this period.

Results from several of these studies can be used to illustrate the situation in recent years. In Kansas, over 100 stream sites have been monitored since 1977 for a variety of pesticides,

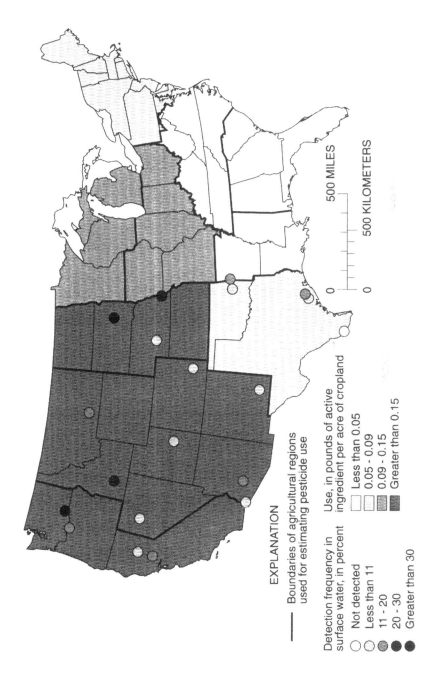

EXPLANATION

— Boundaries of agricultural regions
used for estimating pesticide use

Use, in pounds of active
ingredient per acre of cropland

☐ Less than 0.05

☐ 0.05 - 0.09

▨ 0.09 - 0.15

▨ Greater than 0.15

Detection frequency in
surface water, in percent

○ Not detected

○ Less than 11

◐ 11 - 20

● 20 - 30

● Greater than 30

500 MILES

500 KILOMETERS

Figure 3.44. Regional agricultural use of 2,4-D in 1971, and detection frequency of 2,4-D in rivers and streams of the western United States, 1967–1971. Use data are from Andrilenas (1974). Detection frequencies are from Manigold and Schulze (1969) and Schulze and others (1973).

including 2,4-D (Butler and Arruda, 1985). Detections of 2,4-D increased during the late 1970's and early 1980's, and 2,4-D was detected in 6 to 8 percent of samples each year from 1981 to 1984 (detection limit unknown). Sampling was done only on an annual basis, however, and no information on concentrations is given in Butler and Arruda (1985). Kansas ranks third in agricultural use of 2,4-D (Gianessi and Puffer, 1991). Figure 3.10 shows national 2,4-D use. In Texas, which ranked first in agricultural use of 2,4-D during 1988–1991, five sites on the Colorado River were monitored from 1973 to 1982, with quarterly sampling (Andrews and Schertz, 1986). The most commonly detected herbicide was 2,4-D, with detections at all five sites and in 31 percent of the samples (detection limit of 0.01 µg/L). Maximum concentrations at the five sites ranged from 0.04 to 0.21 µg/L. One of the sites (Wharton, Texas) was also a site in the USGS studies of western streams previously discussed (Brown and Nishioka, 1967; Manigold and Schulze, 1969; Schulze and others, 1973). 2,4-D was detected in 6 percent of samples from 1966 to 1971 in these studies (detection limit of 0.02 µg/L), but in 26 percent of samples from 1973 to 1982 in the Texas study. Whether the increase in detections at this site is due to the lower detection limit in the later study is not known. Concentrations of pesticides, including 2,4-D, were monitored at 20 sites on 12 streams from 1976 to 1978 in Wyoming, which also has relatively high agricultural use of 2,4-D (Butler, 1987). Samples were taken in early summer and autumn at each site. The most commonly detected herbicide was 2,4-D, with detections at 13 sites in 29 percent of the samples. The maximum concentration was 1.2 µg/L, but 90 percent of the concentrations were 0.08 µg/L or less. The early summer samples had more detections and higher concentrations of 2,4-D than the autumn samples. Finally, in Pennsylvania, where agricultural use of 2,4-D is relatively low, concentrations of pesticides, including 2,4-D, were measured in the Susquehanna River at Harrisburg, Pennsylvania, from April 1980 to March 1981 (Fishel, 1984). Concentrations of 2,4-D were detected in approximately 75 percent of the samples. The concentrations were low, with a maximum of 0.41 µg/L, and varied throughout the year. Highest concentrations occurred from August through January, in contrast to atrazine, which showed the more familiar pattern of highest concentrations in spring and early summer (Figure 3.45). The lack of a seasonal pattern in 2,4-D occurrence may have been due to the mixture of land uses in the Susquehanna drainage basin. Land use in the basin at the time of the report was divided among forested land–62 percent, cropland–10 percent, pasture–7 percent, urban–5 percent, and highways, public buildings, and recreational areas–16 percent (Fishel, 1984). 2,4-D has applications in each of these land-use categories (Tables 3.3 and 3.4), and the distribution of concentrations observed at Harrisburg probably was affected by inputs from each. Variable application times would result in a relatively undefined seasonal pattern of occurrence. This situation may be common in other areas of the nation with mixed land use because of the widespread use and the variety of applications for 2,4-D.

In summary, it appears that the most widely used phenoxy herbicide, 2,4-D, is a relatively common contaminant in surface waters of the United States, although recent monitoring data are sparse. Levels of 2,4-D are low, however, with most observed concentrations much less than 1 µg/L. The relation between use of 2,4-D and its occurrence in surface waters is not well defined by the available data. This is due to the widespread use and variety of applications of 2,4-D, the lack of detailed data on non-agricultural applications, and the relatively few recent studies in which both application and occurrence data have been documented. Little information has been published on monitoring for MPCA, the other phenoxy compound with significant agricultural use.

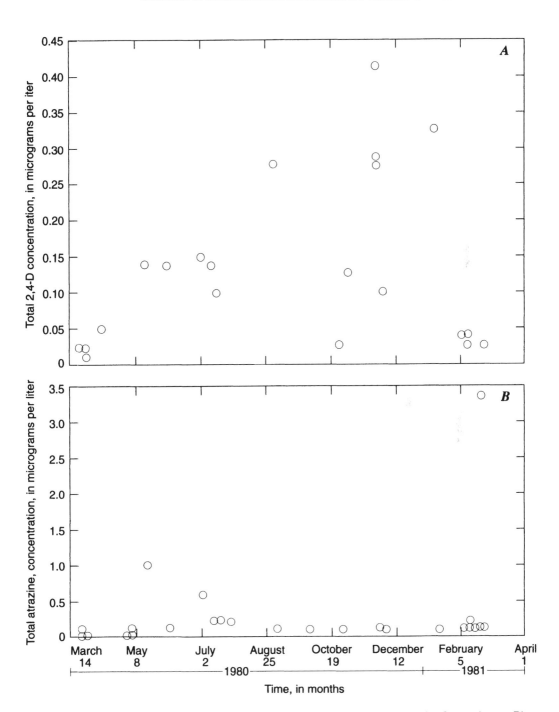

Figure 3.45. Seasonal patterns of 2,4-D (*A*) and atrazine (*B*) concentrations in the Susquehanna River at Harrisburg, Pennsylvania, from March 1980 to April 1981. Data are from Fishel (1984).

OTHER HERBICIDES, INSECTICIDES, AND FUNGICIDES

A number of high-use pesticides do not fall into the insecticide and herbicide categories discussed previously. Some of these compounds were analytes in the Mississippi River Basin study described previously (Section 3.1). Several of the herbicides also were included in the long-term study of Lake Erie tributaries by Baker and others (Richards and Baker, 1993). In the following discussion, these two studies are used to describe what is known about the occurrence and distribution of these miscellaneous pesticides in surface waters. For a number of these compounds, agricultural use in the area covered in these studies represents a major part of the nationwide use (maps showing agricultural use of many of these compounds are included in Section 3.2). For compounds not included in these studies, or for compounds whose major use is in other areas of the United States, applicable studies from Tables 2.2 and 2.3 are used. For pesticides with little occurrence data, the physical and chemical properties may be used to assess whether contamination of surface waters is likely.

Herbicides

Several of the most commonly used agricultural herbicides are not included in the triazine and acetanilide or phenoxy acid groups discussed earlier. These include butylate and EPTC (thiocarbamates), dicamba (benzoic acid derivative), linuron (substituted urea), and pendimethalin and trifluralin (dinitroaniline compounds). Except for dicamba, all of these were included as analytes in the Mississippi River Basin study as shown in Table 3.5 (Goolsby and Battaglin, 1993; Larson and others, 1995). Butylate, EPTC, pendimethalin, and trifluralin were detected in at least seven of the eight subbasins sampled, and in 16, 33, 9, and 35 percent, respectively, of samples collected from May through September 1991. Linuron was detected at only one site in one sample. Detection limits ranged from 0.002 to 0.01 mg/L for these five compounds. Concentrations were very low, with 95 percent of samples containing less than 0.03 µg/L. The total mass transported in the rivers represented 0.05 percent or less of the EPTC and butylate and 0.01 percent or less of the trifluralin, pendimethalin, and linuron applied in each subbasin. Agricultural use of these herbicides in the Mississippi River Basin represents a large part of national agricultural use: butylate–77 percent, EPTC–77 percent, pendimethalin–65 percent, trifluralin–69 percent, and linuron–53 percent (Gianessi and Puffer, 1991). Similar results have been obtained for butylate, EPTC, and linuron in the much smaller river basins of Lake Erie tributaries studied by Richards and Baker (1993). In this study, covering 1983 through 1991, butylate, EPTC, and linuron were detected at low concentrations in all seven rivers monitored. The percentages of samples with detections were 4 to 22 percent for butylate and 6 to 24 percent for EPTC (the percentage for linuron was not reported, but was less than 50 percent). The maximum concentrations were considerably higher in these smaller rivers (drainage basin sizes ranged from 4.4 to 6,000 mi^2), but most samples still had very low concentrations. In each basin, 95 percent of all samples analyzed had butylate concentrations of 0.05 µg/L or less, linuron concentrations of 0.68 µg/L or less, and EPTC concentrations of 0.07 µg/L or less. These three herbicides were among the top 20 pesticides used in Ohio during this period.

Dicamba was not an analyte in any of the large-scale studies reviewed, but was included in several of the state and local studies. In five studies conducted in the late 1970's and mid-1980's (Butler, 1987; Wnuk and others, 1987; Knapton and others, 1988; Lambing and others, 1988; Rinella and Schuler, 1992), dicamba was detected at 13 of 59 sites and in 12 of 79 samples from rivers in Iowa and several western states. Detection limits in these studies ranged from 0.01 to 0.1 µg/L. Concentrations were low, with most below 0.1 µg/L. The maximum concentration detected was 1.4 µg/L in an Iowa river (Wnuk and others, 1987). Dicamba has a medium potential for movement in surface runoff (Goss, 1992) and has been observed in runoff from field plots under simulated rainfall conditions (Trichell and others, 1968). Laboratory studies of the persistence of dicamba in water suggest that it is fairly resistant to biodegradation and hydrolysis, and unlikely to volatilize (Muir, 1991), although field studies do not indicate a long aquatic lifetime. Dicamba is very mobile in soil (Norris and Montgomery, 1975) and may undergo leaching before significant runoff occurs.

Insecticides

Some of the most commonly used insecticides (as of 1989) are not included in the organochlorine or organophosphorus groups discussed earlier (Tables 3.1 and 3.3). These include several carbamates (carbaryl, carbofuran, aldicarb, methomyl, and oxamyl), a thiocarbamate (thiodicarb), propargite, and the pyrethroids (permethrin and a variety of compounds primarily for home and garden use). Several of these were included as analytes in the Mississippi River Basin study as shown in Table 3.5 (Goolsby and Battaglin, 1993; Larson and others, 1995). Carbofuran and carbaryl were detected at seven of eight sites, in 35 percent and 9 percent, respectively, of samples collected from May to September 1991. The detection limit was 0.002 µg/L for both compounds. Concentrations were very low: 95 percent of all samples contained 0.08 µg/L or less of carbofuran and 0.01 µg/L or less of carbaryl. The total mass transported in the rivers represented from less than 0.01 to 0.4 percent of the carbofuran applied in each subbasin and from less than 0.01 to 0.08 percent of the carbaryl. Propargite was detected at three of eight sites (1 percent of samples) with a detection limit of 0.01 µg/L. Permethrin was detected at two of eight sampling sites (2 percent of samples) with a detection limit of 0.01 µg/L and a maximum concentration of 0.03 µg/L. Water samples were filtered before analysis in this study, and any permethrin sorbed to suspended sediment would not have been included in the analysis. Permethrin has a relatively high sorption coefficient (see section on Phase-Transfer Processes in Chapter 4) and very low water solubility (0.006 µg/L) (Wauchope and others, 1992), and it is likely that a significant portion would remain sorbed to particles if present in one of these rivers. Agricultural use of these four insecticides in the Mississippi River Basin represents nearly half of their national agricultural use: carbofuran–56 percent, carbaryl–43 percent, propargite–46 percent, and permethrin–37 percent (Gianessi and Puffer, 1992a). Although studies show that these compounds are used substantially on other crops throughout the United States, only a few studies actually targeted the compounds in surface waters. It is not known whether these compounds would behave in other environmental settings as they do in the Mississippi River Basin.

Aldicarb (Figure 3.18), another commonly used carbamate insecticide, was not targeted in any of large-scale studies reviewed and was an analyte in only one of the studies included in Table 2.2. It was not detected at any of four sites in a USGS reconnaissance study in Colorado during 1988–1989, which had a detection limit of 0.5 µg/L (Butler and others, 1991). Aldicarb has received much attention as a ground-water contaminant (U.S. Environmental Protection Agency, 1992), but apparently very little as a potential surface water contaminant. The half-life of aldicarb in surface water is estimated to be 20 to 360 days (Howard and others, 1991), and its physical and chemical properties suggest a medium potential for movement in surface runoff from fields (Goss, 1992). Aldicarb is normally incorporated into soil when applied (U.S.

Environmental Protection Agency, 1988), however, and is oxidized rather quickly in soil to the sulfoxide and more slowly to the sulfone (Howard and others, 1991), greatly reducing the potential for surface water contamination.

Two other carbamates with wide use, methomyl (Figure 3.26) and oxamyl (Figure 3.28), also were not targeted in any of the large-scale studies reviewed. Methomyl was an analyte in three studies included in Table 2.2. It was not detected in two USGS reconnaissance studies in California and one in Colorado in the late 1980's (detection limits ranged from 0.01 to 2.0 µg/L). The physical and chemical properties of methomyl suggest a medium potential for movement in surface runoff from fields (Goss, 1992). A long-term field study of pesticide losses from irrigated fields (Spencer and Cliath, 1991) has shown that the amount of methomyl lost from fields, as a proportion of the amount applied, is similar to that of some of the triazine and acetanilide herbicides observed in the Mississippi River Basin study. In addition, methomyl has been shown to be resistant to both biotic and abiotic degradation in water (Walker and others, 1988). Thus, there does appear to be a potential for contamination of surface waters by methomyl, but existing studies are not sufficient to assess whether this has actually occurred. Oxamyl was an analyte in the Colorado reconnaissance study, and was not detected (detection limit of 0.5 µg/L). The estimated half-life of oxamyl in soil is very short (4 days), and it is not considered likely to be transported in surface runoff from fields (Goss, 1992).

Fungicides

Fungicides, representing approximately 10 percent of all pesticide use (excluding fungicidal use of sulfur and copper), have significantly different use patterns than the herbicides and insecticides (Gianessi and Puffer, 1992b), as shown in Figure 3.5. It should be noted that applications of sulfur are not included in the map of total fungicide use. Sulfur accounted for approximately 61 percent of total fungicide use in the United States during 1988–1991. Studies of the effects of fungicidal use of sulfur on surface water quality are not included in this text, since natural sources should far exceed contributions from use of sulfur as a fungicide. Copper, accounting for approximately 8 percent of total fungicide use, also is not included in Figure 3.5 and is not discussed in this text. The heaviest use of fungicides, in terms of the total mass applied, is in California and Florida, which account for 22 and 17 percent of nonsulfur fungicide use in the United States, respectively. Other states with relatively high agricultural use of fungicides include Georgia, Michigan, New York, Alabama, and North Carolina. Crops with highest fungicide use include grapes, various fruits, tomatoes, sugarbeets, peanuts, potatoes, cotton, and almonds. Compounds with the highest agricultural use include chlorothalonil, mancozeb, captan, maneb, ziram, and benomyl (Table 3.1). Maps showing agricultural use of several of these compounds are included in Section 3.2. Several fungicides have significant noncrop uses (seed treatments or postharvest application), which are not accounted for in Table 3.1 or Figure 3.5. These include PCNB (pentachloronitrobenzene), carboxin, captan, thiram, and benomyl. In addition, some fungicides have significant non-agricultural uses (Table 3.3).

In general, the major agricultural fungicides have low water solubility and moderate sorption characteristics, and their persistence in soil is variable. Their potential for transport from fields in runoff is estimated as medium to large (Goss, 1992). Captan and benomyl hydrolyze in water, but other high-use fungicides are relatively stable in water (Howard, 1991; Howard and others, 1991; U.S. Environmental Protection Agency, 1991). Several of these compounds are of concern from a human-health standpoint. Mancozeb and maneb both degrade to ethylene thiourea (ETU). ETU is classified as a probable human carcinogen by the USEPA (Nowell and Resek, 1994) and is relatively stable in water (Howard and others, 1991). Benomyl is classified as a probable human carcinogen, and captan and chlorothalonil are classified as possible human carcinogens by the USEPA (Nowell and Resek, 1994). Despite these classifications, little data have been published on the occurrence of these compounds in surface waters. None of the large-scale studies in Table 2.1 included any of the high-use fungicides. Captan was included as an analyte in the extensive monitoring of Illinois surface waters conducted by the IEPA from 1985 to 1989 (Moyer and Cross, 1990) and was not detected in any of the 580 samples. However, Illinois is not a high-use area for captan (Figure 3.33), and it is hydrolyzed more readily in water than several of the other high-use fungicides. No published studies were found in which occurrence of these compounds in surface waters was investigated in areas of high agricultural use.

3.4 LONG-TERM TRENDS IN PESTICIDE OCCURRENCE IN SURFACE WATERS

From the studies reviewed, little can be concluded about long-term trends in pesticide occurrence in surface waters of the United States for several reasons. The major reason is the general lack of consistent long-term studies in which the same sites are sampled over a number of years. The agricultural-use data in Tables 3.1 and 3.2 can be used to make broad generalizations about the groups of pesticides for which occurrence has possibly increased or is likely to have decreased in surface waters over the last 25 years. Thus, it can be expected that OC contamination has decreased, whereas contamination by herbicides probably has increased. Beyond that, however, not much can be said on the basis of use data because these data represent aggregated national use and may not reflect use trends in a particular area. In addition, many factors other than the amount applied determine whether individual pesticides will be found in surface waters (see Section 4.1). For rivers, simply comparing pesticide concentrations observed in recent studies with concentrations observed in previous studies may not be valid, even for data from the same site, unless many years of data obtained with a fairly intensive sampling schedule are available. The normal seasonal variations in concentration, combined with year-to-year variations caused by differences in weather and agricultural practices, make comparison of recent and past concentration data tenuous. For large lakes, comparison of recent concentration data with older data may be more useful, as seasonal and year-to-year variations in pesticide concentrations are likely to be less significant. For certain pesticide classes, limited data from specific areas serve as examples of the probable underlying national trends in occurrence in surface waters. These classes are discussed in the following sections.

ORGANOCHLORINE INSECTICIDES

The national-scale studies of the 1960's provided a glimpse of trends in the occurrence of the major OCs in rivers. The trends during this period, however, became a moot point when use of most organochlorines was banned or severely restricted in the early 1970's. The question now is whether levels of organochlorines in surface waters in recent years are significantly different from levels observed before, and shortly after, use of these compounds was curtailed. That is, has the situation improved, or do current inputs from contaminated soil, contaminated bed sediments, and the atmosphere continue to pose a threat?

At a few sites, relatively long-term records exist for certain OCs. Water and bed-sediment samples from the Yakima River at Kiona, Washington, were analyzed for DDT from 1968 to 1982 by the USGS (Rinella and others, 1993) and again in 1985 by Johnson and others (1988). Fish samples from further upstream in the Yakima River were analyzed for DDT from the mid-1960's to the mid-1980's by the U.S. Fish and Wildlife Service (Johnson and others, 1988).

During 1968–1970, total DDT (sum of DDT, DDD, and DDE) was detected in about half of the water samples at concentrations ranging from 0.01 to 0.06 µg/L. Concentrations declined after 1971, and in 27 samples collected from 1974 to 1985, the total DDT concentration was always less than 0.01 µg/L. Results from the fish analyses also show a decline in total DDT residues. Samples collected from 1970 to 1974 had concentrations ranging from 1,400 to 2,600 µg/kg, whereas samples collected at the same site in 1980 had a mean concentration of 380 µg/kg. Fish collected in 1985 from the Kiona site (farther downstream), however, had mean concentrations of 1,100 to 3,000 µg/kg for two species. Total DDT continued to be detected in more recent sampling of the Yakima River by the USGS, using methods with a lower reporting limit of 0.001 µg/L. Concentrations of total DDT measured at Kiona, near the mouth of the river, were generally in the 0.001 to 0.01 µg/L range, equaling or exceeding the USEPA chronic-toxicity criterion for aquatic organisms of 0.001 µg/L in 9 of 10 samples collected year-round during 1988–1989 (Rinella and others, 1993). Concentrations were higher in upstream areas where return flows from agricultural drains enter the Yakima River. Analysis of fish samples collected during 1989–1990 also reveal continued contamination. Concentrations of total DDT in resident fish from agricultural return flows and the lower Yakima River exceeded NAS guidelines established for the protection of fish predators (Rinella and others, 1993). Current levels of DDT in fish from the Yakima River system are among the highest in the nation, apparently a result of the intensive agricultural activity and the heavy use of irrigation in the basin. Main stem flow in the lower river is dominated by return flows from irrigated agricultural land, which carry large amounts of eroded, DDT-contaminated soil. The Yakima River provides an excellent case study of the continuing problem of surface water contamination by OCs. While the Yakima River Basin represents somewhat of a worst-case scenario in terms of the potential for contamination, it is not unique. In other intensively farmed areas with past use of DDT, erosion of contaminated soil continues to serve as a source of DDT to surface waters, resulting in concentrations in water and fish above USEPA and NAS guidelines, many years after DDT use ceased (Cooper, 1991). A 1985 study in California (Mischke and others, 1985) concluded that erosion and mechanical movement of DDT-contaminated soil into waterways was responsible for the continued presence of DDT in fish from California rivers. In this study, 99 soil samples from 32 counties were analyzed for *p,p'*- and *o,p'*-DDT, DDD, and DDE. All sampling sites were in areas where past agricultural use of DDT was confirmed. Every sample had residues of DDT or degradation products of DDT, with total DDT (the sum of both isomers of DDT, DDD, and DDE) concentrations in many samples in the micrograms per gram (parts-per-million) range.

It should be kept in mind that for the persistent, hydrophobic OCs, trends may be more easily identified by examining concentration data for sediments and biota. Sediment and tissue concentrations are the focus of a companion review, in which trends in organochlorine concentrations are discussed in much more detail (Nowell, 1996).

Organochlorine insecticide concentrations in the Great Lakes have received considerable attention, primarily because of their bioaccumulation in fish and the resulting negative effects on the commercial and recreational fishing industry. Numerous organochlorines have been detected in the water column of the Great Lakes (Oliver and Charlton, 1984; Biberhofer and Stevens, 1987; LeBel and others, 1987; Poulton, 1987; Stevens and Neilson, 1989), but at very low levels (nanograms per liter or lower), as shown in Tables 2.1, 2.2, and 2.3. The existing water column data do not allow an evaluation of temporal trends, but, as with rivers and streams, long-term trends in organochlorine concentrations are better assessed from data on concentrations in bed sediments and in tissues of fish and other biota.

ORGANOPHOSPHORUS AND OTHER INSECTICIDES

Although many studies have targeted OPs over the past 20 years, including the long-term study of Lake Erie tributaries (Richards and Baker, 1993), detections have been infrequent and sporadic, making it difficult to infer any long-term trend in occurrence or concentrations. In the USGS/USEPA nationwide study of rivers conducted from 1975 to 1980 (Gilliom and others, 1985), detections of OPs were too infrequent to test for trends in concentrations at most sites. At seven sites where sufficient data did exist, no significant trends were evident over the study period. Other insecticides, such as the carbamates and pyrethrins, have been targeted infrequently, and none were targeted in any of the long-term studies reviewed. Insufficient data are available to determine whether any trend exists in either concentrations or occurrence of these compounds.

TRIAZINE AND ACETANILIDE HERBICIDES

Use of triazine and acetanilide herbicides has risen dramatically since the early 1970's (Table 3.2). Monitoring for these compounds in rivers and lakes has increased with use, but very few studies included them as analytes before the mid-1980's. Recent studies have shown that concentrations of these compounds in rivers are very seasonal, with a sharp increase in concentration shortly after application followed by a relatively rapid decline to near or below detection limits for the remainder of the year (Goolsby and Battaglin, 1993; Richards and Baker, 1993; Schottler and others, 1994; Larson and others, 1995). These seasonal peaks in concentration are influenced strongly by the timing of rainfall with respect to application, so that year-to-year variability at a specific sampling site can be quite large (Figures 3.46 and 3.47) This variability makes detection of long-term trends difficult.

The study of Lake Erie tributaries (Richards and Baker, 1993) is the longest and most complete continuous record of triazine and acetanilide concentrations, and can be used to show this variability more clearly. Figure 3.47 shows the monthly mean concentrations of the three most heavily used and most frequently detected herbicides in this study—alachlor, atrazine, and metolachlor—for 1983 to 1991 in one of the smaller basins studied. The strong effect of weather is evident in the much lower concentrations observed in the drought year of 1988. The authors of this study (Richards and Baker, 1993) found no statistically significant trend in the concentrations of these three compounds over the 9-year period. For the other triazine and acetanilide herbicides in this study, agricultural use and detection frequencies are lower, making trend detection even less feasible.

Atrazine has been monitored at a number of sites in the central United States for as many as 17 years, although there are gaps in the data at most sites (Ciba-Geigy 1992a,c,d,f, 1994a). Though the sampling frequency at most locations was not high enough to provide information on long-term trends, the sampling frequency at some sites was high enough to determine annual mean concentrations for a number of different years. These data can be used to examine the trends in atrazine concentrations in these regions during the sampling periods. Annual mean concentrations of atrazine at four sites on the Mississippi River (Figure 3.48) and three of its tributaries (Figure 3.49) are shown for various years during 1975–1991. These plots clearly show that annual mean atrazine concentrations in rivers vary considerably from year to year. Much of this variation is undoubtedly due to weather, but it also may be due to low sampling frequency at some sites. For example, the sampling frequency at the Keokuk, Iowa, site was one sample per month in most years and two samples per month in 1977. With monthly sampling, the timing of

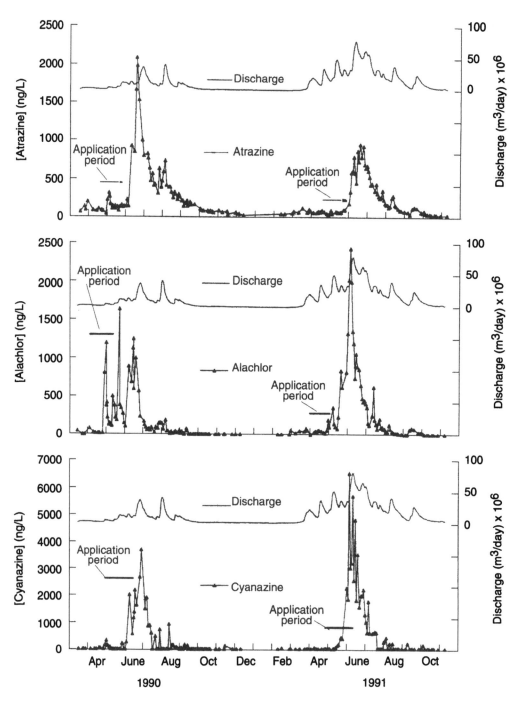

Figure 3.46. Seasonal patterns of atrazine, alachlor, and cyanazine concentrations, and river discharge in the Minnesota River at Mankato, Minnesota, from April 1990 to October 1991. Redrawn from Schottler and others (1994), with permission from *Environmental Science and Technology*. Copyright 1994, American Chemical Society.

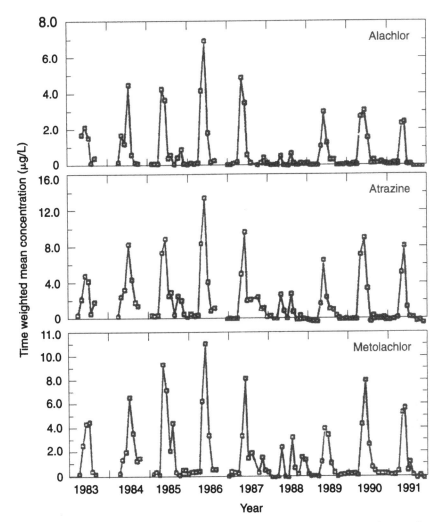

Figure 3.47. Monthly, time-weighted mean concentrations of alachlor, atrazine, and metolachlor in Honey Creek, Ohio, 1983–1991. Redrawn from Richards and Baker (1993), with permission from *Environmental Toxicology and Chemistry.* Copyright 1993, Elsevier Science Ltd.

sampling during the spring and early summer can have a large effect on the calculated annual mean, depending on whether the river was sampled during the peak atrazine concentrations. Peak herbicide concentrations in rivers of the Midwest can occur over days to weeks (see Section 5.1), and may or may not be caught by a single monthly sample. The most reliable mean concentrations from the plots in Figure 3.48 are probably those from the Vicksburg, Mississippi and the St. Gabriel, Louisiana sites on the Mississippi River. At these sites, the sampling frequencies are higher and peak atrazine concentrations in the lower Mississippi River are spread out over a longer time. Elevated atrazine concentrations at the Vicksburg site lasted from 1 to 4 months each year during the monitoring period, and at least five samples were taken during each period of elevated concentrations (Ciba-Geigy, 1992d). The data from these two sites,

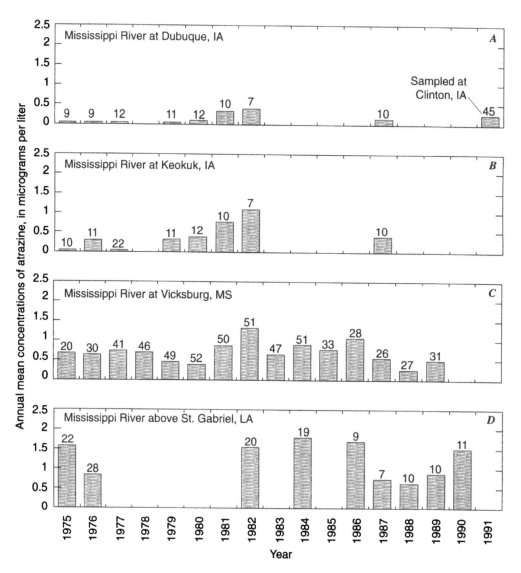

Figure 3.48. Annual mean concentrations of atrazine at four sites on the Mississippi River, 1975–1991. Number of data points included in mean is indicated above bar for each year. For annual mean concentrations reported as less than 0.1 µg/L, a value of 0.05 µg/L was arbitrarily assigned for plotting. A blank column indicates that no samples were collected that year or that sampling frequency was too low to calculate an annual mean concentration. Mean concentration for 1991 in plot (*A*) is from samples collected at Clinton, Iowa, approximately 50 miles downstream of Dubuque. All annual mean values were calculated as the simple mean of all observations, except for the site at Clinton, Iowa, for which a monthly, time-weighted mean concentration was calculated. Data are from Ciba-Geigy (1992d), except for the site in Clinton, Iowa, which is from Larson (1995).

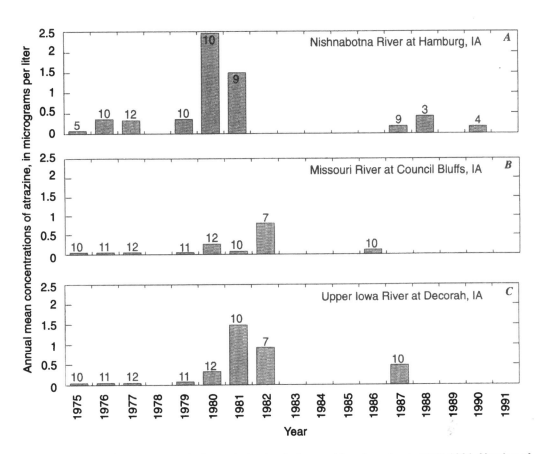

Figure 3.49. Annual mean concentrations of atrazine in three midwestern rivers, 1975-1991. Number of data points included in mean is indicated above bar for each year. For annual mean concentrations reported as less than 0.1 µg/L, a value of 0.05 µg/L was arbitrarily assigned for plotting. A blank column indicates that no samples were collected that year or that sampling frequency was too low to calculate an annual mean concentration. All annual mean values were calculated as the simple mean of all observations. Data are from Ciba-Geigy (1994a).

which serve to integrate the inputs from the entire Mississippi River Basin, indicate that no significant trend in atrazine concentrations occurred during this period. The plots for the sites on the upper Mississippi River (Figure 3.48A,B) and the Missouri, Nishnabotna, and Upper Iowa Rivers (Figure 3.49) all show increased annual mean concentrations of atrazine from 1980 to 1982. Unfortunately, mean concentrations for 1983–1986 are not available for most of these sites. Without data from these periods, it is difficult to say whether the higher annual means during 1980 to 1982 represented a general trend toward increased atrazine concentrations in the upper Midwest, or whether the higher annual means were due to weather patterns or inadequate sampling frequencies. Data from later years at these sites imply that there has been no significant overall trend during the 17-year period.

In summary, there are insufficient data to assess trends in occurrence in surface waters for most of the triazine and acetanilide herbicides. Data from a limited number of sites in the

Mississippi River Basin imply that atrazine concentrations have been relatively stable over the last 15 to 20 years. However, data from the existing studies show that, as use of triazine and acetanilide herbicides increased during the 1970's and 1980's, several of these compounds (such as atrazine, cyanazine, alachlor, and metolachlor) have become the most commonly observed pesticides in rivers and streams in agricultural areas, particularly in the central United States. When these compounds are used as preemergent herbicides, as they are in large parts of the United States, a fairly predictable fraction—0.3 to 2 percent of the amount applied—may be lost to surface waters (Goolsby and Battaglin, 1993; Richards and Baker, 1993; Schottler and others, 1994; Larson and others, 1995). The amount will vary somewhat with weather, topography, and soil characteristics. The trend in a particular basin, then, can be inferred from use data for specific compounds. If, for example, atrazine use in an area decreases while cyanazine use increases, a corresponding change in surface water occurrence of the two compounds would be expected.

CHAPTER 4

Factors Controlling the Behavior and Fate of Pesticides in Surface Waters

4.1 SOURCES OF PESTICIDES TO SURFACE WATERS

INTRODUCTION

As discussed in Section 3.3, pesticides are applied in a variety of agricultural and non-agricultural settings throughout the United States. In each of these applications, a fraction of the applied pesticide may be transported from the site of application and enter the broader environment, where it is perceived as an environmental contaminant rather than as a useful chemical. Pesticides may enter surface waters directly through runoff, spills, or various effluents. Contamination also may be indirect, with pesticides first entering the atmosphere or ground water, and then transported to surface waters. Once in surface water, some pesticides can be deposited in sedimentation areas, which can then act as a long-term source to the water column through resuspension, biotic uptake, and diffusion. In the following sections, the various sources of pesticides to surface waters are discussed. The methods of application or routes of delivery of pesticides to each source, the important processes involved in transport of pesticides from each source to surface waters, and finally, the relative importance of each source of pesticides to surface waters are examined.

PESTICIDES FROM AGRICULTURAL APPLICATIONS

Generally, the major source of most pesticides to surface waters is agricultural use. Agricultural use accounts for about 75 percent of total pesticide use in the United States (Aspelin, 1994). The compounds used (Table 3.1) vary tremendously in chemical structure, application rate, and potential for movement to surface waters. Any of the pesticides listed in Table 3.1 potentially could be transported from the point of agricultural application to streams and other surface waters. Table 2.5 lists the compounds observed in surface waters in the studies reviewed. A comparison of these two lists shows that many of the pesticides used in agriculture have not been detected in surface waters. Specific pesticides have not been detected in surface waters for several possible reasons, including low potential for transport in surface runoff because of pesticide properties or application techniques, low application rates (grams per hectare), and a lack of studies targeting the chemical in surface waters.

Agricultural application practices include aerial spraying, near-ground spraying from a tractor, soil incorporation, chemigation, and direct application to plant foliage. In almost all cases, the target for the pesticide is either the soil or the plant surface. Once a pesticide has been applied,

217

a number of physical and chemical processes can diminish its presence. These processes either destroy the chemical structure of the pesticide through transformation processes (Section 4.2) or move the pesticide among environmental compartments through phase-transfer and transport processes (see Transport of Pesticides in Surface Waters, Section 4.2). A pesticide potentially can leave the agricultural field in its molecular form and enter the atmosphere, vadose zone, ground water, or surface water. It also can leave the field in a particle-associated form in runoff or through plant uptake followed by harvest. The specific combination of environmental conditions, agricultural management practices, and pesticide properties determines if, when, and how the pesticide will leave the field and move into the broader environment, including surface waters.

Pesticides generally move from fields to surface waters in runoff or in drainage induced by rain or irrigation. Runoff of pesticides from agricultural fields can occur by overland flow, interflow (water that enters the shallow subsurface and then returns to the soil surface), and flow through tile-drainage networks. Generally, the water is routed to drainage ditches or natural topographic drains and, ultimately, to a surface water system. Leonard (1990) suggests that at the field microscale, "pesticide extraction into runoff may be described as mechanisms of (i) diffusion and turbulent transport of dissolved pesticide from soil pores to runoff stream; (ii) desorption from soil particles into the moving liquid boundary; (iii) dissolution of stationary pesticide particulates; (iv) scouring of pesticide particulates and their subsequent dissolution in the moving water. Pesticides are also entrained in runoff attached to suspended soil particles."

Pesticides can leave the field either as a dissolved chemical or associated with soil particles, depending largely on the properties of the compound (Wauchope, 1978). Most pesticides observed in runoff from agricultural fields are predominately in the dissolved form, except for pesticides with very low water solubilities (less than 1 mg/L) or strong ion-exchange capabilities with clay minerals (ionic compounds like paraquat and MSMA).

Leonard (1990) summarized four dominant factors that affect pesticide transport in runoff. The first factor is climate, including duration, amount, and intensity of rainfall, timing of rainfall with respect to pesticide application, and rainwater temperature. The second factor is soil characteristics, including soil texture and organic matter content, surface crusting and compaction, antecedent water content (before rainfall), slope and topography of the field, and degree of soil aggregation and stability. The third factor is the physical and chemical properties of the pesticide. Properties that control the runoff of pesticides include water solubility, acid/base and ionic properties, sorption properties, and persistence. The fourth factor is agricultural management practices. Included in this factor are pesticide formulation, application rate, application placement (soil surface, soil incorporation, or foliar), erosion control practices, plant residue management, use of vegetative buffer strips, and irrigation practices.

Pesticides removed in runoff from a treated agricultural area constitute only a small percentage of the total application of the compound. Leonard (1990) and Wauchope (1978), in their reviews of the literature, concluded from field plot studies that normal runoff losses are 2 to 5 percent of application for pesticides formulated as wettable powders, approximately 1 percent of application for foliar-applied organochlorine insecticides (OCs), and less than 0.3 percent of application for the remaining pesticides. Larson and others (1995) observed the same range of percentages for losses of 26 pesticides into large, integrating rivers into the Mississippi River Basin (Table 3.5).

For most surface waters downstream from agricultural areas, runoff from agricultural fields is the major source of their pesticide load. As an example, Figure 3.46 shows the time and concentration profile of the herbicide atrazine in the Minnesota River, which drains a large area of intensive row-crop agriculture (over 80 percent of the land is cultivated) (Schottler and others, 1994). The peaks in atrazine concentration for 1990 and 1991 occur soon after its application. The riverine concentrations of atrazine then decline until a relatively constant concentration is achieved during the low-flow period. The elevated concentrations in late spring and early summer are attributed to inputs of atrazine from rain-induced runoff from agricultural fields. The relatively constant low-level concentrations (about 0.5 to 1 percent of maximum concentrations) observed during the low-flow period are thought to be due primarily to inputs from ground water, although discharge from reservoirs, surface runoff from fields, and discharge from tile drains also may add low levels of pesticides to streams during this period. The other source of atrazine to this site is atmospheric deposition, but the estimated mass delivered by this route directly to the river is relatively unimportant compared to runoff processes in this intensively farmed basin (Capel, 1991).

PESTICIDES FROM FORESTRY APPLICATIONS

Pesticides serve a number of purposes in silviculture. Herbicides are used primarily for site preparation and conifer release. In site preparation, herbicides are used to reduce competing vegetation in areas where replanting is to take place. Conifer release involves application of an herbicide several years after planting to release the growing trees from competing, overtopping vegetation. With decreased competition for light and water, conifers can normally outgrow competing vegetation without further treatment. Thus, herbicides usually are applied to replanted areas only once or twice in the 25 to 50 years between planting and harvesting. Minor uses of herbicides include manipulation of wildlife habitat and maintenance of rights-of-way. Insecticides are used for control of outbreaks of specific pests, such as the gypsy moth, the spruce budworm, bark beetles, cone and seed insects, and grasshoppers (on rangeland). Fungicides and fumigants are used primarily on nursery stock.

The discussion of pesticide use in Section 3.2 implies that pesticide use in forestry may be insignificant compared to agricultural use, both in terms of the mass of pesticide applied and the acreage involved. Pesticide use in forests, however, needs to be considered for several reasons. Forested lands in the United States are often relatively pristine and are highly valued for their aesthetic and recreational uses. Forested land also serves as a habitat for wildlife and supports a number of important fisheries. Many of the national parks and wilderness areas border forested land, which may be treated with pesticides. The headwaters of most of the nation's major river systems are in forested areas. Finally, pesticide use in forestry represents a significant portion of the total use of a number of pesticides, such as triclopyr, hexazinone, and diflubenzuron. Forestry applications, therefore, must be considered when evaluating the results of monitoring studies and research on occurrence of these pesticides in surface waters.

Several methods of pesticide application are used in forestry. Aerial application has been used with liquid formulations of insecticides, including *Bacillus thuringiensis* var. *kurstaki* (Bt), and with liquid and pellet formulations of herbicides. Several variations of ground-based application are used for liquid or pellet formulations of herbicides, including broadcast spraying from vehicles, manual spot spraying, single-stem injection, and banded spraying along tree rows in commercially owned forests. Typical application rates for herbicides, inferred from national forest use data, range from approximately 1 lb a.i. per acre for triclopyr, glyphosate, and 2,4-D, to about 2 lb a.i. per acre for hexazinone (U.S. Forest Service, 1992, 1993, 1994a).

The forest environment has a number of characteristics that can affect movement of pesticides to surface waters. Forested land in the United States generally has higher slope and receives more precipitation than agricultural land. Forests generally have relatively shallow soils with high infiltration capacity, low pH, and high organic carbon content. The presence of year-round vegetation in forests is another important difference from agricultural land (Norris, 1981). Pesticides applied in forests may reach surface waters by several different routes. Direct input of herbicides to streams or lakes can result from aerial application of liquid or pellet formulations or from spray drift from ground spraying. Pellets or liquids aerially applied to dry ephemeral stream channels have a high probability of entering surface waters, especially if rainfall occurs shortly after application. Pesticides may move in overland flow (surface runoff), although this is uncommon due to the high infiltration rate of forest soils. Pesticides also may be leached and move to surface waters in subsurface (downslope) flow (Norris, 1981).

A number of studies have monitored the levels of pesticides in forest streams in the southeastern United States and in Canada. In nearly all cases, these studies can be described as field experiments, in which a known amount of a pesticide was applied to a section of a watershed, with subsequent sampling of stream water ranging from weeks to years. Pesticides routinely used in forestry typically have not been included as analytes in monitoring of ambient levels of pesticides in surface waters. For this reason, most of the reviewed studies of pesticide occurrence in forest streams are tabulated in Table 2.3. Results from these studies are discussed in Section 5.4 and will be described here only briefly. In nearly all the studies reviewed, the authors conclude that use of the studied pesticide should have no adverse impacts on surface water quality or aquatic life, provided that appropriate safeguards, such as buffer strips, are used during application. This conclusion is based on the fact that elevated concentrations of the pesticides appear as short-lived pulses in these small streams and are often quickly diminished by dilution. Adverse ecological effects, when observed, have been short-term and reversible. As mentioned earlier, normal forestry practices require pesticide applications only once or twice in the 25- to 50-year growth period. This low frequency also helps to minimize impacts. It also should be pointed out that mechanical alternatives to herbicide use for site preparation generally are regarded as having much larger negative impacts on stream water quality because of increased erosion and nutrient losses (Neary and others, 1993).

PESTICIDES FROM ROADWAYS AND RIGHTS-OF-WAY

Pesticides used on roadways and other rights-of-way are discussed in Section 3.2. The general method of application to roadsides is spraying from a moving truck. Spraying by hand also is used for small areas such as under guard rails and near bridges and overpasses.

Movement of pesticides used along roadsides to the greater environment can occur through spray drift, volatilization, runoff, or leaching, although the relative contribution of each route is largely unknown. Volatilization may occur before or after spray droplets reach the target location. Surface runoff into the roadside drainage system is a potentially important route by which these chemicals may reach surface waters. Depending on the soil characteristics, leaching of the chemicals into the subsurface, followed by lateral movement in the ground water to a discharge point, also could be a route of introduction into surface waters at some locations.

Only a few investigations on the environmental fate of pesticides applied to roadsides have been published. McKinley and Arron (1987) investigated the behavior of 2,4-D and picloram applied to a right-of-way in eastern Ontario, Canada. After 8 months, they found low levels of residues in soil up to 36 m from the application site. They also detected low levels in samples from an adjacent lake. All concentrations were well below any level of environmental concern. The appearance of the pesticides off site and in the lake suggests at least minimal atmospheric drift or surface runoff. Watson and others (1989) studied the fate of picloram applied to roadsides in a mountain valley bottom and on a mountainside. They found no evidence of the movement of picloram from either of the application sites.

With the controlled use of pesticides along roadsides, these chemicals are probably not a large contributor to surface waters in most areas. In some remote areas, the only source of these chemicals to surface waters could be the roadside use of pesticides, perhaps along with atmospheric deposition.

PESTICIDES FROM URBAN AND SUBURBAN APPLICATIONS

Pesticides are introduced into the urban environment in a variety of ways. Homeowners and professional applicators apply herbicides, insecticides, and fungicides to lawns and gardens as liquid sprays, dusts, and granular solids. Golf courses, cemeteries, and some parks are treated similarly. In many parts of the United States, building foundations and the soil surrounding them are routinely treated with insecticides for controlling termites or other destructive insects. Controlling mosquitoes also has a high priority in some parts of the country for both public health and aesthetic purposes. Some lakes and reservoirs in urban areas are treated with herbicides for controlling algae or undesirable weeds (such as Eurasian watermilfoil). Controlling specific insect pests for agricultural purposes (such as the medfly in California) has included aerial spraying of insecticides in urban areas.

Processes affecting the movement of pesticides to surface waters in urban areas are the same as in agricultural areas, but some important differences between the two environments could affect this movement. Urban areas have large expanses of impermeable surfaces, such as concrete and asphalt roads and sidewalks, from which pesticides can be removed easily by runoff water from rain or sprinklers. These surfaces also provide a more or less continuous pathway along which pesticides may be transported by water with virtually no loss from sorption. Thus, if pesticides applied in urban areas reach impervious surfaces (by spray drift, direct aerial application, or runoff from lawns and gardens), there is a relatively high probability—compared to pesticides applied in agricultural areas—that they will be transported to surface water bodies. In studies done with turf plots (Harrison and others, 1993), however, it has been found that very little runoff of water occurs from well-maintained grass, even with large amounts of precipitation. So, at least for applications to lawns, the limiting step may be reaching the impervious surfaces. This is discussed further in Section 5.3. Storm sewer systems also provide a direct pathway for movement of pesticides to lakes or rivers. Similarly, effluent from sewage treatment plants may contain pesticides, particularly in urban areas where storm sewers and sanitary sewers are combined. Effluents from sewage treatment plants often flow directly into rivers.

There have been relatively few studies of the effects of urban pesticide use on surface water quality. Results from selected studies are discussed in Section 5.3. Recent studies in the Mississippi River Basin have shown rather clearly that urban use of diazinon is resulting in measurable concentrations of this pesticide in several major rivers (Larson and others, 1995). For the most part, however, the significance of urban areas as a source of pesticides to surface waters is difficult to determine. Many of the compounds used in urban areas also are used in agriculture (such as 2,4-D, dicamba, trifluralin, diazinon, and pendimethalin), so that the source of certain pesticides detected in surface waters is often undetermined. In addition, some pesticides (such as isazofos, isophenphos, oryzalin, and MCPP) used almost exclusively in urban areas have not been targeted in most published studies of surface water quality. For surface waters receiving runoff from both urban and agricultural areas, it is likely that the urban contribution of pesticides is a small percentage of the total pesticide input because of the much greater use of pesticides in agriculture. The limited data available, however, indicate that surface waters within or downstream from urban areas are likely to contain measurable residues of pesticides from urban applications (see Section 5.3).

PESTICIDES FROM AQUATIC APPLICATIONS

Pesticides used in aquatic applications are discussed in Section 3.2. Pesticides are applied to surface waters by a variety of methods, depending on the type of pesticide and the target organism. Herbicides for macrophyte control are commonly introduced directly into the water in shallow areas. An older application method—the total water column treatment method or parts-per-million system—entailed covering the entire surface of the water body with the pesticide formulation. More recently, a number of application techniques have been developed that allow more efficient and safe use of aquatic herbicides. Examples include a variety of controlled-release formulations that can provide the required concentration of herbicide for a longer time and allow the most efficient placement of the herbicide (Murphy and Barrett, 1990). For herbicides taken up by the roots of macrophytes, such as dichlobenil, the most efficient placement is often the sediment-water interface. Use of controlled-release formulations and bottom-placement introduces less of the chemical into the rest of the water column, minimizing effects on fish and other nontarget organisms. With contact herbicides, such as diquat and glyphosate, the chemical may be sprayed directly onto emergent or floating vegetation. Some chemicals are applied aerially to surface waters if the area to be treated is large (Wang and others, 1987a). The choice of which herbicide compound, formulation, and application technique to use is determined by a number of factors, including the plant species to be controlled, type of water body (static or flowing), and water characteristics such as turbidity, pH, and temperature (Murphy and Barrett, 1990). A practical manual for application and use of aquatic pesticides has been prepared by Hansen and others (1983).

Aquatic pesticides are introduced directly into surface water bodies, and the processes that control their behavior and fate are those processes that are specific to surface waters. The two groups of controlling processes are transformation and phase transfer. Important transformation processes, which remove the parent chemical from the environment, include hydrolysis, photolysis, and biodegradation. Phase-transfer processes include the transfer from the dissolved phase to the vapor phase (i.e., volatilization) and transfer from the dissolved phase to the particulate phase (i.e., sorption), with possible subsequent deposition to the bed sediments. These two phase-transfer processes are important in the behavior of acrolein and copper, respectively. The mechanisms of the transformation and transfer processes are described in detail in Section 4.2.

Each of the compounds used in aquatic applications has its target flora, and each behaves differently in water. Generally, most are short-lived in water and do not have a long-term impact on surface water ecosystems. For example, acrolein, used to control macrophytes, has a half-life of 4 to 5 hours in flowing surface waters (Bowmer and Saintly, 1977). Fluridone applied to experimental ponds in several geographic regions of the United States had a mean half-life of 5 days in the water column (West and others, 1979). The aquatic behavior of simazine (Hawxby and Mehta, 1979) and 2,4-D and related compounds (Hoeppel and Westerdahl, 1983) have been studied with respect to their use as aquatic herbicides. Extensive monitoring was conducted after large-scale applications of 2,4-D to reservoirs along the Tennessee River for control of Eurasian watermilfoil in the late 1960's (Smith and Isom, 1967; Wojtalik and others, 1971) and to a Georgia lake in 1981 (Hoeppel and Westerdahl, 1983). The authors of these studies report that the applications had no adverse effects on nontarget macrophytes, phytoplankton, zooplankton, benthic invertebrates, or fish. Concentrations of 2,4-D were very high immediately after application—up to 4,800 mg/L at one location 8 hours after application (Wojtalik and others, 1971). One month after application, concentrations had declined to background levels. Changes in abundance and species composition of some organisms were noted, but the authors suggest that these changes were caused more by the reduction in Eurasian watermilfoil, which served as a substrate and food source, than from toxic effects of the 2,4-D. Decreases in dissolved oxygen and pH, resulting from breakup and dissolution of the watermilfoil, lasted for 1 to 2 months after application in these studies.

PESTICIDES FROM MANUFACTURING WASTE AND ACCIDENTAL SPILLS

All manufactured pesticides potentially can be released into the environment as part of an industrial waste stream. Although available data do not indicate that this a widespread phenomenon, a few studies have reported the presence of pesticides in surface waters and attributed them to manufacturing waste disposal. In the 1970's, a series of investigations for the insecticide mirex in Lake Ontario showed that its occurrence was due to either inputs from the manufacturing waste stream or to disposal of unused chemicals from a secondary industrial user (Kaiser, 1974; Scrudato and DelPrete, 1982). The presence of mirex in Lake Ontario could not be accounted for on the basis of any legitimate agricultural use of the compound. It had a narrow registration and was used primarily to control fire ants in the southeastern region of the United States. In another study, the concentrations of alachlor and two related compounds were measured in several transects across the Mississippi River (Pereira and others, 1992). The fluxes of alachlor and the other compounds (one of which is used as a starting material in the manufacturing process) were significantly higher on one side of the river, suggesting direct inputs from an alachlor manufacturing facility in St. Louis. Oliver and Nicol (1984), investigating the Niagara River over a 2-year period, observed constant low-level inputs of organochlorine compounds, presumably from waste disposal sites, and numerous unpatterned concentration spikes, indicating direct discharges from industrial sources. Several reports of pesticide inputs from manufacturing facilities overseas may be looked at as examples of potential problems in the United States. In Spain, pesticides (trifluralin, atrazine, and simazine) and pesticide precursors have been measured in the waste effluent of a pesticide-manufacturing facility discharging directly into the Llobregat River, Barcelona, Spain (Rivera and others, 1985). There also have been reports from other countries describing the occurrence and effects of accidental discharges of pesticides to surface waters. In 1986, a fire at a chemical manufacturing facility in Switzerland resulted in 20 compounds with a combined total mass of about 1.5 metric tons entering the Rhine River (Capel and others, 1988). The immediate result was damage to the biotic community extending approximately 400 km downstream. Other examples of pesticides accidentally spilled into the Rhine River over the years, such as the insecticide endosulfan in 1969 (Greve and Wit, 1971), also are known. Wherever pesticides are manufactured or stored near surface water bodies, the possibility of direct inputs of waste and of spill discharges exists.

It should be noted that there may be more data on pesticide contamination resulting from manufacturing wastes and spills that was not accessed in this book. Much of this type of data would be collected by local, state, and federal regulatory agencies, and may not be published in the open literature. While the authors reviewed many agency reports, this book concentrated primarily on the published scientific literature.

PESTICIDES FROM GROUND WATER

The U.S. Environmental Protection Agency (USEPA) has compiled available data on pesticides in ground water of the United States and has identified 133 compounds (117 parent compounds and 16 transformation products) that have been detected in at least one well (U.S. Environmental Protection Agency, 1992). The pesticides detected in more than 100 wells in this database were alachlor, aldicarb and two transformation products, atrazine, bromacil, carbaryl, carbofuran, cyanazine, 2,4-D, DBCP, DDT, 1,2-dichloropropane, diuron, ethylene dibromide (EDB), linuron, methomyl, metolachlor, metribuzin, oxamyl, and simazine. The data in this report are from 68,824 nonstatistically chosen wells. For some of the compounds on this list, such as aldicarb, the frequent detections are partially the result of intensive sampling in a relatively small geographic area, and they may not represent a significant potential source to surface waters in general. In a statistically based sampling of 1,300 drinking-water wells across the United States, DCPA (dacthal) and metabolites, and atrazine were the most commonly detected pesticides (U.S. Environmental Protection Agency, 1990). Pesticides in these two groups have the greatest potential for surface water contamination by ground water.

Pesticides enter the subsurface by a number of mechanisms. In alluvial aquifer systems, the water and pesticide can move from the stream to the aquifer during periods of high flow, which often correspond to the periods of high pesticide concentrations in surface water (Squillace and others, 1993). Pesticides also can enter ground water through leaching and spills. Whenever a pesticide is applied to the ground, it potentially can move through the subsurface by advective flow with water from rain or irrigation. The rate of this leaching process is dependent on the properties of the pesticide (particularly its water solubility and extent of sorption to the soil), the rate of transformation of the pesticide in soil, and the characteristics of the soil itself (particularly the particle size, mineral composition, and organic carbon content). Pesticides can also reach the ground water through spills at distribution centers and mixing areas and through back-siphoning down a well during tank cleaning. A detailed description of the movement of pesticides to ground water is included in a companion book on pesticide occurrence in ground water of the United States (Barbash and Resek, 1996).

Pesticides can enter the surface water system at points where ground water is released to surface waters. During periods of low flow, which in the midwestern United States corresponds to the period of minimal farming activity (October to March), the majority of the pesticides observed in streams are assumed to be coming from ground water (Klaseus and others, 1988; Squillace and Thurman, 1992; Squillace and others, 1993; Schottler and others, 1994). This was illustrated in the Minnesota River (Figure 3.46), where frequent samples were obtained over a 19-month period. Atrazine was observed all year, with very low concentrations during the base-flow period. The authors attribute the source of atrazine during the base-flow period primarily to inputs from ground water, since the land surface was frozen and covered with snow for much of this period. They also note, however, that water derived from tile drains, which collect leachate from fields during at least part of this period, also may be a source of atrazine during base flow. Squillace and others (1993, 1996) have investigated inputs of atrazine during the base-flow period in the Cedar River in Iowa. They documented the seasonal movement of pesticides from the river to the alluvial aquifer during the spring (termed bank storage) and the subsequent movement of pesticides from the alluvial aquifer back to the river during the autumn and winter months. They reported that the majority of atrazine detected in the river during base-flow periods in 1989 and 1990 was derived from ground water discharged from the alluvial aquifer adjacent to the river. Inputs from tributaries, which aggregate most of the water collected in tile drains in this basin, were reported to account for 17 and 40 percent of the total atrazine inputs during base flow in 1989 and 1990, respectively, with the remainder coming from ground water (Squillace and others, 1996). They further state that the atrazine entering the river during extended periods of base flow is derived from ground water recharged at some distance from the river, rather than bank-storage water.

The importance of ground water contributions of pesticides to surface waters varies both geographically and seasonally. It can be important only in areas where the ground water is released to surface waters and in lakes when a large fraction of a lake's water budget is due to ground water inflows. In rivers, the input of pesticides by ground water is minimal or negative in periods of high flow, but often is significant and perhaps dominant in periods of low flow. Only pesticides with certain chemical and biological characteristics are likely to move from ground water to surface waters. To enter and readily move through the ground water system, pesticides must be relatively water soluble and have little affinity for solid surfaces. These characteristics allow the pesticide to enter and readily move through the ground water system. Pesticides also must have relatively slow transformation rates, because the residence time of the compound in ground water is at least a few months in the case of bank storage (the time between spring discharge and autumn base flow) and perhaps a number of years for movement through larger aquifer systems. Pesticides that undergo relatively fast chemical or biological transformation (half-lives of days to weeks) will largely disappear before being released to a surface water body. Given these constraints, only a few pesticides have a strong potential to be delivered to surface waters from ground water in appreciable quantities. The most common example in the midcontinental United States is atrazine (Squillace and Thurman, 1992; Squillace and others, 1993; Schottler and others, 1994).

PESTICIDES FROM THE ATMOSPHERE

Numerous pesticides have been observed in various atmospheric matrices (air, aerosols, rain, snow, and fog). Majewski and Capel (1995) have reviewed the existing observations of pesticides in the atmosphere. The authors report that 63 pesticides and pesticide transformation products have been identified in the atmosphere. One of the conclusions of the book is that "nearly every pesticide that has been analyzed for has been detected in one or more atmospheric matrix throughout the country at different times of the year." In general, the more volatile pesticides and those that are applied aerially have a greater chance of entering the atmosphere.

Pesticides enter the atmosphere through a variety of processes during and after application. Pesticides can be (and often are) released into the atmosphere during agricultural application. Some are applied aerially, some are applied as a spray from a few centimeters above the soil surface, and others are incorporated directly into the soil. With both aerial and ground-based spraying, it is very likely that some fraction of the applied pesticide will not reach the field, but rather remain in the atmosphere. Pesticides also can enter the atmosphere after reaching the soil surface through vapor desorption (release of vapor-phase pesticides from the soil, often termed volatilization from soil) and through wind erosion of soil particles with associated pesticides. The magnitude of the movement of any particular pesticide into the atmosphere is dependent on the pesticide's physical and chemical properties, application method, and formulation. Pesticides also can be released into the atmosphere from plant surfaces. If a pesticide has been applied to, or transported to, a surface water body, it can be released into the atmosphere through direct air-water partitioning to an extent based on its Henry's Law constant. Pesticides also can enter the atmosphere during the manufacturing process and industrial uses. Once in the atmosphere, the pesticide can be transported by wind currents, undergo photochemical and hydrolytic degradation, and be deposited to aquatic and terrestrial surfaces.

The movement of pesticides from the atmosphere to surface waters can occur by several mechanisms. A pesticide in the vapor phase can undergo direct air-water partitioning to an extent based on its Henry's Law constant. Pesticides associated with atmospheric particles (aerosols) can undergo dry deposition (dryfall) to a surface water body. Both vapor-phase and particulate-phase pesticides can be scavenged from the atmosphere by rain, snow, and fog, and deposited in surface waters. Pesticides also can undergo the same depositional mechanisms to soil and plant surfaces and enter the terrestrial pool of pesticides. Some unknown fraction of these pesticides eventually may be transported to surface waters.

The relative importance of atmospheric inputs of pesticides to surface waters is directly dependent on the magnitude of the other sources of pesticides to that water body. Atmospheric deposition of pesticides occurs globally. OCs have been detected in the Arctic (Hargrave and others, 1988; Gregor, 1990; Muir and others, 1990), and atrazine has been detected in remote Alpine lakes in Switzerland (Buser, 1990), although the reported water concentrations were very low. If atmospheric deposition of pesticides to surface waters in active agricultural areas is of the same order of magnitude as deposition to remote areas, the atmospheric contribution to surface waters of currently used agricultural pesticides may be overwhelmed by the amount entering surface waters directly from agricultural fields. As an example, Glotfelty and others (1990) have shown that less than 3 percent of the atrazine found in the Wye River, a tributary to Chesapeake Bay whose drainage basin is heavily agricultural, was contributed by atmospheric deposition. Atmospheric deposition of atrazine to Chesapeake Bay itself was about 10 percent of the total loading. Another example is the DDT contamination in the Great Lakes. Strachan and Eisenreich (1990) estimated that more than 97 percent of the total DDT (sum of DDT, DDD, and DDE) and metabolite burden in Lakes Superior, Huron, and Michigan is due to atmospheric deposition, whereas in Lakes Erie and Ontario, whose basins are more heavily agricultural, only 22 and 31 percent, respectively, of total DDT residues are estimated to be the result of atmospheric deposition.

PESTICIDES FROM BED SEDIMENTS

Pesticides present in bed sediments of lakes and streams, often termed in-place pollutants, are an important and continual source of some chemicals to the overlying water. The specific pesticides of most interest are the recalcitrant, hydrophobic, OCs that were commonly used in the United States from the 1950's through the 1970's and are now banned or have severe use restrictions. The OCs used most heavily in the past are listed in Table 3.1. Nowell (1996) has reviewed the existing literature concerning pesticides in bed sediments. Rinella and others (1993) present an excellent case study of a common in-place pesticide—DDT and its metabolites—in the Yakima River in Washington (see Section 3.4).

For the most part, these pesticides are in the bed sediments of lakes and streams due to sedimentation of particle-associated pesticides from the water column to long-term depositional areas. These particles will be transported near the sediment-water interface or resuspended and redeposited in and out of the bed sediments until they reach a long-term depositional area. The long-term depositional areas are particle-size dependent. Fine-grained sediments (silt and clay), to which most hydrophobic pesticides tend to associate, accumulate in the low-energy portions of surface water systems, such as the deepest areas of lakes and reservoirs, the shallow, back-water areas of streams, and behind dams in reservoirs. Sediment-deposited pesticides are buried slowly in the sediments by continual fresh sedimentation.

The in-place pesticides act as a continual source of contamination of the water column through a variety of processes. Resuspension of the bottom materials, driven by energy inputs into the system such as strong wind-induced currents, lake turnover, and unusually large water discharges in rivers, can erode the long-term sedimentation areas and move the particle-associated pesticides into the water column. Biota, such as benthic feeding fish and benthic worms, also can disrupt the bed sediments and introduce the particle-associated pesticides back into the water column, although this is on a much more limited areal scale than the physical erosion of sediments. Finally, diffusion of the pesticides from the sediment porewater to the overlying water column also can be a release mechanism. Once a pesticide has reentered the water column from the bed sediments, regardless of the mechanism, it will undergo sorption or desorption in the drive to reach equilibrium. It may remain in the water column, be taken up by biota, enter the atmosphere through volatilization, undergo transformation, or be redeposited to the surficial sediment.

The movement of pesticides from bed sediments to the water column contributes to measurable concentrations of these compounds in the water column and biota of many surface water systems. Although some of the present organochlorine contamination of surface waters can be attributed to atmospheric deposition and fresh additions of historically contaminated soil particles, a large fraction should be attributed to the release of the in-place pesticides in many surface water systems (Baker and others, 1985; Gilliom and Clifton, 1990). Because most in-place pesticides are distributed widely and are present at low concentrations (micrograms per kilogram), remediation is not practical. Sediments will continue to contribute recalcitrant, hydrophobic pesticides to surface waters, albeit at a slow and diminishing rate over time.

4.2 BEHAVIOR AND FATE OF PESTICIDES IN SURFACE WATERS

INTRODUCTION

The behavior, transport, and fate of an organic chemical in surface waters is controlled by the properties of the chemical and the environmental conditions in the water. The structure of the organic chemical determines its physical, chemical, and biological properties. The surface water environment that surrounds the organic chemical consists of physical, chemical, and biological components. The interaction of the chemical structure and environmental conditions controls the chemical's behavior and ultimately its effect on the environment. The environmental processes that control an organic chemical's behavior and fate in surface water can be classified into three types: (1) transformation processes, which change its chemical structure; (2) phase-transfer processes, which control its movement between water, biota, suspended sediments, bed sediments, and the atmosphere; and (3) transport processes, which move it away from its initial point of introduction to the environment and throughout the surface water system.

TRANSFORMATION PROCESSES

The transformation of a pesticide results in changes in its chemical structure. One or more new chemicals are produced, and the original pesticide disappears. These new chemicals can be organic or inorganic molecules and ions. From an environmental-effects point of view, the ideal fate for a pesticide is ultimate transformation to inorganic species, such as water, carbon dioxide, and chloride ions (termed mineralization). Unfortunately, in many instances, the chemicals formed from transformation reactions are long-lived intermediates, which themselves can have a negative impact on the environment. Often the initial transformation products undergo subsequent transformation reactions before mineralization. By this process, a large number of transformation products can potentially be formed, some of which may retain pesticidal properties. Some pesticide formulations are applied as inactive agents and gain pesticidal properties only after transformation (see Section 5.5).

Chemical transformations can be mediated by chemical, biological, or physical means. In surface waters, chemically induced abiotic hydrolysis and oxidation-reduction reactions often occur. Biodegradation is the general term for biologically mediated reactions. Microorganisms can induce pesticides to undergo both hydrolytic and oxidation-reduction reactions. Photolysis is a chemical reaction induced by the energy from sunlight.

Generally, first-order or pseudo-first-order kinetic expressions are adequate to describe transformation processes. Unfortunately, the actual effect of the environmental conditions on the kinetics of transformation processes can be difficult to determine (Macalady and others, 1986). Only with very detailed laboratory and field studies can the exact transformation mechanism(s) be identified. For most purposes, transformation reactions are grouped together in a kinetic expression to describe the disappearance of a pesticide with a lumped, pseudo-first-order reaction rate constant. This has been the approach generally used in interpreting or predicting the fate of pesticides in surface water systems. Howard and others (1991) have tabulated measured or estimated rate constants for various transformation reactions for a number of pesticides.

Hydrolysis is the chemical (sometimes biologically mediated) reaction of a pesticide with water, usually resulting in the cleavage of the molecule into smaller, more water-soluble portions and in the formation of new C–OH or C–H bonds. This process is important for many organophosphorus and carbamate pesticides. The hydrolysis rate of a given organic compound is dependent on the characteristics of the solution. The strongest factor is pH. Hydrolysis reactions can be a result of direct attack by the water molecule (H_2O), the hydronium ion (H_3O^+), or the hydroxide ion (OH^-). These are termed neutral, acid, and base hydrolysis, respectively. At low pH, reactions are dominated by acid-catalyzed hydrolysis, whereas at high pH, reactions are dominated by base-catalyzed hydrolysis. At intermediate pH values, both neutral and acid, or neutral and base-catalyzed reactions, can be important to the overall rate of hydrolysis. It should be noted that acid or base catalysis does not necessarily occur in all hydrolysis reactions, and that neutral catalyzed reactions alone sometimes may govern the overall rate of reaction. In these cases, the rate will not depend on pH. Temperature also is an important factor. Generally, a temperature rise of 10°C increases the reaction rate twofold to fourfold. The presence of certain metal ions, humic substances, and particles can catalyze hydrolysis for some compounds (Armstrong and Chesters, 1968; Mabey and Mill, 1978; Burkhard and Guth, 1981). The structure of the pesticide determines which of these processes, if any, are important in its hydrolysis (Mabey and Mill 1978).

Organic chemicals that can undergo hydrolysis on time scales important for consideration of this process in surface water systems (half-lives of days to years) include alkyl halides, aliphatic and aromatic esters, carbamates, phosphoric esters, and phosphoric acid esters (Vogel and others, 1987). Some pesticides, such as dichlorvos, undergo hydrolysis at rates too fast (half-lives of minutes) to ever be present at significant concentrations in surface waters. Other pesticides, such as DDT and chlordane, undergo hydrolysis at rates too slow (half-lives of years to decades) to warrant consideration of this transformation process. Others, such as pentachlorophenol (PCP) and benfluralin (benefin), contain no hydrolyzable functional groups (Howard, 1991).

Oxidation-reduction reactions are chemically or biologically mediated reactions that involve a transfer of electrons. The process requires two chemical species to react as a couple: one chemical undergoes oxidation (loses one or more electrons) while another undergoes reduction (gains one or more electrons). Many oxidation reactions of pesticides in surface waters are biologically or photolytically induced. In reduced environments, such as bed sediments and the hypolimnion of lakes, abiotic reduction reactions can occur when organic or inorganic reducing agents are present, such as certain transition metals (iron, nickel, cobalt, chromium), extracellular enzymes, iron porphyrins, or chlorophylls. The rates of reduction reactions are dependent on pH and the magnitude of the reduction potential. The reduction half-life of the organophosphorus insecticide (OP) parathion, for example, is on the order of minutes in strongly reducing environments.

Biodegradation is the transformation of pesticides mediated by living organisms using enzymes. Chemical transformation reactions can cause structural changes in an organic chemical, but biodegradation is the only transformation process able to completely mineralize the pesticide (Alexander, 1981). Microorganisms degrade (transform) organic chemicals as a source of energy and carbon for growth, although most of their degradative enzymes are not used directly for growth and energy processes, but rather are part of a metabolic sequence that terminates in energy release (Dagley, 1983). All naturally produced organic compounds can be biodegraded, though this is a slow process for some chemicals. On the other hand, some synthetically produced organic chemicals, including most pesticides, have structures totally unfamiliar to microorganisms, which may not have the enzymes needed for degradation of these compounds. This is the primary reason why some pesticides, such as DDE, hexachlorobenzene (HCB), and mirex, are recalcitrant (very long lived) in the environment. However, even these synthetic compounds are observed to slowly biodegrade, probably owing to a process called cometabolism. In cometabolism, the microorganisms are using other substrates (carbon sources) for growth and energy, and the unfamiliar synthetic compound enters into the process and is transformed. The microorganisms derive no particular benefit from the degradation of this compound. The rate of biodegradation of a pesticide is dependent on chemical structure, environmental conditions, and the microorganisms present. The structure of the organic chemical determines the types of enzymes needed to cause its transformation. Given this strong dependence of degradation rate on chemical structure, some progress has been made in finding predictive relations between the two (see Section 5.6). The concentration of the chemical also can affect its rate of degradation. At high concentrations, a chemical may be toxic to microorganisms; at very low concentrations, it can be overlooked by the organisms as a potential substrate. The environmental conditions (temperature, pH, moisture, oxygen availability, salinity, and concentration of other substrates) determine the species and viability of the microorganisms present. Finally, the microorganisms themselves control the rate of biodegradation depending on their species composition, spatial distribution, population density and viability, previous history with the compound of interest, and enzymatic content and activity (Scow, 1990).

Photolytic transformations of pesticides are caused by the addition of energy from sunlight. The earth's atmosphere filters out light with wavelengths shorter than 290 nm; only wavelengths greater than this reach the earth's surface. Pesticides can undergo a direct reaction with sunlight (direct photolysis) or a secondary reaction with a photoactivated, sunlight-induced, short-lived reactive chemical species (indirect photolysis). The type of photoinduced reaction is dependent on the structure of the pesticide and specific environmental conditions.

Direct photolysis is the result of absorption of sunlight by a pesticide, causing a chemical transformation, such as cleavage of bonds, dimerization, oxidation, hydrolysis, or rearrangement. This reaction will occur only if the pesticide absorbs light at wavelengths present in solar radiation. The light absorption spectrum of most pesticides falls outside or near the fringes of the solar spectrum; therefore, direct photolysis is not an important transformation process for many pesticides. Notable exceptions to this are DNOC, fenitrothion, and metoxuron.

Indirect photolysis is usually a photoinduced oxidation reaction. Sunlight excites a photon absorber, such as nitrate or dissolved organic matter, which in turn reacts with dissolved oxygen to form potential photoreactants such as singlet oxygen ($^1O_2$), hydroxy radical ($^\bullet OH$), superoxide anion (O_2^-), peroxy radical ($ROO^\bullet$), and hydrogen peroxide (H_2O_2). These highly reactive species randomly attack water, dissolved organic matter, dissolved oxygen, or pesticides, if present. Another type of indirect photolysis, triplet photosensitization, occurs when a photon absorber, such as humic acid, transfers excess energy to a pesticide molecule, which then photodegrades. The two most important indirect photolysis reactions are singlet oxygen and nitrate-induced photooxidation. Singlet oxygen is a very efficient photoreactant for specific types of chemical structures, including many OPs. The more general reaction is the nitrate-induced photooxidation that proceeds through the hydroxy radical intermediate and affects all organic molecules. For any specific surface water, the rate of this reaction is a function of the nitrate concentration (Hoigné and others, 1989).

PHASE-TRANSFER PROCESSES

Phase-transfer processes involve the movement of a pesticide from one environmental matrix to another. The important processes that can occur in surface water environments include water-to-solid transfer (sorption), water-to-biota transfer (bioaccumulation), and water-to-air transfer (volatilization from water). In addition to these processes, air-to-solid transfer (vapor sorption) is important in soil environments. Although the physical movement of the chemical is involved, these transfer processes should not be confused with transport processes. Transfer processes are important on the scale of molecular distances (nanometers to micrometers). Once the organic chemical has passed through the physical interface (environmental compartment boundary), it may undergo transport over much larger distances. The phase-transfer processes of sorption and volatilization largely control the overall transport of many pesticides in surface waters.

Pesticides are distributed between particle surfaces and the water to varying degrees. This process, termed sorption, can play a pivotal role in the environmental behavior, transport, and fate of a pesticide in surface water. An organic chemical sorbed to a particle surface behaves differently than it does in the dissolved phase. Chemicals associated with particles generally are less available for biodegradation and are not available for volatilization to the atmosphere. Some particle-associated pesticides, such as atrazine (in soil), undergo sorbent-catalyzed hydrolysis (Armstrong and Chesters, 1968). The extent of sorption of a pesticide is a function of its physical-chemical properties and the properties of the particle and the solution. Relevant aspects for the solution include pH (especially for organic chemicals having a pK_a from 4 to 8), ionic strength, concentration of dissolved organic carbon (DOC), and, to a lesser extent, temperature. The ionic strength of the solution affects the activity coefficient of the organic chemical in water. As the ionic strength of an aqueous solution increases, the chemical's solubility decreases and the extent of sorption slightly increases. The presence of DOC in the water also can affect the activity coefficient of the pesticide, decreasing the extent of sorption. Because sorption is a surface process, the characteristics of the particles that have the greatest influence on the extent of sorption are surface area and surface coverage by organic films. In most surface waters, the majority of the particulate surface area is contributed by silt, clay, and colloidal size particles. In addition to having large surface areas, these sizes of particles are generally the most enriched with organic surface films. It has been shown that organic coatings on particles essentially control the extent of sorption for many organic chemicals (Karickhoff, 1984; Chiou, 1990). The hydrophobicity of an organic chemical, which can be quantified to some extent by its water solubility or octanol-water partition coefficient, also controls the extent of sorption. The extent of sorption at equilibrium is commonly defined in terms of a distribution coefficient, K_d, defined as the ratio of the concentrations of the pesticide between the suspended sediments and the water. An organic carbon-normalized distribution coefficient, K_{oc}, defined as K_d divided by the fractional content of organic carbon in the suspended sediment, also is used widely. Tabulations of K_d and K_{oc} values for pesticides can be found in Howard (1991) and Wauchope and others (1992). For a wide range of pesticides and other organic chemicals, their sorption distribution coefficients have been shown to be correlated strongly with their water solubilities. This provides a tool for predicting the extent of sorption of a particular chemical in a particular environment where the organic coatings of the particles dominate the sorptive process. A number of structure-activity relations have been derived for these types of predictions (Lyman, 1990). Sorption is an extremely complex process. With limited information on sorption and relatively few environmental observations of the process, researchers have often assumed that sorption of organic chemicals is completely reversible, linear (with respect to chemical concentration), and at equilibrium in surface waters. Studies have shown, however, that sorption and desorption are not completely reversible, at least in the laboratory (DiToro and Horzempa, 1982; Karickhoff and Morris, 1985), and that chemical equilibrium may not be reached for biotic particles in surface waters (Swackhamer and Skogland, 1993). For most environmental situations, organic contaminants are present at concentrations low enough that a linear K_d value adequately describes its sorptive behavior. Given the numerous uncertainties in environmental observations of organic chemicals, the quantification of nonequilibrium, nonlinear, and nonreversible sorptive behavior has been difficult.

Some pesticides concentrate in the living tissues of aquatic organisms, such that the concentration in the organism is greater than in the water. The pesticide can accumulate in tissues by two routes. One route is through the process called bioconcentration, which is direct water/tissue partitioning governed by the same mechanisms as sorption of pesticides to organic matter on particles. The second route is through the organism's diet. When one organism eats another that has accumulated pesticides in its tissue, some fraction of that pesticide burden is available for accumulation by the consumer. The combination of these two routes, both of which are thought to be important in the environment, is termed bioaccumulation. It has been observed by many investigators that bioconcentration can be related to hydrophobicity for many persistent chemicals. Thus, numerous structure-activity relations have been developed that relate bioconcentration to a chemical's water solubility or octanol-water partition coefficient (Bysshe, 1990). The two most important parameters determining the extent of bioconcentration for a particular compound are the lipid content of the organism and the rate at which the chemical is metabolized in the organism. Differences of up to two orders of magnitude in bioconcentration can be expected for a single compound because of variations in biotic species, sex, life stage, and size (Seiber, 1987; Bysshe, 1990). Kenaga and Goring (1980) suggest that bioconcentration is not an important process in the overall environmental behavior of chemicals with water solubilities greater than about 1 mg/L.

Pesticides can be transferred from the dissolved aqueous phase to the vapor phase in the atmosphere as a result of volatilization from water. This transfer is controlled by the chemical nature of the air-water interface and the mass transfer (advective) rates of the chemical in water (velocity of water flow, etc.), the pesticide's molecular diffusion coefficients in air and water, and its Henry's Law constant. Thomas (1990) has suggested that the importance of volatilization for a given chemical can be generalized from its Henry's Law constant alone. For pesticides that have a Henry's Law constant less than 3×10^{-7} atm-m^3/mole, volatilization from surface water is unimportant. For pesticides that have a Henry's Law constant greater than 1×10^{-5} atm-m^3/mole, volatilization is significant for all waters. Many of the volatile pesticides used as fumigants have Henry's Law constants in the range where volatilization can be a significant process in their environmental behavior. In contrast to this, only a few of the high-use herbicides exhibit any tendency toward volatilization from surface water. The organochlorine pesticides fall between these two extremes and may or may not volatilize from water, depending on the environmental conditions and the relative concentrations of the compound in the water and in the atmosphere.

Just as pesticides distribute themselves between the water and particle surfaces in water, they also distribute themselves between the air and particle surfaces in soils. The extent of this vapor sorption (i.e., air-to-solid transfer) is a function of the chemical's properties, the soil particle's properties, and the water content of the soil. Chiou (1990) has shown that in dry soils, vapor sorption interactions are stronger between the pesticide and the inorganic surface of the particle (particularly clay surfaces) than with the organic matter on the particle surface. As the water content of the soil increases, the inorganic surface becomes hydrated and the water out-competes the pesticides for the inorganic sorption sites. The extent of vapor sorption decreases and the interactions with the organic carbon surface coatings become the dominant mechanism. Pesticides sorbed by dry soils are released by the addition of water. At about 90-percent relative humidity, the extent of vapor sorption is close to that of sorption in aqueous systems, if the same chemical and particle are compared. It has been suggested that the K_{oc} concept, when based on vapor concentration, can be used to describe vapor sorption (Spencer, 1987). The process of vapor sorption is important in determining the pesticide's dominant environmental matrix in the soil and the routes by which it can leave the soil (into the air through volatilization or into water or solid phases during a runoff event).

TRANSPORT OF PESTICIDES IN SURFACE WATERS

The transport of a pesticide in surface waters depends on the form in which the compound exists in the water and the hydrodynamics of the system. As discussed in the preceding section, a pesticide molecule can exist either in the dissolved phase or it can be associated with a particle or colloid. In the dissolved phase, transport of the pesticide will be governed essentially by water flow. In the associated phase, transport will be governed by the movement of the particle or colloid. Pesticides in the associated phase can undergo a variety of transport processes, depending on the type of substrate with which it is associated. Transport of pesticides associated with dissolved organic matter or colloids is primarily governed by water flow, similar to that of dissolved pesticides. Pesticides associated with particles (sands to clays), fecal pellets, or coagulations of very fine particles tend to settle out in lakes and reservoirs, and in low-energy sections of streams, such as backwaters and behind large objects. Because of the propensity of hydrophobic organic pesticides to associate with natural organic matter, they tend to accumulate in bed sediments with a relatively high organic matter content (more than 1 percent). As discussed in Section 4.1, these sediment deposition areas can serve as long- or short-term sinks for pesticides until the sediments are disturbed by the hydrodynamics of the system.

In streams, high-energy events, such as spring runoff and large storms, can transport bed sediments and their associated pesticides downstream. When particle-associated pesticides are reintroduced into the water column, they may become redistributed between the dissolved and particulate phases in the attempt to reach sorptive equilibrium. The newly dissolved pesticides will be transported with the flow of water, while the pesticides still associated with particles can again return to the bed sediments, once the energy level of the stream subsides to a point where the particles settle out of the water column. Thus, pesticides with low water solubility and a high affinity for surfaces, such as the OCs, tend to remain in the bed sediments of streams for long periods of time and are transported slowly from the system. Pesticides with relatively high water solubility and little affinity for solids, such as triazine and acetanilide herbicides, are transported in a flowing stream at a rate approximating the river's velocity.

In slow-moving surface water systems, such as lakes and reservoirs, the hydrodynamic conditions controlling the transport of pesticides are different from those of faster moving systems such as streams, although the sorptive interactions in slow- and fast-moving waters are essentially the same. Lakes and reservoirs generally lack the strong one-dimensional flow of a river. In a lake or reservoir, the ratio of water inputs (i.e., tributary inflow, overland runoff, direct precipitation, and ground water inputs) to water outputs (i.e., outflow, evaporation, and ground water outflow) determines the hydraulic residence time of the system. Lakes and reservoirs commonly have hydraulic residence times of months to tens of years. Pesticides that exist primarily in the dissolved-phase will be transported with the lake currents and have an average residence time similar to the hydraulic residence time in the lake, if they are not chemically or biologically transformed. Pesticides associated with particles also may be transported with the lake currents, but also can undergo sedimentation just as in flowing water systems. The dynamics of particle and pesticide interactions at the sediment-water interface (benthic nephloid layer) of a lake or reservoir are very complex. A settling, organic-rich particle may be degraded as it falls through the water column or at the sediment-water interface, releasing any associated pesticides to the water column. If the particle is deposited to the sediment-water interface, it can undergo physical and biological mixing into the sediments and be buried. As in streams, these buried sediments can act as a long-term, low-level source of pesticides to the water column. The sediments of lakes and reservoirs also can be resuspended by strong bottom currents or storm events, resulting in the reintroduction of particle-associated pesticides into the water column in a short timespan. These reintroduced pesticides can undergo new sorptive processes, redistribute themselves between the dissolved and associated phase, and be transported through and out of the lake or reservoir accordingly.

CHAPTER 5

Analysis of Key Topics—Sources, Behavior, and Transport

5.1 SEASONAL PATTERNS OF PESTICIDE OCCURRENCE

Most agricultural pesticides, particularly herbicides, are applied during distinct and relatively short seasonal periods. Preemergent herbicides are applied just before planting, for example, and postemergent herbicides are applied a few weeks after the crop begins to sprout. Some crops receive an autumn application of herbicides to kill the plant before the crop is harvested. Some insecticides also are applied at certain times of the year to control specific pests. Sometimes pesticides not routinely used are applied to control an unexpected pest. Seldom in agricultural applications is the same pesticide used continually for long periods of time (i.e., months) during a growing season on the same crop. The seasonal application of a pesticide is the primary source for transport to surface waters, if residues in soil from applications in previous years are minimal when compared to the amount being applied.

The first runoff-inducing rain or irrigation event after application of a pesticide can potentially move significant quantities of the pesticide to surface waters. This has been observed for numerous compounds, especially the preemergent herbicides, in many river systems in the midwestern United States. Schottler and others (1994) observed a strong seasonality in the occurrence of herbicides in the Minnesota River (Figure 3.46), as did Larson and others (1995) and Goolsby and Battaglin (1993) for a number of herbicides in a wide range of stream sizes in the Mississippi River Basin. The seasonal pattern of occurrence for herbicides, such as atrazine and alachlor in midwestern rivers, is well known and somewhat predictable. In late winter and early spring, the concentrations of pesticides are low, often below the detection limit. The source of compounds detected during this time is primarily ground water (Squillace and others, 1993), although discharge from reservoirs, surface runoff from fields, and discharge from tile drains also may add low levels of pesticides to streams. Application of herbicides in the Midwest starts in late April to mid-May, depending on weather conditions. Elevated herbicide concentrations are observed in streams draining agricultural areas for a few days to a few weeks, depending on the timing and number of rain events and the size of the drainage basin. During this period, about 0.2 to 2 percent of the applied chemical may be moved to surface waters. As the crops grow and the rains subside, the movement of pesticides to surface waters is diminished and riverine concentrations decline throughout the summer. For some compounds, such as atrazine, a low-level, relatively constant concentration is reached and maintained throughout much of the autumn and winter. For others, such as metribuzin, alachlor, and EPTC, the concentration drops below detection levels and remains there until the chemicals are applied again the following spring. The low-level herbicide concentrations observed during the low-flow period (autumn through winter) may result from inflow of ground water from alluvial aquifers that were filled up during the high-flow period when pesticide concentrations were also relatively high. The cycle then repeats itself the next spring.

The seasonal cycle of herbicide concentrations in midwestern reservoirs is somewhat different than in rivers. Many reservoirs in the Midwest receive much of their water from surface water sources during the spring runoff period, when concentrations of herbicides in tributary streams are relatively high. The water is stored for use during the remainder of the year. For compounds that are relatively stable in water, concentrations may remain elevated in reservoirs much longer than in streams, since they are not flushed from the system as quickly. Thus, concentrations of pesticides in reservoirs can remain relatively high long after inputs from agricultural fields have declined or ceased. This effect was observed in the 1992 study of midwestern reservoirs (Goolsby and others, 1993) described earlier (Section 3.3). In Figure 5.1, detection frequencies for herbicides and selected degradation products in reservoirs and streams are compared. The number of reservoirs with detections was nearly constant for most of the analytes from the June-July sampling period through the October-November sampling period. In contrast, the number of streams with detections dropped considerably between the early summer sampling and the late autumn sampling for most analytes. The same contrast was seen in the concentrations of the analytes. In Figure 5.2, concentrations of atrazine, alachlor, and several transformation products in midwestern streams and reservoirs are compared. The stream concentrations follow the pattern described above, with low levels in the preplanting and postharvest periods, and elevated concentrations during the postplanting period. The concentrations in the reservoirs, on the other hand, were much more stable from the early summer period through late autumn, except for alachlor. Alachlor apparently degraded more quickly in the water column of the reservoirs than the other compounds. The seasonal pattern in reservoirs has implications for users of drinking water derived from reservoirs in this region. Compliance with the Safe Drinking Water Act (SDWA) requires that the annual average concentration of a number of pesticides, obtained with quarterly sampling and analysis, remain below a maximum contaminant level (MCL) established for each specific chemical. For most streams supplying drinking water, the normal seasonal pattern in this region results in annual average concentrations below the various MCLs. For reservoirs, the longer period of elevated concentrations increases the likelihood that at least two of the four quarterly samples may have elevated concentrations of some pesticides.

The storage of water with relatively high levels of herbicides in reservoirs also can affect the seasonal pattern of herbicide concentrations in rivers downstream from the reservoir. Depending on the timing of releases of water from the reservoir, downstream concentrations of herbicides would be expected to remain elevated for a longer time than in an unregulated stream. In some cases, the low-level concentrations observed during autumn and winter for certain pesticides, such as atrazine, may be partially attributed to release of water from reservoirs filled during the spring runoff period. Peak concentrations in streams downstream from reservoirs, however, would be expected to be lower because of dilution in the large volume of water in the reservoir (Goolsby and others, 1993). For some compounds with relatively short aquatic lifetimes, such as alachlor, both the duration and magnitude of elevated concentrations downstream from reservoirs may be decreased, due to degradation within the reservoir. For the most part, the effect of reservoirs on seasonal pesticide concentration patterns in streams has not been specifically addressed in published studies.

Seasonal patterns of pesticides in streams may be different in different parts of the nation, depending on the timing of pesticide application and significant rainfall or irrigation. For example, the streams draining the Central Valley of northern California have a strong seasonal

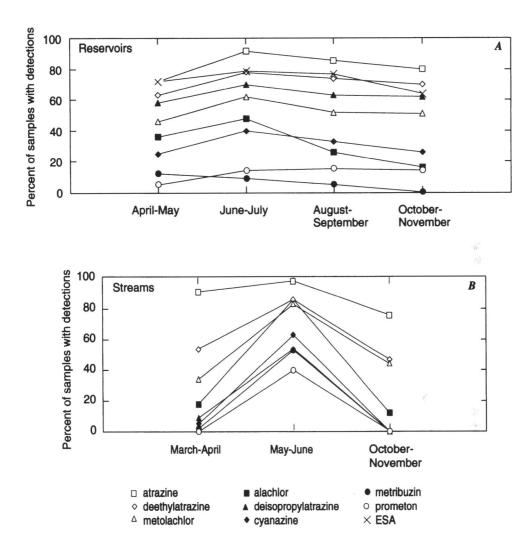

Figure 5.1. Detection frequencies for herbicides and selected degradation products in 76 midwestern reservoirs in 1992 (*A*), and in 147 midwestern streams in 1989 (*B*). Data are from Goolsby and others (1993) and Goolsby and Battaglin (1993).

appearance of methidathion and diazinon—organophosphorus insecticides (OPs) used on orchards—in January and February during the rainy season (Kuivila and Foe, 1995), as shown in Figure 5.3. Herbicides and insecticides used on rice in California also have a distinct seasonal pattern of occurrence in surface waters because of release of irrigation water at specific times (Crepeau and others, 1996), as shown in Figure 5.4. In the Yakima River in Washington, concentrations of 2,4-D followed a distinct seasonal pattern from 1967 to 1971, with elevated concentrations generally occurring from May to September (Manigold and Schulze, 1969; Schulze and others, 1973), as shown in Figure 5.5. In general, available data show that the seasonal input of pesticides into surface waters is dependent on the combination of the timing of pesticide application and subsequent rainfall or irrigation, or release of water in regulated

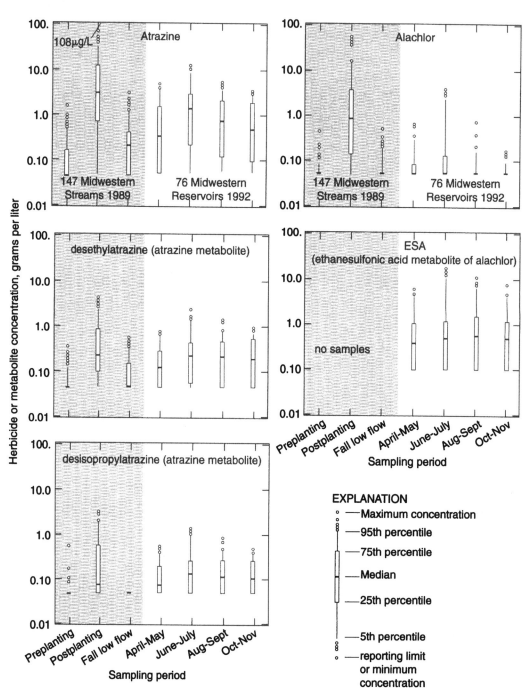

Figure 5.2. Temporal distribution of concentrations of atrazine, alachlor, and selected degradation products in 147 midwestern streams in 1989, and in 76 midwestern reservoirs in 1992. Redrawn from Goolsby and others (1993).

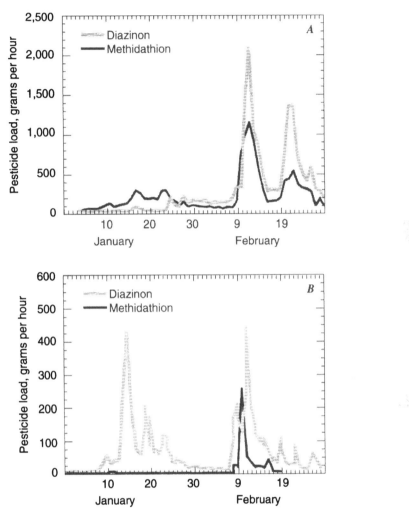

Figure 5.3. Loads (fluxes) of diazinon and methidathion in the Sacramento River at Sacramento (*A*) and the San Joaquin River at Vernalis (*B*) in January and February 1993. Redrawn from Kuivila and Foe (1995).

systems. This is probably true for agriculturally applied pesticides throughout the United States, although there is less published data on the seasonal concentration patterns of pesticides in surface waters outside the midwestern and western United States.

The seasonal pattern in urban areas differs from that of agricultural areas because of differences in the timing of pesticide application. Urban runoff in Minneapolis, Minnesota, recently has been shown to contain the herbicides 2,4-D, MCPP, and MCPA during April through October (Wotzka and others, 1994), as shown in Figure 5.6. The low-level appearance of the herbicides in early spring and late autumn was attributed to use on lawns and gardens by commercial applicators. During mid-summer, significantly higher concentrations of herbicides were detected in runoff and attributed to applications by individual homeowners. During this

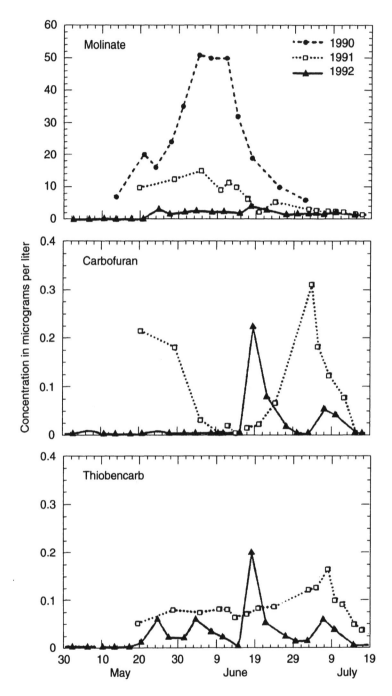

Figure 5.4. Concentrations of three rice pesticides (molinate, 1990–1992; carbofuran, 1991–1992; and thiobencarb, 1991–1992) in the Colusa Basin Drain in the Sacramento Valley, California. Modified from Crepeau and others (1996).

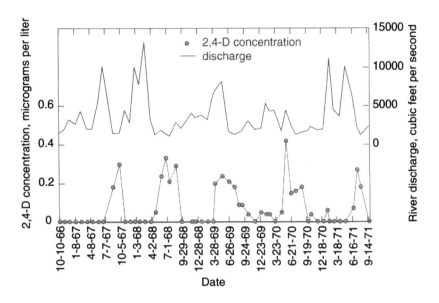

Figure 5.5. Concentrations of 2,4-D and river discharge in the Yakima River at Kiona, Washington, 1966–1971. Data are from Manigold and Schulze (1969) and Schulze and others (1973).

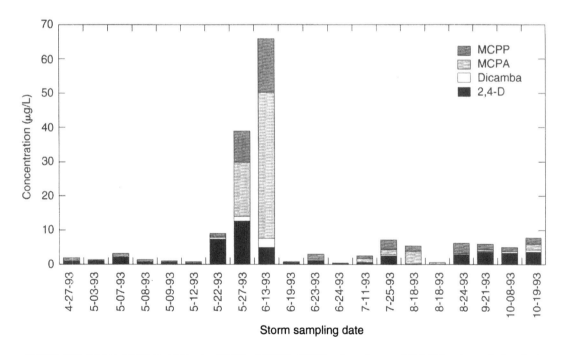

Figure 5.6. Concentrations of the herbicides MCPP, MCPA, dicamba, and 2,4-D in storm drains that drain a residential watershed in Minneapolis, Minnesota, from April to October 1993. Data are from Wotzka and others (1994).

period, inputs of the pesticides were spread out over time with no distinct seasonal pattern. The same observations were made for the insecticide diazinon in the study of the Mississippi River and major tributaries (Larson and others, 1995). In the three river basins with the highest population densities and significant urban centers (the White, Illinois, and Ohio River Basins), the observed flux of diazinon was much greater than would be expected, on the basis of known agricultural use, and had a different seasonal pattern than exclusively agricultural pesticides, such as atrazine, in the same rivers (Figure 5.7). The authors attributed this lack of a seasonal pattern to continual urban use throughout the spring, summer, and autumn. These studies indicate that seasonal patterns of occurrence for urban-use pesticides in surface waters are less distinct and occur over a longer time than for agricultural-use pesticides.

A study of the Susquehanna River in Pennsylvania examined the concentrations of 2,4-D and atrazine over a 12-month period (Fishel, 1984), as shown in Figure 3.45. In the Susquehanna River Basin, there are a variety of land uses, including urban, forested, and agricultural areas (see Section 3.3). Each of these could provide inputs of 2,4-D to the river at various times of the year. Atrazine, on the other hand, has exclusively agricultural uses, and inputs to the river occur mainly in the spring and early summer. Atrazine concentrations in the river show the typical seasonal pattern observed in agricultural areas, whereas 2,4-D concentrations lack strong seasonal patterns, probably from the multiple sources of this compound in the basin.

Resuspension of bed sediments can provide a seasonal source of hydrophobic, recalcitrant pesticides, such as DDT and other organochlorine insecticides (OCs), to surface waters. Bed-sediment particles can be scoured from the bottom and reintroduced into the water column when streamflow is high enough. Pesticides sorbed to these particles may be released to the water column in the dissolved phase before equilibrium is reestablished (see Section 4.2). Resuspension can occur during periods of high flow resulting from spring or autumn rains, extremely large single-storm events, or large releases of irrigation or reservoir waters. In Chesapeake Bay, increases in organochlorine concentrations in the water column (sorbed to suspended sediments) have been attributed to resuspension of bottom sediments by strong currents in parts of the bay (Palmer and others, 1975). Some of these high-energy events in surface waters have a distinct seasonal pattern.

Seasonal patterns in surface-water contamination also have been observed in areas where soil still contains residues from past use of OCs. In the Yakima River Basin in Washington, where irrigation is used to support intensive agricultural activity, total DDT (sum of DDT, DDD, and DDE) concentrations in agricultural drains entering the Yakima River have been shown to be proportional to the suspended-sediment concentration (Johnson and others, 1988; Rinella and others, 1993). Suspended sediment and total DDT concentrations in the river increase during the irrigation season as soil contaminated with DDT is washed into the agricultural drains. The same pattern has been observed in the Moon Lake watershed in Mississippi, where increased total DDT concentrations in the water column occurred during the winter and spring rainy seasons (Cooper, 1991). Soil in this watershed contained significant amounts of DDT (as of 1985), and analysis of sediment cores from Moon Lake showed that recently deposited sediment contained higher amounts of DDT than sediments deposited during the time of heavy DDT use. The authors concluded that DDT in the older sediments was slowly degrading, and the DDT in the recent sediments was coming from eroded soil entering the lake each rainy season. The presence of substantial residues of DDT in soil has been documented in a 1985 study in California (Mischke and others, 1985), and it is likely that seasonal inputs of DDT and other recalcitrant pesticides are occurring in other areas with past use of these compounds (see Section 3.4).

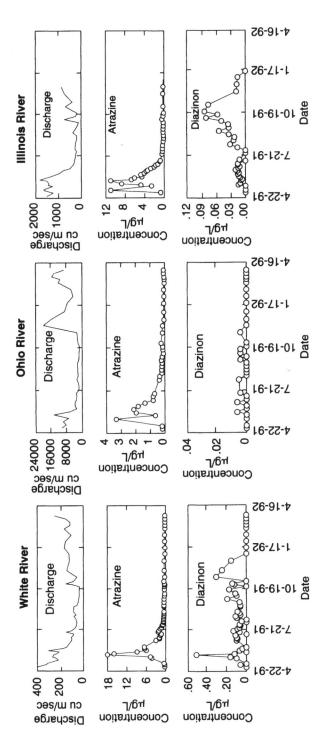

Figure 5.7. Comparison of river discharge, atrazine concentrations, and diazinon concentrations in the White (Indiana), Ohio, and Illinois Rivers, 1991–1992. Data are from Coupe and others (1995).

5.2 SOURCES AND CONCENTRATIONS OF PESTICIDES IN REMOTE WATER BODIES

On a national scale, the dominant source of pesticides to surface waters is agricultural use, with additional inputs from use in urban areas. Sources in more remote areas, such as forests and roadsides, are much more limited in both area and amount of pesticides applied. The compounds currently used for these purposes—such as 2,4-D, picloram, triclopyr, glyphosate, diflubenzuron, and bacterial agents—generally have short environmental lifetimes, and studies suggest that contributions to surface water contamination from these sources are minimal (see discussion in Sections 4.1 and 5.4.).

Thus, in remote non-agricultural areas, atmospheric deposition of relatively long-lived pesticides to surface waters is probably more important than local use. The relative contribution of atmospheric pesticides to a specific surface water body depends on how much of the water budget is derived from drainage, runoff, and precipitation, and how close the water body is to the sources of the pesticides. The magnitude of direct aerial deposition to surface waters is directly proportional to the surface area of the body of water. Generally, lakes are more likely to be affected by atmospheric deposition than streams because the surface areas of lakes represent a much greater proportion of their drainage area than do the surface areas of streams. The significance of the atmospheric input of pesticides to remote lakes and streams is not well known, largely because of the lack of available atmospheric concentration data.

The best understanding of the atmospheric inputs of pesticides to surface water comes from years of study of OCs in and around the Great Lakes. One of the earliest observations of pesticides and other chlorinated hydrocarbons in surface waters in a remote area was from Siskiwit Lake on Isle Royale in Lake Superior (Swain, 1978). Residues of numerous organochlorine compounds were detected in the water, sediment, biota, and precipitation on this island, which is hundreds of miles from the nearest intensive agricultural or industrial activity. The conclusion was that all the organochlorine residues found in the lake had come from atmospheric deposition. This conclusion was supported by observations of the same compounds in precipitation. This finding provided the impetus for many research projects investigating the atmospheric inputs of pesticides and other organic chemicals into the Great Lakes ecosystem. Strachan and Eisenreich (1990) estimated that atmospheric deposition is the greatest source of DDT into Lakes Superior, Michigan, and Huron, where the concentrations range from subnanogram to nanogram per liter. Murphy (1984) used precipitation concentration data from Strachan and Huneault (1979) to estimate the loadings of eight organochlorine pesticides into four of the Great Lakes from 1975 to 1976. The depositional amounts ranged from 112 kg/yr for hexachlorobenzene (HCB) to nearly 1,800 kg/yr for α-HCH, roughly the same as reported by Eisenreich and others (1981). Strachan (1985) reported that precipitation at two locations at opposite ends of Lake Superior contained a variety of organochlorine pesticides. The calculated average yearly loadings ranged from 3.7 kg/yr for HCB to 860 kg/yr for α-HCH. Voldner and Schroeder (1989) estimated that 70 to 80 percent of the toxaphene input to the Great Lakes was derived from long-range atmospheric transport and wet deposition.

The OCs also have been observed in remote surface waters other than the Great Lakes. A number of researchers have reported these chemicals in open ocean areas in the Atlantic and Pacific (Risebrough and others, 1968; Tanabe and others, 1982; Krämer and Ballschmiter, 1988; Iwata and others, 1993). Duce and others (1991) have reviewed the literature on the atmospheric deposition of trace chemical species, including OCs, to the world's oceans. As an example, they estimated atmospheric deposition of the HCHs at 2 and 30 mg/m^2/yr for the South Atlantic and

North Pacific, respectively. They also estimated that 4.8 million kg of HCHs are deposited yearly to ocean surfaces from the atmosphere. DDTs, toxaphene, HCHs, and HCB have been detected in remote wetlands throughout North America (Rapaport and others, 1985; Rapaport and Eisenreich, 1986, 1988). These researchers have used the depth profile of the accumulated pesticides in peat to elucidate the historical atmospheric deposition of these chemicals and have shown that the historical fluxes correlate well with historical-use patterns. In the Antarctic, Tanabe and others (1983) quantified DDTs and HCHs in pack ice, fresh water, and seawater, and Desideri and others (1991) quantified DDTs, HCHs, heptachlor, aldrin, and dieldrin in the same matrices. In the Arctic Ocean, Hargrave and others (1988) quantified DDTs, HCHs, HCB, dieldrin, endrin, heptachlor, and chlordane in either pack ice or seawater. The concentrations were in the subnanogram per liter range. Gregor (1990) observed these compounds and endosulfan in Canadian arctic snow. Hargrave and others (1992) and Lockerbie and Clair (1988) also quantified these OCs in the biota of the Arctic Ocean and noted their accumulation up the food web. Muir and others (1990) linked the presence of toxaphene in arctic water and fish to atmospheric deposition. From these and other data, Richards and Baker (1990) made the following observation:

> ...[atmospheric] transport of toxaphene and other persistent, bioaccumulating compounds has produced dangerous concentration levels in arctic fish and marine animals. These compounds have never been used within 1,000 miles of the Arctic and have not been extensively used in the United States in the last decade. It is ironic that they may represent more of a threat to arctic Native American populations (through dietary intake) than drinking water, with its burden of widely used modern pesticides, does to those living in the corn belt [of the midwestern United States].

Very little research has been done on atmospheric inputs of pesticides to inland surface waters of the United States outside the Great Lakes. Even less has been done on pesticides other than organochlorine compounds. Glotfelty and others (1990) studied the inputs of atrazine, alachlor, and other pesticides in Chesapeake Bay and one of its tributaries (the Wye River). They estimated that about 3 percent of the atrazine load in 1982 and 20 percent of the alachlor load in 1981 in the Wye River was attributable to precipitation inputs. They also estimated that the average summer wet depositional inputs to Chesapeake Bay for atrazine, simazine, alachlor, metolachlor, and toxaphene were 910, 130, 5300, 2500, and 820 kg, respectively, between 1981 and 1984. This area is not remote from agricultural activity, and the bulk of these atmospheric inputs occurred during the time of local use of the compounds (April through June). However, elevated concentrations of simazine and atrazine in rain were observed as early as January, and concentrations continued to rise through the early spring, before any applications of these compounds in the Chesapeake Bay area. The authors hypothesized that the increase in concentrations in rain during this time was due to regional atmospheric transport from agricultural areas farther south, in Florida, Georgia, and North and South Carolina. The timing of planting and herbicide applications in these areas corresponds to the start of the increased concentrations in rain in the Chesapeake Bay area. This suggests that atrazine and simazine can be transported in the atmosphere as much as 600 mi from the point of application. Concentrations of alachlor and metolachlor in rain did not show the same pattern, being present in rain only during the time of local use. The authors conclude that these compounds degraded more quickly in the atmosphere and that regional transport probably does not occur. Buser (1990) quantified atrazine, simazine, and terbuthylazine in rain, snow, and remote Alpine lakes in Switzerland. The concentrations of the herbicides in six mountain lakes, far from agricultural activities, were in the subnanogram per liter range (0.08 to 1 ng/L), whereas rain and snow had concentrations of up to 193 ng/L. The author suggests that for some of the remote lakes, roadside applications may have contributed some of the herbicides, but atmospheric deposition was probably the major source.

If atmospheric deposition contributes pesticides to the Great Lakes, to the Arctic Ocean, to Chesapeake Bay, and to the mountain lakes in Switzerland, then it probably contributes pesticides to most remote surface water environments. The nature of atmospheric deposition of pesticides to remote surface waters is very different from contributions of pesticides used on forests and roadsides. The atmospheric contribution is probably low level (nanogram-per-liter concentrations) and occurs over a long timespan (decades for the OCs), whereas inputs from forest and roadside applications may have higher concentrations (perhaps microgram-per-liter concentrations) and occur over one or more shorter timespans. The continuous atmospheric input is probably of more environmental concern because of the potential for bioaccumulation of the organochlorines and the global nature of the sources and deposition of these compounds. However, relatively little is known about the atmospheric contribution of pesticides to surface waters. More than a decade ago, Eisenreich and others (1981) listed several reasons for this that still hold true today.

1. Inadequate data on atmospheric concentrations of pesticides,
2. Inadequate knowledge of the distribution of pesticides between vapor and particulate phases in the atmosphere,
3. Lack of understanding of the dry deposition process,
4. Lack of appreciation for the episodic nature of atmospheric deposition, and,
5. Inadequate understanding of the temporal and spatial variations in atmospheric concentrations and deposition of pesticides.

5.3 IMPACT OF URBAN-USE PESTICIDES ON SURFACE WATER QUALITY

The only nationwide study of urban runoff—the National Urban Runoff Program (NURP)—was conducted during 1980–1983 (Cole and others, 1984) by the U.S. Environmental Protection Agency (USEPA). In this study, 121 water samples were collected from 61 residential and commercial sites across the United States and analyzed for 127 of the 129 priority pollutants. Of the 20 organochlorine pesticides included in the priority pollutants, 13 were observed in at least one sample. The pesticides observed most often were α-HCH (20 percent of samples), α-endosulfan (19 percent), γ-HCH, or lindane (15 percent), and chlordane (17 percent). Concentrations were generally less than 0.2 µg/L, except for chlordane, which had a maximum concentration of 10 µg/L.

Several smaller-scale studies during the late 1970's and early 1980's monitored the occurrence of pesticides in runoff from urban areas. Water samples from storm sewers draining residential and commercial areas of San Diego, California, from 1976 to 1977 (Setmire and Bradford, 1980), Fresno, California, from 1981 to 1983 (Oltmann and Schulters, 1989), and Denver, Colorado, during 1976 (Ellis, 1978), were analyzed for OCs, OPs, and chlorophenoxy herbicides. The insecticides chlordane, diazinon, and malathion, and the herbicide 2,4-D were detected frequently in all three studies. Concentrations in most samples ranged from 0.1 to 4 µg/L. For comparison, concentrations of pesticides in runoff from agricultural plots commonly exceed 10 µg/L and can reach several hundred micrograms per liter, especially for herbicides (Leonard, 1990).

A more recent (1993) study in Minneapolis, Minnesota, analyzed water in storm sewers draining a residential area for 26 pesticides currently used in urban and agricultural areas (Wotzka and others, 1994). While most samples contained very low or undetectable levels of most of the pesticides, storm-runoff water in June contained the herbicides MCPP, MCPA, and 2,4-D at concentrations up to 40 μg/L (Figure 5.6). These compounds are commonly used on lawns in the Minneapolis area by homeowners and professional applicators. Maximum concentrations of MCPP and 2,4-D in the lake receiving the storm runoff were both 0.2 μg/L. MCPP and 2,4-D were detected in 30 and 40 percent, respectively, of runoff samples analyzed in this study.

There have been few studies of pesticide movement in runoff from grass lawns. In the studies that have been reported, very little runoff (of water) occurred with natural or simulated rainfall on well-maintained turf, even from plots with considerable slope (Harrison and others, 1993). In this study, turf plots (9 and 14 percent slopes) were treated with fertilizers and pesticides for 2 years in a manner typical of that employed by professional lawn care services. Pesticides applied were pendimethalin, 2,4-D ester, 2,4-DP ester, dicamba, and chlorpyrifos. Pesticides were applied in spring, early summer, late summer, and autumn, and the plots were irrigated 1 week before and 2 days after each application. Runoff water was collected during each irrigation event and analyzed for pesticides. No residues of pendimethalin, chlorpyrifos, or the esters of 2,4-D or 2,4-DP were detected in any samples of runoff water. Dicamba, and the acid forms of 2,4-D and 2,4-DP (formed by hydrolysis of the esters), often were detected in runoff water from the first irrigation event following application of the pesticides. Concentrations often were quite high in these samples, with 2,4-D and 2,4-DP concentrations generally in the 10 to 100 μg/L range, but occasionally reaching 200 to 300 μg/L. Dicamba concentrations generally were lower, but reached 252 μg/L in at least one sample. These curbside concentrations agree fairly well with the concentrations observed in the Minneapolis storm sewers mentioned above. Several important problems with this study, however, make generalizing these results to real lawns questionable. First, the amount of irrigation water applied in each event had to be raised to extremely high levels to produce runoff from the plots. The amount of water applied (150 millimeters per hour for 1 to 1.5 hours) corresponds to a storm with a return frequency of much more than 100 years for this region. The lack of runoff at more reasonable rainfall levels was attributed to the high capacity of the thickly grassed plots—maintained in virtually ideal conditions—to hold water. Harrison and others (1993) state that it is not clear how well the results reflect the response of turfgrass subject to normal use and of lawns less well-maintained. The authors also mention that the underlying soil may have contained a zone of highly permeable weathered limestone, which would allow infiltration of large amounts of water and lessen the likelihood of surface runoff. Second, the detection limits for all the pesticides were somewhat high, ranging from 2.4 to 20 μg/L. It is possible that more of the pesticides would have been detected, or that 2,4-D, 2,4-DP, and dicamba would have been detected for a longer time after application, if detection limits had been lower. Third, not all the data from the study are presented in the paper. From the data shown, it appears that irrigation did not always follow application of the pesticides by the same amount of time. In one case, runoff samples were not collected until 38 days after application (with positive detections of all three compounds), whereas in others, samples were collected within 3 days of application. It is difficult to determine the persistence and runoff potential of these compounds in lawns from the data shown.

Recent studies in the Mississippi River Basin have shown rather clearly that urban use of diazinon is resulting in measurable concentrations in several major rivers, as shown in Figure 5.7 (Larson and others, 1995). Concentrations of diazinon in the White (Indiana), Illinois, and Ohio

Rivers were low, generally in the 0.01 to 0.05 µg/L range, but measurable throughout the summer and autumn. The seasonal pattern was similar to the pattern described for urban-use herbicides in Section 5.1, and the estimated riverine flux indicated that much of the diazinon observed originated from non-agricultural applications.

For the most part, however, the significance of urban areas as a source of pesticides to surface waters is difficult to determine. As can be seen from a comparison of Tables 3.1 and 3.3, many of the compounds used in urban areas also are used in agriculture (such as 2,4-D, dicamba, trifluralin, and pendimethalin), so the source of certain pesticides detected in surface waters is often unclear. In addition, pesticides used almost exclusively in urban areas (such as isazofos, isophenphos, oryzalin, and MCPP) have been targeted in very few published studies of surface water quality. For surface waters receiving runoff at least partly from agricultural areas, it is likely that the urban contribution of pesticides is a small proportion of the total pesticide input, because of the much greater agricultural use of pesticides. For example, the estimated agricultural use of pesticides was 811 million lb a.i. in 1993 versus 73 million lb a.i. used in the home and garden sector, and 197 million lb a.i. used in the industrial, commercial, and government sector (Aspelin, 1994), an unknown portion of which was in urban areas. From the limited data available, however, it does appear that surface waters within urban areas, or downstream from urban areas, are likely to contain measurable residues of pesticides from urban applications.

Insecticides are used frequently in urban areas for control of specific pests, such as mosquitoes, termites, and Mediterranean fruit flies (medflies). The chemicals used to control mosquitoes include methoprene, the bacteria *Bacillus thuringiensis* var. *israelensis* (Bti), temephos, resmethrin, malathion, naled, fenthion, and chlorpyrifos (Zoecon Corporation, 1990). In recent years, the OPs and pyrethroid insecticides have been largely replaced by methoprene, an insect growth regulator, and Bti, a selective biological larvicide. These two mosquito-specific agents are less toxic to nontarget organisms than the more traditional insecticides listed above (Hester and others, 1980). Chlordane was the major chemical used to control termites until it was banned in the United States in 1988 (Howard, 1991). Chemicals used in recent years for termite control include the OPs chlorpyrifos, isofenphos, and fenvalerate; and the pyrethroids, permethrin and cypermethrin (Kard and McDaniel, 1993). Medflies have been controlled in California predominantly with aerial applications of malathion (Segawa and others, 1991). In a monitoring study of residues of malathion following aerial spraying in California, concentrations in surface waters within the sprayed area were highly variable, but exceeded water-quality criteria for the protection of aquatic organisms in some cases. Concentrations of malathion in rain runoff ranged from less than 0.1 to 44 µg/L, and the authors conclude that the impact of this runoff on surface waters may be the most significant water quality problem resulting from the spraying program. Other than in this study, the insecticides used to control specific pests generally have not been studied in urban surface waters, and the water quality impacts associated with these applications are largely unknown.

5.4 IMPACT OF FORESTRY-USE PESTICIDES ON SURFACE WATER QUALITY

Several studies have monitored the levels of pesticides in forest streams, mostly in the forests of the southeastern United States and in Canada (Table 2.3). In nearly all cases, these studies can be described as field experiments, in which a known amount of a pesticide was applied to a section of a watershed, with subsequent sampling of stream water for a period of weeks to years. In some studies, forest soil, foliage, aquatic biota, ground water, or runoff water also were sampled. In a few cases, ecological effects or toxic effects on specific organisms were studied (Kreutzweiser and others, 1989; Neary and others, 1993). Studies have been conducted with most of the commonly used silvicultural herbicides (2,4-D, hexazinone, triclopyr, picloram, and glyphosate), as well as with imazapyr, sulfometuron-methyl, amitrole, dicamba, and three of the most commonly used insecticides (*Bacillus thuringiensis* var. *kurstaki* [Bt], diflubenzuron, and carbaryl). See Tables 2.2 and 2.3. No studies of forest streams were found in which the most common silvicultural fungicides or fumigants (dazomet and methyl bromide) were included as target analytes. Laboratory experiments have shown that the half-life of methyl bromide in moving surface waters is very brief (less than 2 hours) because of volatilization (Gentile and others, 1989).

For the herbicides, results of most of the studies were quite similar. In general, contamination of surface waters consisted of pulses of herbicide at relatively high concentrations for short periods after application (hours to a few days), quickly diminishing to low or nondetectable levels. Peak concentrations were below levels known to cause acute effects in most organisms. In some cases, low-level contamination continued for long periods after application—up to 3 years for hexazinone (Lavy and others, 1989). In several of the studies, the pulses of highest concentrations of the herbicide were attributed to direct application to the stream (accidently or by design). The authors of several studies stress that the use of buffer strips along streams, which is the normal practice, is important in preventing direct inputs to streams. More difficult to avoid, especially with aerial application, is application of the pesticide to ephemeral stream channels within the forest. This also was seen as a cause of pulses of higher concentrations in streams. These channels, which may be dry during application, may contain flowing water during subsequent rains and, depending on the time between application and rain, can serve as a relatively direct source of pesticides to streams. The long-term, low-level contamination observed in some of the studies was attributed to both subsurface (downslope) movement in ground water and subsequent release to surface waters (Neary and others, 1983) and to residues remaining on leaf litter (Lavy and others, 1989).

Studies in the southeastern United States indicate that application technique is an important factor in determining the amount of herbicide entering streams. Contamination was most likely with aerial application because of the inadvertent direct inputs to streams or application to ephemeral channels. Contamination was less likely with ground-based broadcast application, and much less likely with stem injection or spot spraying (Michael and Neary, 1993). The peak concentrations of herbicides occasionally have been relatively high, with levels in the 10 to 500 µg/L range. This is, to a certain extent, due to the small size of most of the streams studied. When samples were taken farther downstream, where discharge was higher, concentrations were much lower, owing to dilution with water flowing from untreated areas. Peak concentrations also were quite short-lived, since most of the streams studied were flushed rather quickly with water from untreated areas upstream. This complicates the assessment of the significance of these residues, since most aquatic toxicity data are based on exposure to pesticides in static systems. The effects of low-level, short-lived pulses of pesticides on aquatic organisms is largely unknown (Michael and Neary, 1993). In nearly all the herbicide studies reviewed, the authors conclude that use of the herbicide studied should have no adverse impacts on surface water quality or aquatic life, provided that appropriate safeguards, such as buffer strips, are used when the herbicide is applied. It also should be pointed out that mechanical alternatives to herbicide use for site preparation generally are regarded as having much larger negative impacts on stream water quality, owing to increased erosion and nutrient losses.

Much less has been written about potential surface water contamination resulting from current forestry use of insecticides. Studies of the effects of diflubenzuron use have been reviewed (Fischer and Hall, 1992). The results suggest that adverse environmental impacts are unlikely because of the short half-life of this compound in water, especially if adequate buffer strips along surface waters are used. Studies of the fate of Bt (a bacterial insecticide) and its potential occurrence in surface waters are scarce. In one Canadian study, viable Bt spores were detected in surface waters (and in treated drinking water) up to 13 days after aerial application for control of spruce budworm (Menon and De Mestral, 1985). Laboratory experiments in the same study showed that the bacteria could survive for well over 50 days in lake water. The authors also found that normal chlorination procedures for drinking water did not significantly reduce the population of Bt in water. The apparent lack of attention given to the monitoring of Bt residues may be a reflection of its low toxicity. Bt has little effect on stream invertebrates, and the 96-hour LC_{50} (the concentration lethal to 50 percent of a test population exposed for 96 hours) for rainbow trout is reported as 300 to 1,000 mg/L (Eidt, 1985; Eidt and others, 1989; Kreutzweiser and others, 1992). No adverse effects of Bt have been reported in humans (Menon and De Mestral, 1985).

The consensus, then, from existing studies is that current practices in silvicultural pesticide use are having minimal adverse effects on human or ecosystem health and on surface water quality. There are some significant gaps in the data, however, at least in the published literature. The most intensively used pesticides in forestry, in terms of mass applied per unit area, are the fungicides and fumigants used on nursery stock. No studies were found in which the most commonly used of these compounds were included as analytes in surface waters. In addition, little has been published on monitoring of Bt in surface waters.

5.5 PESTICIDE TRANSFORMATION PRODUCTS IN SURFACE WATERS

Like all organic compounds, pesticides undergo a series of reactions that eventually transform the original compound (parent) into carbon dioxide and other inorganic species (mineralization). Transformation products are the various intermediate chemicals created, then destroyed, during the mineralization process. Transformation products sometimes are referred to as metabolites or degradation products. The latter two terms describe transformation products created through biological processes (biodegradation), but not through chemical transformation processes (e.g., hydrolysis, oxidation and reduction, and photolysis). The rates at which these transformation reactions occur in aquatic systems vary considerably for individual chemicals and under different environmental conditions. For some compounds, such as dichlorvos (an OP), complete mineralization can occur in minutes to hours, whereas complete mineralization of mirex or DDT (OCs) may take decades to millennia. Some transformation products, such as DDE (from DDT) and paranitrophenol (from methyl parathion), can exhibit longer environmental lifetimes than their parent compounds. Different transformation products may be created from the same parent compound by various transformation processes.

A complete understanding of pesticide transformation and of the occurrence of the transformation products in surface waters does not exist. For a few compounds—such as DDE, the aerobic metabolite of DDT—considerable data are available to provide a relatively thorough understanding of its formation and aquatic occurrence, behavior, and fate (Callahan and others, 1979). For the vast majority of pesticides, however, knowledge of the occurrence and behavior of transformation products is much less complete or altogether lacking. Table 2.5 lists the transformation products that have been targeted in surface waters in the studies listed in Tables 2.1 and 2.2. Of the 136 studies listed in Tables 2.1 and 2.2, 84 studies included at least one pesticide transformation product as a target analyte. However, only 10 of these studies targeted transformation products of pesticides other than OCs. Of the 101 studies listed in Table 2.3, only 15 studies addressed pesticide transformation products. An extensive research effort has been undertaken to elucidate the structure and toxicity of pesticide transformation products, but this effort, for the most part, has not been extended to the occurrence and behavior of these compounds in surface waters.

Of the thousands of possibilities, only 20 transformation products, representing 15 parent pesticides, have been targeted in surface waters in the studies reviewed (Table 2.5). Only a few transformation products have been targeted in surface waters for several reasons. First, not all transformation products created from the various processes have been identified, even in an ideal laboratory setting; this is true for both the newer pesticides and those that have been used for years. Second, many transformation intermediates are very transient species and may exist only for seconds or minutes. Only transformation products with relatively long lifetimes in aquatic environments (usually greater than days) can be isolated and detected. Third, with a few exceptions, the presence of pesticide transformation products in surface waters has been perceived only recently as an important concern. In the last few years, interest in pesticide transformation products has increased due to increased awareness of possible environmental impacts and the realization that data on metabolites in surface waters can help researchers understand behavior and fate processes. Finally, most transformation products tend to be more polar than the parent compounds because most transformation processes add oxygenated functional groups to the molecule (carboxylic acid and hydroxyl groups). These more polar compounds generally are more difficult to isolate from water and many cannot be analyzed by gas chromatography—the most common analytical method for pesticides. As a result, these transformation products are not commonly included as target analytes in monitoring studies.

Historically, the transformation products most commonly targeted in surface waters have been those included in the USEPA priority pollutant list from the early 1970's (Callahan and others, 1979). These include DDD and DDE (from DDT), endosulfan sulfate (from endosulfan), endrin aldehyde (from endrin), heptachlor epoxide (from heptachlor), α-HCH (a component of the technical mixture and a transformation product of γ-HCH, or lindane), and dieldrin (a pesticide itself and a transformation product of aldrin). In the studies reviewed (Table 2.5), DDD and DDE have been found at about the same percentage of sites as DDT, and heptachlor epoxide was found at nearly twice the percentage of sites as heptachlor. Endosulfan sulfate and endrin aldehyde were not detected in any of the studies reviewed. Dieldrin and the various isomers of HCH were the most common organochlorines detected in the studies reviewed, but their presence undoubtedly is due to both the application of the compounds themselves and to the transformation of their parent compounds. Similar results were obtained in a review of data on priority pollutants retrieved from the USEPA's STORET (STOrage and RETrieval) water quality database (Staples and others, 1985), as shown in Figure 5.8. In this study, data on detections of priority pollutant pesticides in ambient waters are summarized from 1980 to 1982. Concentrations of transformation products of the organochlorines, like the parent compounds, are very low in the water column. Most of these compounds are still quite hydrophobic and are more likely found in bed sediments. Water column concentrations reported in the studies reviewed were nearly all less than 1 μg/L, and most were well below 0.1 μg/L (Tables 2.1 and 2.2).

More recently, two of the transformation products of atrazine—deethylatrazine and deisopropylatrazine—have been the focus of a number of studies, and a basic understanding of their occurrence in surface waters is developing. These two metabolites were among the most commonly detected compounds in the studies reviewed. Deethylatrazine and deisopropylatrazine were detected at 87 and 64 percent, respectively, of the sites at which they were targeted

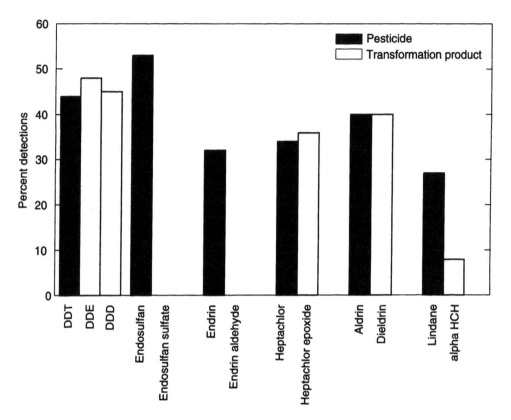

Figure 5.8. Detection frequencies of selected organochlorine pesticides and their transformation products in ambient waters, 1980–1982. Frequencies are from the U.S. Environmental Protection Agency's STORET (water quality database). Number of samples ranged from 800 to 9,000 for the compounds. Dieldrin is a transformation product of aldrin, but was also used as an insecticide prior to 1974; α-HCH is a transformation product of lindane, but is also a major ingredient of technical-grade HCH. Data are from Staples and others (1985).

(Table 2.5). It should be noted that the studies targeting these compounds are all quite recent and were primarily conducted in areas with high use of atrazine. Two recent regional-scale studies have investigated the occurrence and temporal variability of these two metabolites in streams and reservoirs of the midwestern United States. In 1989 and 1990, 147 streams were sampled in spring (preplanting), summer (postplanting), and autumn (postharvest) seasons (Thurman and others, 1992; Goolsby and Battaglin, 1993). In 1992, 76 reservoirs in the same area were sampled four times from April to November (Goolsby and others, 1993). Results from both regional-scale studies are shown in Figure 5.2. Stream concentrations followed the typical seasonal pattern observed in the Midwest, with highest concentrations of both parent compounds and metabolites occurring shortly after pesticide application in early summer. Concentrations of the metabolites generally were lower than atrazine concentrations during all three sampling periods. The median concentration of atrazine in the streams was 3.8 mg/L in the postplanting period, whereas the median concentrations of deethylatrazine and deisopropylatrazine were 0.28 and 0.09 µg/L, respectively (Thurman and others, 1992). Maximum concentrations during the postplanting period were 108, 4.4, and 3.2 µg/L for atrazine, deethylatrazine, and deisopropylatrazine, respectively. Concentrations for both parent compounds and metabolites in the reservoirs had much less seasonal variability than stream concentrations. Concentrations of deethylatrazine, deisopropylatrazine, and the ethanesulfonic acid metabolite of alachlor (ESA) were relatively stable during the entire sampling period in the reservoirs, with the result that metabolite concentrations may be significantly higher in reservoirs than in streams for extended periods of time. Concentrations of the atrazine metabolites in the reservoirs generally were in the 0.1 to 1.0 µg/L range throughout the period, while atrazine concentrations in the reservoirs generally were in the 0.1 to 2.0 µg/L range. Reservoir concentrations of the alachlor metabolite ESA generally were in the 0.1 to 1.0 µg/L range and were consistently higher than alachlor concentrations. ESA was not monitored in the stream study.

Some pesticide transformation products may be more stable in surface waters than their parent compound. This was shown in a recent study of the Sacramento River in California, in which concentrations of two herbicides (molinate and thiobencarb) and two insecticides (carbofuran and methyl parathion) commonly used on rice fields in this area were monitored, along with several transformation products (Domagalski and Kuivila, 1991). By law, water used for irrigation of rice fields in this area must be held on the fields for a prescribed period before being released to agricultural drains to allow degradation of residual molinate to occur. Concentrations of the parent compounds and transformation products were measured as the released irrigation water was transported down the river. Transformation products of carbofuran and molinate were detected in the river and appeared to behave conservatively (i.e., concentrations did not decline) for up to 90 hours. Methyl parathion was not detected in the river, apparently having degraded while still on the fields. Paranitrophenol, a transformation product of methyl parathion, was detected in all samples of river water, however, and appeared to be stable over the study period. This study, along with the atrazine metabolite studies described above, suggests that pesticide transformation products may be much more prevalent in surface waters than the data in Table 2.5 would suggest. The likelihood of finding a particular transformation product in surface waters depends on the relative stabilities of the parent and transformed compound and on their individual physical and chemical properties. The importance of the occurrence of transformation products in surface waters depends on their toxicological properties. This topic is discussed further in Section 6.3.

5.6 MODELING OF PESTICIDES IN SURFACE WATERS

Conceptual and mathematical models are used to understand or predict the environmental behavior or fate of pesticides. The purposes, complexities, and usefulness of models vary considerably. Donigian and Carsel (1992), in arguing the value of models, observed the following:

> Several methods, including both monitoring and mathematical modeling, can be used to evaluate exposures from pesticide use in various environmental media (i.e., air, soil, water). Monitoring can be used to determine the presence or absence of chemicals in the specific media of interest; however, monitoring is a de facto (or after-the-fact) evaluation. Also, it cannot be used in a priori type evaluations to predict potential exposures, and monitoring over the

typically large spatial scales of interest is extremely costly and usually not practical or feasible. Moreover, the national scale of the problem, the number of pesticides in use (approximately 600 active ingredients), chemical properties (e.g., transport and transformation), and the multimedia requirements (e.g., surface water, ground water, air, etc.) dictates comprehensive monitoring designs, costly implementation, and significant logistical problems. The use of mathematical models of pesticide environmental fate and transport processes is one alternative to comprehensive monitoring programs.

On the other hand, models have real limitations, not the least of which is their credibility (Wauchope, 1992).

Models can be classified in various ways. One method is by level of complexity (Decoursey, 1992), ranging from screening to research models. Simple screening models usually have minimal input data, are relatively easy to use, and produce fast results. The underlying assumptions are usually simple, and the quality of the output varies considerably depending on the underlying theories and assumptions. The primary function of these models is the preliminary assessment of specific geographic environments, or pesticide behavior and fate processes for future study. At the other end of this classification are complex research models. These models attempt to simulate the actual physical, chemical, or biological processes accurately. Research models require a large amount of input data, are relatively difficult to use correctly, and may require large amounts of computer time. The output of research models often describes pesticide concentrations as a function of time, space, or environmental media. There is somewhat of an inherent conflict between the scientific community that develops models—usually research models, to help understand mechanisms of behavior and fate—and many end-users, such as regulators, who would like to use models for planning purposes (Donigian and Carsel, 1992). Wauchope (1992) notes that complex systems require complex models, but that the "...realistic model has less credibility. The less the user knows of the subtleties involved, the more suspicious he/she becomes of the models as their difficulty increases. The reverse is also true. The more one knows about the system (specific environment and(or) chemical) the more one recognizes the vast simplification of reality and the extremity of the assumptions used in the simplest models."

For the topic of pesticides in surface waters, a number of different types of models, based on chemical or hydrologic perspectives, provide some understanding of the observed behavior of pesticides, or of some capability for predicting their behavior. The types of models pertinent to pesticides in surface waters include structure-activity models that predict chemical properties or the equilibrium state of a transfer process, field runoff models for understanding the delivery of pesticides to surface waters, surface water transport models for simulating rivers and lakes, and regional multimedia models for predicting the equilibrium state of pesticides among land, air, and surface water. Each of these types of models will be discussed separately, with examples of their utility and limitations for pesticides in surface waters.

STRUCTURE-ACTIVITY MODELS

Accurate chemical property values, such as water solubility, vapor pressure, and hydrolysis and biodegradation rate constants, are critical to the utility of most environmental models that describe or predict the behavior or fate of a pesticide. Predictive structure-activity models for physical, chemical, and biological properties of organic chemicals are estimation tools for use with new pesticides or for properties that are difficult to measure experimentally (such as water solubility of very hydrophobic compounds). These models usually are based on one of two approaches. One approach is based on relations between known properties of similar compounds. As an example, there have been numerous correlations reported in the scientific literature that relate a chemical's water solubility to its octanol-water partition coefficient (Lyman and others, 1990). An equation derived from the regression of these two properties then can be used to predict the octanol-water partition coefficient from the measured water solubility for a new compound. Reviews of various correlation methods have been included in Lyman and others (1990). Included in this review are methods for estimating octanol-water partition coefficient, water solubility, rate of hydrolysis, diffusion coefficient in water, and liquid density.

A second approach is based on summing molecular fragments to yield predictive chemical properties. The premise is that different functional groups on organic molecules affect a chemical property in a consistent, predictable manner. Thus, one could predict that the addition of a hydroxyl or a carboxylic acid group will increase the water solubility of an organic chemical, whereas addition of a methyl group or halogen will decrease its water solubility. A review of various predictive fragmentation methods is in Lyman and others (1990). Included in this review are predictive fragmentation methods for octanol-water partition coefficient, water solubility, acid dissociation constant, rate of hydrolysis, aqueous activity coefficients, boiling point, heat of vaporization, vapor pressure, diffusion coefficients in water, and liquid density.

Structure-activity relations also have been used to model phase transfer processes in aquatic, atmospheric, and terrestrial environments. For transfer processes purely chemical in nature, such as the sorption of a pesticide to the organic carbon coating on an aquatic particle, fundamental chemical properties are linked closely to environmental behavior and provide strong predictive modeling tools over a broad range of environmental conditions. For transfer processes that have a strong mass transport component linked with phase transfer, such as volatilization of a pesticide from water, structure-activity models tend to be less robust. One structure-activity approach for modeling environmental behavior is based on a linear regression of known properties of similar compounds. As an example, there have been numerous correlations reported in the scientific literature relating a chemical's water solubility to its bioaccumulation factor (Lyman and others, 1990). These regression equations then can be used to predict the bioaccumulation factor from a measured water solubility for a new compound. A review of various correlation methods is included in Lyman and others (1990). Included are correlation methods for estimating bioaccumulation from octanol-water partition coefficient and water solubility, sorption to aquatic particles (normalized for organic carbon) from octanol-water partition coefficient and water solubility, volatilization from water from Henry's Law constant, and volatilization from soil from water solubility and octanol-water partition coefficient.

Both uses of structure-activity relations (to estimate chemical properties and to estimate environmental behavior through phase-transfer processes) are used widely by the scientific and industrial communities and by regulatory agencies in initial attempts to model the environmental transport, behavior, and fate of new pesticides. Physical and chemical properties and environmental fate constants usually are included as inputs to the models discussed below. A considerable amount of on-going research is being undertaken to increase the usefulness of structure-activity relations and to better understand the underlying principles upon which they are founded.

RUNOFF MODELS

The primary mechanism for the movement of most agricultural pesticides to surface waters is runoff from agricultural fields. There has been considerable work in understanding and modeling the processes involved in runoff. Leonard (1990) has outlined the conceptual model for runoff (Figure 5.9). The general factors that need to be considered include the following:

1. Climatic conditions (rainfall duration, amount, and intensity; timing of rainfall after pesticide application; and time to runoff after inception of rainfall);

2. Soil conditions (soil texture, organic matter content, surface compaction and crusting, antecedent water content, slope, and degree of aggregation);

3. Pesticide characteristics (water solubility, sorption properties, polarity and ionic nature, persistence in soil, formulation, and application rate); and

4. Agricultural management practices (pesticide placement in or on soil; erosion control practices; residue management; and irrigation duration, amount, intensity, and timing after pesticide application).

These four broad considerations involve a large number of complex, individual processes that are physical, chemical, or biological in nature. Numerous mathematical models, with varying degrees of complexity, have been used to quantify pesticide losses in runoff. Prediction of pesticide concentrations in runoff—over time and with varying amounts of rainfall—is a particularly important, but mathematically difficult, problem.

Models described in the scientific literature for simulation of pesticide behavior in agricultural runoff vary in approach and purpose. Simulation models specifically designed for pesticides in runoff include an early model, PRT, or Pesticide Runoff Transport (Crawford and Donigian, 1973), which estimated soil and pesticide losses from field-sized areas. PRT was extended to include other water quality components and was called ARM, or Agricultural Runoff Model (Donigian and Crawford, 1976). This model then was linked to HSPF, or Hydrologic Simulation Program—FORTRAN (Johanson and others, 1980; Donigian and others, 1983), a simulation model for stream transport. Bruce and others (1975) developed an event model to estimate runoff of pesticides, water, and eroded particles from field-sized areas. ACTMO, or Agricultural Chemical Transport Model (Frere and others, 1975), simulated runoff losses of soil and pesticides from field- to basin-sized areas. CPM, or Cornell Pesticide Model (Steenhuis and Walter, 1980), includes simulations of both pesticide runoff and leaching. CREAMS, or Chemicals, Runoff, and Erosion from Agricultural Fields Management Systems (Knisel, 1980), is the result of the U.S. Department of Agriculture's national Agricultural Research Service Project. This model is useful for making relative comparisons of runoff from various management practices for field-sized areas. CREAMS was later modified to incorporate vertical movement of chemicals and renamed GLEAMS, or Ground Water Loading Effects of Agricultural Management Systems (Leonard and others, 1987). The USEPA sponsored the development of PRZM, or Pesticide Root Zone Model (Carsel and others, 1985). PRZM was developed initially to simulate root-zone processes and leaching, but later included a simulation of pesticide runoff. This model is used by the USEPA in pesticide registration evaluations (Zubkoff, 1992). GLEAMS and PRZM also have been used to simulate pesticide losses from forested watersheds and turf (Lin and Graney, 1992; Dowd and others, 1993; Nutter and others, 1993).

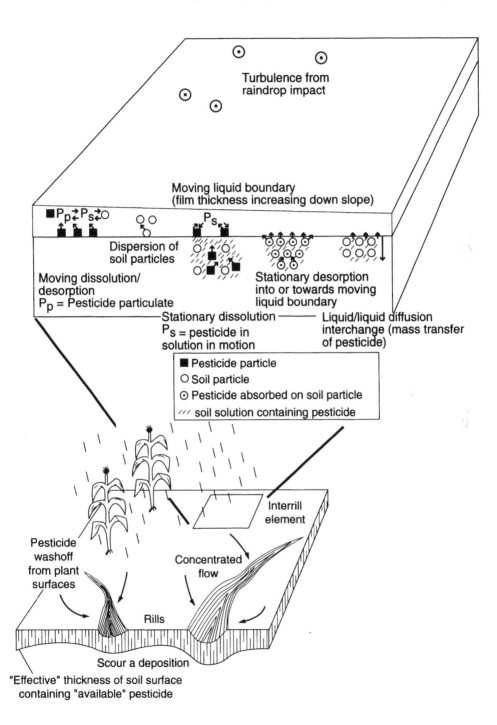

Figure 5.9. Conceptual model for runoff from agricultural fields. Redrawn from Leonard (1990), with permission from the Soil Science Society of America.

SURFACE WATER TRANSPORT MODELS

The simulation of pesticide behavior in streams and rivers generally is done in the context of a modified advection-dispersion model. Models specifically designed for pesticides in streams include HSPF (Johanson and others, 1980; Donigian and others, 1983), STREAM, or Stream Transport and Agricultural Runoff of Pesticides for Exposure Assessment (Donigian and others, 1986), and SURFACE (a mathematical model used for prediction of concentrations of pesticides in surface waters), which is used in conjunction with PRZM (Gustafson, 1990). Models often are developed to simulate behavior in a specific river (Schnoor and others, 1982; Wanner and others, 1989; Schnoor and others, 1992). The following are examples of modeling of pesticides in rivers and streams. Bicknell and others (1985) used HSPF to model the behavior of pesticides, nutrients, and sediment in a large river in Iowa and estimate changes in water quality resulting from various agricultural management practices. Wanner and others (1989) and Capel and others (1988) modeled a spill of pesticides into the Rhine River by modifying a hydrologic model of the river to include pesticide loss and retardation processes. Gustafson (1990) calibrated SURFACE with the 1985 concentrations of atrazine, alachlor, cyanazine, and metolachlor at 17 sites, and then predicted their annualized mean concentrations in 48 streams in the midwestern United States. The predictions, compared with the measured data collected in 1986, gave an average error of 68 percent. Schnoor and others (1992) developed a dynamic, environmental-processing-based, one-dimensional model for large rivers and used it in conjunction with the data from the pesticide spill on the Rhine River (Capel and others, 1988; Wanner and others, 1989).

Models used for simulation of the behavior of pesticides in lakes also vary in their approach and purpose. Some researchers use a simple mass-balance approach, whereas others have attempted to model the in-lake processes in detail. Models specifically developed to model behavior in lakes and reservoirs are SLSA, or Simplified Lake and Stream Analyzer (DiToro and others, 1982), TOXIWASP/WASTOX/WASP4 (Ambrose and others, 1983; Burns 1983), EXAMS II, or EXposure Analysis Modeling Systems (Burns and Cline, 1985), and TOXIC (Schnoor and McAvoy, 1981). The following are a few examples of how pesticides were modeled in lakes and reservoirs. Schnoor and others (1982) modeled the time-dependent fate and transport of atrazine into, within, and out of a reservoir. Crossland and others (1986) examined the behavior of two pesticides—methyl parathion and pentachlorophenol (PCP)—in an outdoor experimental pond and modeled their compartmental distributions and transformation rates. Halfon (1986, 1987) modeled the behavior and transport of mirex into and within Lake Ontario and used the mirex concentration preserved in a sediment core to calibrate a model of in-lake processing of the chemical. O'Connor (1988) presented the conceptual and mathematical approaches for modeling persistent, hydrophobic pesticides (such as DDE) in lakes and reservoirs, with an emphasis on sorption and sedimentation, the most important fate processes for this class of chemicals. Sato and Schnoor (1991) evaluated three fate models for pesticides in lakes and reservoirs with a 16-year data set of dieldrin in a reservoir in Iowa. They concluded that each of the models has its advantages and its limitations. Ulrich and others (1994) simulated the seasonal pattern of atrazine concentrations by using its behavior and fate processes in a one-box model of a small lake and compared the simulations to field observations.

MULTIMEDIA MODELS

The use of regional multimedia models for predicting the behavior of organic chemicals has been suggested for almost 2 decades (Baughman and Lassiter, 1978; Neely and Mackay, 1982; Mackay, 1991; Mackay and others, 1992). The premise of these models is that a chemical's fate is controlled by two groups of factors. One is the set of properties of the chemical, such as water solubility, vapor pressure, and hydrolysis reaction rates. The other is the set of properties of the environment into which the chemical is introduced, such as organic carbon content of soil and aquatic particles, temperature of air and water, and pH of water. In the real world, the environment changes spatially, but the chemical properties always remain the same. The ultimate value of the regional multimedia models is that a common evaluative environment can be created, and the behavior of many chemicals can be compared on a normalized basis to understand behavior and fate processes and to assess generic behavior and fate as a "prelude to region-specific environments" (Mackay and others, 1992). This approach has become accepted by the scientific community, industry, and regulatory agencies as an initial, inexpensive tool in evaluating the behavior and fate of pesticides in the combined atmospheric, terrestrial, and hydrologic environment.

Three levels of complexity have been suggested for these regional multimedia models. The simplest level describes the equilibrium distribution among various environmental media— air, soil, surface water (dissolved), suspended sediment, bed sediment, and fish—of an arbitrary mass of chemical introduced at a point in time. The relative volumes and characteristics of the environmental compartments are chosen to simulate a generic environment. Mackay and others (1992) suggest an example of a model environment as an area of 10^{11} m^2 (about the size of the state of Ohio) with an active atmospheric height of 1,000 m, and a surface water coverage of 10 percent with a mean depth of 20 m. The soil compartment is assumed to be 10 cm thick and contains an organic carbon content of 2 percent. The bed sediment has the same area as the surface water with an active depth of 1 cm and an organic carbon content of 4 percent. Aquatic particles are present at a concentration of 5 mg/L and have an organic carbon content of 20 percent. Finally, the fish are present at a quantity that represents a volume fraction of 10^{-6} of the total water volume and a lipid content of 5 percent. The model output is concentration and mass of the chemical in each of the compartments. The absolute values in the output do not have particular significance. However, the relative magnitudes of concentration and mass among the compartments can be compared for a given compound. In addition, the relative magnitudes of concentration and mass in the same compartments can be compared when the model is applied to different compounds. The environmental compartments with the smallest volumes—aquatic particles and fish—often have the highest relative concentrations for hydrophobic pesticides and are of most concern from the perspective of human and aquatic health. The regional multimedia models with the simplest level of complexity only consider how the chemical will distribute itself at equilibrium. The models do not consider attainment of equilibrium or losses of the chemical from transformation. This type of model has been used to study the distribution of pesticides in forested environments in Canada (Zitko and McLeese, 1980), in the Po River watershed in northern Italy (Del Re and others, 1989), and on Prince Edward Island, Canada (Burridge and Haya, 1989).

The second level of complexity of regional multimedia models considers the equilibrium and steady-state distribution of a chemical introduced continually into the environment. When the input and output rates are equal, rates of loss of the chemical by reaction and advection can be calculated. The same chemical properties and generic environmental compartments are used as in the simple model.

The most complex form of the present multimedia models does not assume that equilibrium has been reached. This form of the model simulates the actual environment most realistically. The advective rates of transfer between adjacent compartments (air–water, soil–water, soil–air, and so forth) are described by generic transport velocity characteristics. From the chemical properties, environmental compartment properties, and the intermedia transport rates, the concentration and mass in each compartment, transformation rates within a compartment, and the transport rates between compartments can be calculated. Although this is a powerful tool for the prediction of the behavior and fate of a chemical with time, the weakest link lies in the ability of the modeler to choose the correct intermedia transport rates. In the case of pesticides, one of the most important intermedia transport rates is soil–water and, as discussed above, a parameter very difficult to model. Nevertheless, the regional multimedia models, particularly the nonequilibrium models, provide the only tools for quickly and inexpensively predicting the behavior and fate of new pesticides in the surface water environment.

USE OF MODELS

Future work in the modeling of pesticide transport, behavior, and fate in surface waters lies in two main areas. The first is the continued refining and development of the kinds of models described above. Each of the types of models discussed has an area of particular utility to scientists and managers concerned with the behavior and fate of pesticides in surface waters. Each type of model also has limitations in its usefulness, and these must be recognized and considered by users of the models. The second aspect of the future of modeling is the linking of various behavior and fate models with environmental effects, toxicological (or socioeconomic) models, and geographic information systems (Zubkoff, 1992).

The use of mathematical models to address regulatory questions and concerns about pesticide transport, behavior, and fate is an area of rapid growth. The USEPA sees multiple uses of models in their registration and reregistration process of pesticides. Zubkoff (1992) outlined seven potential areas of usefulness.

1. Helping to determine whether additional studies on the fate and distribution of a candidate chemical in the environment and/or ecological effects may be needed when full chemical characterization is complete.
2. Helping to more fully integrate data submissions of laboratory and field observations.
3. Estimating probable fate and distribution of an agrochemical after a severe runoff event.
4. Comparing alternative chemical application rates and methods for the same chemical/soil/crop/environmental combinations.
5. Comparing different soil/crop/environmental combinations representing different geographical areas with the same chemical.
6. Evaluating preliminary designs of proposed field studies.
7. Gaining insight into the environmental fate of modern chemicals that are applied at 1 to 2% of the rates of older chemicals when sampling designs and analytical methods are not available.

The USEPA Office of Pesticide Programs commonly uses the SWRRB, or Simulator for Water Resources in Rural Basins, and PRZM models to evaluate chemical runoff from fields and EXAMS II to evaluate pesticide fate and transport in surface waters. Zubkoff (1992) discussed the regulatory use of simple screening models and reviewed a number of existing models, and more complex models under development, used by regulatory and academic scientists.

The viewpoint of the pesticide industry on the use of models in the regulatory process is somewhat different. Russell and Layton (1992) have suggested the following:

...uses for modeling to address regulatory concerns include: (a) evaluating the rate of off-target movement of both existing and potential products in response to changes in application rates, application dates, and environmental scenarios; (b) assisting in the design of field studies by predicting off-target movement in order to refine sampling schedules or methods, or to provide loading information for aquatic toxicity tests (such as microcosms or mesocosms); (c) assisting in the evaluation of field data and extending the usefulness of field data through interpolation and/or extrapolation; and (d) substituting for field studies when the results clearly indicate environmental safety even in worst case settings.

They also see several problems in using models for regulatory purposes. The first problem is that most models currently used in this way were developed as research tools, not with the intention of being used for regulatory purposes. Second, models can be used incorrectly, such as the use of PRZM, a model developed to predict the downward movement of pesticides in soil, to evaluate potential for surface runoff. Third, the absolute accuracy of models is limited, and many models have not been validated against field data. Finally, models should not be used without clearly defined objectives, standard procedures, and thoroughly trained personnel.

In general, most scientists—academic, regulatory, and industrial—would agree that models provide a valuable tool in understanding and predicting the behavior and fate of pesticides in the environment in general, and in surface waters in particular. Regulatory decisions should not be based on results from modeling efforts alone, but modeling should be used to provide one of the pieces of evidence in the entire puzzle of pesticide concerns.

CHAPTER 6

Analysis of Key Topics—Environmental Significance

6.1 IMPLICATIONS FOR HUMAN HEALTH

Under provisions of the Safe Drinking Water Act (SDWA), the U.S. Environmental Protection Agency (USEPA) has established an enforceable maximum contaminant level (MCL) allowed in drinking water for certain pesticides with past or present use in the United States (Table 6.1). The MCLs are health-based standards and are results of chronic toxicity tests conducted with animals. The MCLs are derived from the highest concentration at which no adverse health effects were observed in the test animals, multiplied by a safety factor of 100, or 1,000 in the case of suspected or probable carcinogens. Considerations of treatment feasibility, cost of treatment, and analytical detection limits also were included in the derivation of the MCLs. The USEPA also has established a maximum contaminant level goal (MCLG) for all chemicals with an established MCL. The MCLG is a nonenforceable concentration of a drinking-water contaminant that is protective of human health and allows an adequate margin of safety (Nowell and Resek, 1994), without regard for economic or analytical constraints. The MCLG is set at zero for known or probable human carcinogens. Pesticides with an MCLG of zero include alachlor, chlordane, dibromochloropropane (DBCP), EDB, heptachlor, heptachlor epoxide, pentachlorophenol (PCP), hexachlorobenzene (HCB), and toxaphene. Of these, alachlor is the only one with significant current use in the United States. These standards apply to finished (treated) drinking water supplied by a community water supply, and require that the annual average concentration of the specific contaminant be below the MCL. As of 1994, the SDWA requires most suppliers of drinking water to monitor for 39 pesticides or pesticide transformation products in finished water, 14 of which are no longer registered for use in the United States. Pesticides with current agricultural use, for which MCLs have been established and monitoring is required, include seven herbicides (alachlor, atrazine, 2,4-D, diquat, glyphosate, picloram, and simazine), four insecticides (carbofuran, lindane, methoxychlor, and oxamyl), and one fungicide (PCP). In addition, monitoring is required for 13 pesticide-related compounds for which MCLs have not been established (U.S. Environmental Protection Agency, 1994b). From a compliance standpoint, the standards (MCLs) do not apply to most water bodies reported on in this book, since most of the studies reviewed were not analyzing finished drinking water.

For many of the pesticides with no established MCL, other (nonregulatory) criteria have been established. The USEPA has issued drinking-water health advisory (HA) levels for adults and children for various exposure periods. The National Academy of Sciences has issued a Suggested No-Adverse-Response Level (SNARL) for many pesticides. Both the HA and SNARL values represent estimates of the maximum level of a contaminant in drinking water at which no adverse effects would be expected. The lifetime HA and the SNARL are derived in the same

Table 6.1. Standards and criteria for protection of human and aquatic organism health for pesticides targeted in surface waters

[All standards and criteria values are from Nowell and Resek (1994). Concentrations are in microgram(s) per liter. Human Health: MCL, Maximum contaminant level for drinking water established by the U.S. Environmental Protection Agency (USEPA); MCLG, Maximum contaminant level goal for drinking water established by the USEPA (equal to zero for known or probable human carcinogens); HA (child, long term), Health advisory level for drinking water established by the USEPA (for a 10-kilogram child over a 7-year exposure period and for a 70-kilogram individual over a 70-year exposure period). SNARL, Suggested No-Adverse-Response Level for drinking water established by the National Academy of Sciences (NAS). Exceeded Values, Number of studies in which a criteria value was exceeded / Number of studies in which an analyte was targeted. All studies from Tables 2.1 and 2.2 that targeted the compound are included in the denominator, regardless of whether a maximum concentration was reported. The number of studies with exceeded values may be an underestimate, because some studies did not report a maximum concentration. Aquatic Organism Health: USEPA, Acute and Chronic, Established concentration below which adverse effects on aquatic organisms are not expected for acute or chronic exposure. National Academy of Sciences and the National Academy of Engineering (NAS/NAE), Concentration established in 1973 below which adverse effects are not expected. Exceeded Values, as defined above. nsg, no standards given]

| | Human Health | | | | | | Studies in which MCL or HA exceeded | Aquatic Organism Health | | | |
| | MCL | MCLG | HA | | SNARL | Exceeded values | | USEPA | | NAS/NAE | Exceeded values |
			Child	Adult				Acute	Chronic		
INSECTICIDES											
Organochlorine:											
Aldrin	nsg	nsg	0.3	nsg	nsg	2/57	Bradshaw and others, 1972 Page, 1981	3	nsg	0.01	13/57
Chlordane	2	0	0.5	nsg	nsg	2/50	Warry and Chan, 1981 Kuntz and Warry, 1983	2.4	0.0043	nsg	17/50
DDT[1]	nsg	nsg	nsg	nsg	nsg	nsg		1.1	0.001	nsg	38/74
Dieldrin	nsg	nsg	0.5	nsg	nsg	4/72	Klaasen and Kadoum, 1975 Warry and Chan, 1981 Leung and others, 1982 Kuntz and Warry, 1983	2.5	0.0019	nsg	37/72
Endosulfan	nsg	nsg	nsg	nsg	nsg	nsg	None	0.22	0.056	nsg	4/45
Endrin	2	2	4.5	1.6	nsg	0/60	None	0.19	0.0023	nsg	16/60
HCH (all isomers)[2]	0.2	0.2	33	2	nsg	6/65	Nicholson and others, 1964 Bradshaw and others, 1972 Page, 1981 Warry and Chan, 1981 Kuntz and Warry, 1983 Takita, 1984	2	0.08	nsg	14/65

Table 6.1. Standards and criteria for protection of human and aquatic organism health for pesticides targeted in surface waters—Continued

| | Human Health | | | | | | | Aquatic Organism Health | | | |
| | | | HA | | | | | USEPA | | | |
	MCL	MCLG	Child	Adult	SNARL	Exceeded values	Studies in which MCL or HA exceeded	Acute	Chronic	NAS/NAE	Exceeded values
Heptachlor	0.4	0	5	nsg	nsg	2/54	Page, 1981; Kuntz and Warry, 1983	0.52	0.0038	nsg	14/54
Kepone	nsg	nsg	nsg	nsg	nsg	nsg	None	nsg	nsg	nsg	nsg
Methoxychlor	40	nsg	500	340	700	1/36	Warry and Chan, 1981	nsg	0.03	nsg	6/36
Mirex	nsg	nsg	nsg	nsg	nsg	nsg	None	nsg	0.001	nsg	3/29
Perthane	nsg	nsg	nsg	nsg	nsg	nsg	None	nsg	nsg	nsg	nsg
Toxaphene	.3	0	nsg	nsg	8.75	1/37	Thurman and others, 1992	0.73	0.0002	nsg	5/37
Organophosphorus:											
Azinphos-methyl	nsg	nsg	nsg	nsg	87.5	0/4	None	nsg	0.01	nsg	0/4
Chlorpyrifos	nsg	nsg	30	20	nsg	0/9	None	0.083	0.041	nsg	2/9
Crufomate	nsg	nsg	nsg	nsg	nsg	nsg	None	nsg	nsg	nsg	nsg
DEF	nsg	nsg	nsg	nsg	nsg	nsg	None	nsg	nsg	nsg	nsg
Diazinon	nsg	nsg	5	0.6	14	1/30	Kuivila and Foe, 1995	nsg	nsg	0.009	15/30
Dichlorvos (DDVP)	nsg	nsg	nsg	nsg	nsg	nsg	None	nsg	nsg	0.001	0/1
Dimethoate	nsg	nsg	nsg	nsg	nsg	nsg	None	nsg	nsg	nsg	nsg
Disulfoton	nsg	nsg	3	0.3	0.7	0/2	None	nsg	nsg	0.05	0/2
Disyston	nsg	nsg	nsg	nsg	nsg	nsg	None	nsg	nsg	nsg	nsg
Ethion	nsg	nsg	nsg	nsg	nsg	nsg	None	nsg	nsg	0.02	1/18
Ethoprop	nsg	nsg	nsg	nsg	nsg	nsg	None	nsg	nsg	nsg	nsg
Fenitrothion	nsg	nsg	nsg	nsg	nsg	nsg	None	nsg	nsg	nsg	nsg
Fensulfothion	nsg	nsg	nsg	nsg	nsg	nsg	None	nsg	nsg	nsg	nsg
Fenthion	nsg	nsg	nsg	nsg	nsg	nsg	None	nsg	nsg	0.006	0/2
Fonofos	nsg	nsg	20	10	nsg	0/6	None	nsg	nsg	nsg	nsg
Imidan	nsg	nsg	nsg	nsg	nsg	nsg	None	nsg	nsg	nsg	nsg
Malathion	nsg	nsg	200	200	140	0/28	None	nsg	0.1	nsg	4/28
Methamidophos	nsg	nsg	nsg	nsg	nsg	nsg	None	nsg	nsg	nsg	nsg
Methidathion	nsg	nsg	nsg	nsg	nsg	nsg	None	nsg	nsg	nsg	nsg
Methyl parathion	nsg	nsg	30	2	30	0/28	None	nsg	nsg	[3]0.02	5/28
Methyl trithion	nsg	nsg	nsg	nsg	nsg	nsg	None	nsg	nsg	nsg	nsg

Table 6.1. Standards and criteria for protection of human and aquatic organism health for pesticides targeted in surface waters—Continued

| | Human Health | | | | | | | Aquatic Organism Health | | | |
| | MCL | MCLG | HA | | SNARL | Exceeded values | Studies in which MCL or HA exceeded | USEPA | | NAS/NAE | Exceeded values |
			Child	Adult				Acute	Chronic		
Parathion	nsg	nsg	nsg	nsg	30	0/27	None	0.065	0.013	nsg	1/27
Phorate	nsg	nsg	nsg	nsg	0.7	0/9	None	nsg	nsg	[3]0.05	0/9
Phosphamidon	nsg	nsg	nsg	nsg	nsg	nsg	None	nsg	nsg	0.03	0/1
Ronnel	nsg	nsg	nsg	nsg	nsg	nsg	None	nsg	nsg	nsg	nsg
Sulprofos	nsg	nsg	nsg	nsg	nsg	nsg	None	nsg	nsg	nsg	nsg
Terbufos	nsg	nsg	1	0.9	nsg	2/6	Petersen, 1990; Baker, 1988b	nsg	nsg	nsg	nsg
Trithion	nsg	nsg	nsg	nsg	nsg	nsg	None	nsg	nsg	nsg	nsg
Other Insecticides:[4]											
Aldicarb	nsg	nsg	nsg	7	7	0/1	None	nsg	nsg	nsg	nsg
Carbaryl	nsg	nsg	1,000	700	574	0/6	None	nsg	nsg	0.02	1/6
Carbofuran	40	40	50	36	nsg	0/9	None	nsg	nsg	nsg	nsg
Deet	nsg	nsg	nsg	nsg	nsg	nsg	None	nsg	nsg	nsg	nsg
Dibutyltin (DBT)	nsg	nsg	nsg	nsg	nsg	nsg	None	nsg	nsg	nsg	nsg
Fenvalerate	nsg	nsg	nsg	nsg	nsg	nsg	None	nsg	nsg	nsg	nsg
Methomyl	nsg	nsg	300	200	175	0/3	None	nsg	nsg	nsg	nsg
Oxamyl	200	200	175	175	nsg	0/1	None	nsg	nsg	nsg	nsg
Permethrin	nsg	nsg	nsg	nsg	nsg	nsg	None	nsg	nsg	nsg	nsg
Propargite	nsg	nsg	nsg	nsg	nsg	nsg	None	nsg	nsg	nsg	nsg
Tributyltin (TBT)	nsg	nsg	nsg	nsg	nsg	nsg	None	nsg	nsg	nsg	nsg
HERBICIDES											
Triazine and Acetanilide:											
Alachlor	2	0	100	nsg	700	10/15	Takita, 1984; Yorke and others, 1985; Wnuk and others, 1987; Baker, 1985, 1988b; Squillace and Engberg, 1988; Moyer and Cross, 1990; Lewis and others, 1992; Goolsby and Battaglin, 1993; Goolsby and others, 1993	nsg	nsg	nsg	nsg

Table 6.1. Standards and criteria for protection of human and aquatic organism health for pesticides targeted in surface waters—Continued

| | Human Health | | | | | | | Aquatic Organism Health | | | |
	MCL	MCLG	HA Child	HA Adult	SNARL	Exceeded values	Studies in which MCL or HA exceeded	USEPA Acute	USEPA Chronic	NAS/NAE	Exceeded values
Ametryn	nsg	nsg	900	60	nsg	0/11	None	nsg	nsg	[3]0.01	1/11
Atratone	nsg	nsg	nsg	nsg	nsg	nsg	None	nsg	nsg	nsg	nsg
Atrazine	3	3	50	3	150	16/41	Wu and others, 1980 Fishel, 1984 Baker, 1985, 1988b Ward, 1987 Wnuk and others, 1987 Fujii, 1988 Squillace and Engberg, 1988 Moyer and Cross, 1990 Lewis and others, 1992 Thurman and others, 1992 Goolsby and Battaglin, 1993 Goolsby and others, 1993 Pereira and Hostettler, 1993 Squillace and others, 1993	nsg	nsg	[3]1.0	21/41
Cyanazine	nsg	nsg	20	10	nsg	4/23	Baker, 1985, 1988b Wnuk and others, 1987 Thurman and others, 1992	nsg	nsg	nsg	nsg
Cyprazine	nsg	nsg	nsg	nsg	nsg	nsg	None	nsg	nsg	nsg	nsg
Hexazinone	nsg	nsg	3,000	200	nsg	0/1	None	nsg	nsg	nsg	nsg
Metolachlor	nsg	nsg	2,000	100	nsg	0/18	None	nsg	nsg	nsg	nsg
Metribuzin	nsg	nsg	300	200	nsg	0/18	None	nsg	nsg	nsg	nsg
Prometon	nsg	nsg	200	100	nsg	0/13	None	nsg	nsg	nsg	nsg
Prometryn	nsg	nsg	nsg	nsg	nsg	nsg	None	nsg	nsg	nsg	nsg
Propachlor	nsg	nsg	100	90	700	0/5	None	nsg	nsg	nsg	nsg
Propazine	nsg	nsg	500	10	325	0/13	None	nsg	nsg	nsg	nsg

Table 6.1. Standards and criteria for protection of human and aquatic organism health for pesticides targeted in surface waters—Continued

| | Human Health | | | | | | | Aquatic Organism Health | | | |
| | | | HA | | | | | USEPA | | | |
	MCL	MCLG	Child	Adult	SNARL	Exceeded values	Studies in which MCL or HA exceeded	Acute	Chronic	NAS/NAE	Exceeded values
Simazine	4	4	50	4	1,505	3/18	Ward, 1987 Baker, 1988b Thurman and others, 1992	nsg	nsg	10	1/18
Simetone	nsg	nsg	nsg	nsg	nsg	nsg	None	nsg	nsg	nsg	nsg
Simetryn	nsg	nsg	nsg	nsg	nsg	nsg	None	nsg	nsg	nsg	nsg
Terbutryn	nsg	nsg	nsg	nsg	nsg	nsg	None	nsg	nsg	nsg	nsg
Phenoxy Acid:											
2,4,5-T	nsg	nsg	300	70	700	0/27	None	nsg	nsg	[3]50	0/27
2,4,5-TP (Silvex)	50	50	70	52	5.25	0/30	None	nsg	nsg	[3]1.4	0/30
2,4-D	70	70	100	70	87.5	0/33	None	nsg	nsg	3	2/33
2,4-D (Methyl ester)	nsg	nsg	nsg	nsg	nsg	nsg	None	nsg	nsg	nsg	nsg
2,4-DP	nsg	nsg	nsg	nsg	nsg	nsg	None	nsg	nsg	nsg	nsg
Other Herbicides:											
Acrolein	nsg	nsg	nsg	nsg	nsg	nsg	None	nsg	nsg	nsg	nsg
Bensulfuron-methyl	nsg	nsg	nsg	nsg	nsg	nsg	None	nsg	nsg	nsg	nsg
Butylate	nsg	nsg	1,000	700	nsg	0/7	None	nsg	nsg	nsg	nsg
Chloramben	nsg	nsg	200	100	1,750	0/1	None	nsg	nsg	[3]25	0/1
Dacthal	nsg	nsg	5,000	4000	nsg	0/1	None	nsg	nsg	nsg	nsg
Dicamba	nsg	nsg	300	200	8.75	0/7	None	nsg	nsg	200	0/7
Dinoseb	7	7	10	7	39	0/2	None	nsg	nsg	nsg	nsg
Diquat	20	20	nsg	20	nsg	0/1	None	nsg	nsg	0.5	0/1
EPTC	nsg	nsg	nsg	nsg	nsg	nsg	None	nsg	nsg	nsg	nsg
Fluometuron	nsg	nsg	2,000	90	nsg	0/1	None	nsg	nsg	nsg	nsg
Linuron	nsg	nsg	nsg	nsg	nsg	nsg	None	nsg	nsg	nsg	nsg
Molinate	nsg	nsg	nsg	nsg	nsg	nsg	None	nsg	nsg	nsg	nsg
Norflurazon	nsg	nsg	nsg	nsg	nsg	nsg	None	nsg	nsg	nsg	nsg
Paraquat	nsg	nsg	50	30	59.5	0/1	None	nsg	nsg	[3]50	0/1

Table 6.1. Standards and criteria for protection of human and aquatic organism health for pesticides targeted in surface waters—Continued

| | Human Health | | | | | | | Aquatic Organism Health | | | |
| | MCL | MCLG | HA | | SNARL | Exceeded values | Studies in which MCL or HA exceeded | USEPA | | NAS/NAE | Exceeded values |
			Child	Adult				Acute	Chronic		
Pendimethalin	nsg	nsg	nsg	nsg	nsg	nsg	None	nsg	nsg	nsg	nsg
Picloram	500	500	700	500	1,050	0/5	None	nsg	nsg	[3]500	0/5
Propham	nsg	nsg	nsg	nsg	nsg	nsg	None	nsg	nsg	nsg	nsg
Thiobencarb	nsg	nsg	nsg	nsg	nsg	nsg	None	nsg	nsg	nsg	nsg
Trifluralin	nsg	nsg	30	2	700	0/11	None	nsg	nsg	[3]25	0/11
					FUNGICIDES						
Captan	nsg	nsg	nsg	nsg	350	0/1	None	nsg	nsg	nsg	nsg
Chlorothalonil	nsg	nsg	200	nsg	nsg	0/1	None	nsg	nsg	nsg	nsg
HCB	1	0	50	nsg	7	1/10	Kuntz and Warry, 1983	nsg	3.7	nsg	1/10
PCNB	nsg	nsg	nsg	nsg	nsg	nsg	None	nsg	nsg	nsg	nsg
PCP	1	0	300	220	21	0/3	None	20	13	nsg	0/3
				TRANSFORMATION PRODUCTS							
Azinphos-methyl oxon	nsg	nsg	nsg	nsg	nsg	nsg	None	nsg	nsg	nsg	nsg
Carbofuran phenol	nsg	nsg	nsg	nsg	nsg	nsg	None	nsg	nsg	nsg	nsg
2-Chloro-2',6'-d ethyl-acetanilide	nsg	nsg	nsg	nsg	nsg	nsg	None	nsg	nsg	nsg	nsg
Cyanazine amide	nsg	nsg	nsg	nsg	nsg	nsg	None	nsg	nsg	nsg	nsg
DDD	nsg	nsg	nsg	nsg	nsg	0/59	None	0.6	nsg	0.006	18/59
DDE	nsg	nsg	nsg	nsg	nsg	nsg	None	1,050	nsg	nsg	nsg
Deethylatrazine	nsg	nsg	nsg	nsg	nsg	nsg	None	nsg	nsg	nsg	nsg
Deisopropylatrazine	nsg	nsg	nsg	nsg	nsg	nsg	None	nsg	nsg	nsg	nsg
Desmethyl norflurazon	nsg	nsg	nsg	nsg	nsg	nsg	None	nsg	nsg	nsg	nsg
Endosulfan sulfate	nsg	nsg	nsg	nsg	nsg	nsg	None	nsg	nsg	nsg	nsg
Endrin aldehyde	nsg	nsg	nsg	nsg	nsg	nsg	None	nsg	nsg	nsg	nsg
ESA (Ethanesulfonic acid)	nsg	nsg	nsg	nsg	nsg	nsg	None	nsg	nsg	nsg	nsg

Table 6.1. Standards and criteria for protection of human and aquatic organism health for pesticides targeted in surface waters—Continued

| | Human Health | | | | | | Aquatic Organism Health | | | |
| | MCL | MCLG | HA | | SNARL | Exceeded values | Studies in which MCL or HA exceeded | USEPA | | NAS/NAE | Exceeded values |
			Child	Adult				Acute	Chronic		
Heptachlor epoxide	0.2	0	0.1	nsg	nsg	4/54	Bradshaw and others, 1972 Page, 1981 Warry and Chan, 1981 Kuntz and Warry, 1983	0.52	0.0038	nsg	20/54
2-Hydroxy-2'6'-diethyl-acetanilide	nsg	nsg	nsg	nsg	nsg	nsg	None	nsg	nsg	nsg	nsg
2-Ketomolinate	nsg	nsg	nsg	nsg	nsg	nsg	None	nsg	nsg	nsg	nsg
4-Ketomolinate	nsg	nsg	nsg	nsg	nsg	nsg	None	nsg	nsg	nsg	nsg
Oxychlordane	nsg	nsg	nsg	nsg	nsg	nsg	None	nsg	nsg	nsg	nsg
Paranitrophenol	nsg	nsg	nsg	nsg	nsg	nsg	None	nsg	nsg	nsg	nsg
Photomirex	nsg	nsg	nsg	nsg	nsg	nsg	None	nsg	nsg	nsg	nsg
Terbufos sulfone	nsg	nsg	nsg	nsg	nsg	nsg	None	nsg	nsg	nsg	nsg

[1]DDT totals include five studies in which only "total DDT" (sum of DDT, DDD, and DDE) was reported.

[2]HCH data includes studies in which any of four isomers (α, β, δ, and γ [lindane]) were targeted. MCL established for lindane only.

[3]Recommended maximum concentration in marine waters (no freshwater value established).

[4]Other insecticides category includes compounds used as acaricides, miticides, and nematocides, as well as insecticides.

manner, so that a lifetime HA value for one compound can be compared with a SNARL value for another. A complete description of the derivation and use of the MCL, HA, and SNARL values can be found in Nowell and Resek (1994). Values of these criteria for the pesticides observed in the reviewed studies are shown in Table 6.1. In some cases, there are large differences between the different criteria values for a particular pesticide. Nowell and Resek (1994) have recommended that, for sources of drinking water, the MCL, if available, should be used for comparison of observed concentrations with criteria values. If no MCL has been established, the HA should be used. If neither MCL nor an HA has been established, the SNARL should be used.

While the MCL, HA, and SNARL values do not directly pertain to ambient concentrations of pesticides in surface waters, they do provide values with which the observed levels can be compared. Since these values are based on the toxicity of the compounds, they can give some idea of the potential significance of the levels observed in surface waters. Pesticides that exceeded a criteria value in at least one of the reviewed studies (Tables 2.1 and 2.2) are noted in Table 6.1. Some compounds with established criteria have not been included in any of the reviewed studies, or have been targeted very infrequently. Five pesticides with established MCLs—dalapon, dibromochloropropane (DBCP), endothall, ethylene dibromide (EDB), and glyphosate—were not included in any of the reviewed studies listed in Tables 2.1 and 2.2. Of these five, only glyphosate (Figure 3.12) and endothall are currently used in United States agriculture. Several important qualifications should be mentioned regarding the data in Table 6.1. First, many of the studies reviewed did not give information on the maximum concentration detected for each analyte. Thus, the number of studies in which criteria values were exceeded may actually be higher than is shown in Table 6.1. Second, studies listed in Table 6.1 cover 1958 to 1993, so that the data do not necessarily reflect the current situation. Finally, in the table, a reported concentration higher than a criteria value in a single sample in one study is counted the same as many samples with concentrations above criteria values in another study. Despite these limitations, several important points are evident.

1. Relatively few of the pesticides targeted in the reviewed studies were detected at a level exceeding a drinking water criteria value. Of the 52 pesticides or transformation products with an established criterion, 15 were detected at a concentration exceeding it in at least one sample. Eight of these were organochlorine compounds or degradation products detected in studies primarily from the 1970's.

2. In recent years, the pesticides most often detected at levels exceeding criteria values have been the triazine and acetanilide herbicides atrazine, alachlor, cyanazine, and simazine. The large number of studies in which these compounds exceeded criteria values is, in part, due to more intensive sampling of midwestern surface waters in recent years. In several of the studies in which criteria values were exceeded, rivers were sampled frequently during the spring runoff, increasing the likelihood of sampling during peak herbicide concentrations. As discussed in Section 5.1, the increased use of these compounds, coupled with their relatively high potential for transport in runoff, results in elevated concentrations in surface waters of the Midwest in spring and early summer. Peak concentrations of these compounds can exceed criteria values, especially in the smaller rivers.

3. Herbicides other than the triazines and acetanilides, including the high-use compounds 2,4-D, dicamba, butylate, and trifluralin, were never detected in surface waters at levels exceeding criteria values in the reviewed studies.

4. Insecticides commonly used in recent years rarely reach levels in surface waters that exceed drinking-water criteria concentrations.

Atrazine concentrations exceeded an established criteria value most often in the reviewed studies, and can be used to more fully explain the human health implications of the levels of pesticides detected in surface waters. As discussed earlier, atrazine concentrations have a seasonal pattern in many rivers throughout the central United States (Figures 3.46, 5.2, and 5.7). In some of these rivers, the MCL of 3 mg/L is exceeded for days to weeks. Peak concentrations generally are higher in smaller rivers, but the duration of elevated concentrations often is longer in larger rivers (Goolsby and Battaglin, 1993). Drinking water for millions of people is obtained from surface water sources in the central United States. Community water supplies drawing water from the three major rivers in this region—the Mississippi, Ohio, and Missouri Rivers— serve approximately 10.5 million people (Ciba-Geigy, 1992d). Smaller rivers and reservoirs provide drinking water to approximately 4.3 million people in Ohio, Illinois, and Iowa (Ciba-Geigy, 1992d), and the situation is similar in other states of the region.

A series of exposure assessments has been done for populations served by these various water sources (Ciba-Geigy, 1993a,b, 1994b; Richards and others, 1995). In these assessments, annual average concentrations of atrazine were estimated for water bodies used as sources of drinking water, using existing monitoring data from Ciba-Geigy, Monsanto, U.S. Geological Survey (USGS), water utilities, and other sources. The number of people exposed to various levels of atrazine in drinking water then was calculated and expressed as a percentage of the total population covered in each of the different assessments. The results of these assessments indicate that the vast majority of people whose drinking water is derived from surface water in the central United States are exposed to annual average atrazine concentrations well below the MCL. In the assessment covering the Mississippi, Missouri, and Ohio Rivers, 85 percent of the population whose drinking water is derived from these rivers was exposed to average concentrations between one-tenth and one-third of the MCL, and approximately 40 percent were exposed to average concentrations slightly over one-third of the MCL. No segment of the assessed population was exposed to average atrazine concentrations above the MCL. In the assessment for Iowa, 97.8 percent of the assessed population using drinking water derived from surface waters was exposed to average atrazine concentrations of less than one-third of the MCL. One lake included in the assessment had an annual average atrazine concentration of over 6 µg/L—more than twice the MCL. This reservoir supplies drinking water to approximately 0.7 percent of the assessed population using surface water sources. Results were similar in the assessments for Ohio and Illinois. In Ohio, no surface water source had an annual average concentration over the MCL, and sources for approximately 8 percent of the assessed population relying on surface water had annual concentrations of slightly over 2 µg/L, or two-thirds of the MCL. In Illinois, no surface water source had an annual average concentration over the MCL, and sources for approximately 4 percent of the assessed population relying on surface water had annual concentrations over 1 µg/L. In both Ohio and Illinois, a large portion of the drinking water derived from surface water comes from the Great Lakes, where atrazine concentrations are quite low. In Illinois, Lake Michigan accounts for nearly 80 percent of the drinking water derived from surface waters; in Ohio, Lake Erie accounts for about 42 percent. Annual mean atrazine concentrations used in the assessments for Lakes Michigan and Erie were 0.1 and 0.07 µg/L, respectively. The results from these assessments probably can be extrapolated to much of the Midwest. Illinois, Iowa, and Ohio ranked first, fourth, and sixth among states, respectively, in atrazine use in the late 1980's, the period in which much of the data for these assessments was collected. The Mississippi, Ohio, and Missouri Rivers drain much of the area of heaviest atrazine use in the United States (Figure 3.7).

Similar results also would be expected for several other herbicides with established criteria values used in the Mississippi River Basin, based on the results of recent studies of rivers and reservoirs in the region (Thurman and others, 1992; Goolsby and Battaglin, 1993; Goolsby and others, 1993; Periera and Hostettler, 1993; Richards and Baker, 1993). In these studies, atrazine was usually present in surface waters at higher concentrations and for longer periods of time than were alachlor, metolachlor, metribuzin, propachlor, cyanazine, butylate, and trifluralin. In addition, criteria values for all these compounds, except alachlor and trifluralin, are considerably higher than atrazine's MCL of 3 µg/L (Table 6.1). It is unknown whether the same results would be obtained for atrazine in other areas of the United States where atrazine use is high, but where cropping patterns and weather conditions are different.

Several assumptions and potential sources of error in the exposure assessments discussed above should be noted. First, in each of the assessments, some water bodies that served as drinking-water sources had insufficient concentration data to estimate an annual average. In some cases, this resulted in a portion of the population not being included in the assessment. In the Illinois and Iowa assessments, approximately 8 and 12 percent, respectively, of the population using surface water sources were excluded for this reason. In most of these cases, the water bodies with insufficient data were small interior rivers, which are likely to have relatively high atrazine levels compared to larger rivers and lakes. Thus, exclusion of these rivers from the assessment probably lowered the overall average concentration. In other cases, average concentrations for some water bodies were inferred from data available on other water bodies, with adjustments made for differences in atrazine use and land use. Richards and others (1995) state that "...in every case where an ambiguous situation could be resolved, the higher estimate of the exposure concentration was used, in order to produce a 'worst possible case' estimate." Second, the annual mean concentrations used in these assessments were based on concentrations of atrazine in samples of raw (untreated) water. Removal of atrazine and other pesticides by water treatment procedures is discussed below. Finally, the validity of the average concentrations used is dependent on many factors, including the number of samples analyzed, the timing of sampling with respect to application, and on how representative the sampling periods were of longer-term conditions. Data from numerous studies of surface waters in the central United States, discussed earlier in Section 3.3, generally agree with the data used in the exposure assessments. It is beyond the scope of this book, however, to evaluate the validity of the assessments as a whole.

The concentrations used in these assessments were from untreated water. While no attempt was made to comprehensively review the literature on removal of pesticides during water treatment, several studies were reviewed that examined the effects of various treatment procedures employed at water treatment plants on levels of pesticides in water (Junk and others, 1976; LeBel and others, 1987; Lykins and others, 1987; Schroeder, 1987; Wnuk and others, 1987; Miltner and others, 1989; Patrick, 1990). For the most part, these studies have found that most routine treatment procedures, including sand filtration, clarification, softening, and chlorination, have little effect on concentrations of many of the pesticides used in recent years. Exceptions include the insecticide carbofuran, which apparently was completely transformed at the elevated pH used in the softening step, and the herbicide metribuzin, which reacted during the chlorination step (Miltner and others, 1989). It was noted in this study that formation of transformation products was the likely reason for the disappearance of both compounds. For carbofuran, conversion to either 3-hydroxycarbofuran or carbofuran phenol was suggested as a possible reaction, but no attempt to identify conversion products was made. For metribuzin, chlorinated transformation products were observed but not identified.

Several treatment procedures have been shown to reduce the concentrations of many pesticides in water. In a 1-year study of several alternative water treatment procedures (Lykins and others, 1987), ozonation of water reduced concentrations of atrazine and alachlor by an average of 83 and 84 percent, respectively. These compounds are not removed during chlorination (Schroeder, 1987; Miltner and others, 1989). An average of 57 percent removal of combined organochlorine insecticides (OCs) also was achieved with ozonation in the same study. It should be pointed out that while ozonation may remove the parent compound, transformation products still may be present in the water. Adams and Randtke (1992) found that deethylatrazine was the primary byproduct formed when atrazine-fortified natural waters were subjected to ozonation. Up to seven other degradation products also were formed, depending on the experimental conditions. Powdered activated carbon (PAC), often used in water treatment plants to control taste and odor problems, removed 70 to 80 percent of alachlor in tests with spiked water (Schroeder, 1987). PAC addition during periods of elevated pesticide levels was suggested as a cost-effective method of removing up to 80 percent of a number of the current high-use herbicides (Miltner and others, 1989). In a Canadian study, however, treatment with PAC did not effectively remove atrazine or deethylatrazine from raw water (Patrick, 1990). Concentrations of both compounds in finished water were essentially unchanged from concentrations in the untreated water, even though PAC was added at levels well above the concentrations normally used for taste and odor control. In addition, the routine use of PAC for taste and odor control may not be sufficient to reduce pesticide levels during the period of highest concentrations, since problems with taste and odor are usually associated with algal blooms, which occur later in the summer than the peak pesticide concentrations in surface waters in agricultural areas. Granular activated carbon (GAC) has been shown to be effective in removing a number of pesticides from water (Lykins and others, 1987; Schroeder, 1987; Miltner and others, 1989). Filtration through GAC removed 94 to 97 percent of both atrazine and alachlor, and 90 to 93 percent of combined OCs from Mississippi River water over a 1-year period (Lykins and others, 1987). More than 94 percent of alachlor added to river water from Kansas was removed by filtration through GAC (Schroeder, 1987). GAC filtration was somewhat less effective in a 1984 study at a treatment plant in Ohio (Miltner and others, 1989). The mean removal of six herbicides ranged from 47 to 72 percent when water was passed through an 18-inch-deep GAC filtration bed. Activated carbon also was reported to be the most efficient means of removing phenoxy herbicides from water (Que Hee and Sutherland, 1981). GAC is the mandated water treatment process for achieving compliance with SDWA regulations for most pesticides, including 2,4-D and triazine and acetanilide herbicides with established MCLs (U.S. Environmental Protection Agency, 1994c).

From the studies discussed above, it can be concluded that drinking water supplied by treatment plants relying on conventional procedures probably contains some pesticides at levels similar to that of the untreated water. Treatment plants using PAC or GAC for taste and odor control may be reducing the concentrations of some pesticides, but are probably not completely removing them. However, data from the exposure assessments, discussed above, and from the concentration data shown in Tables 2.1, 2.2, and 6.1, indicate that surface waters used as sources of drinking water seldom contain pesticides at levels above the MCLs established by the USEPA. It also should be noted, however, that several pesticides with MCLGs of zero were detected in a number of the reviewed studies. These include the commonly detected herbicide, alachlor, which has been shown to be unaffected by conventional water treatment, as previously discussed.

Some have expressed concern that the existing standards and criteria do not adequately protect against potential effects of low-level exposure to a number of pesticides with past and present use in the United States (Colborn and Clement, 1992). In particular, a number of pesticides have been shown to affect the endocrine systems of various organisms. Included are many of the OCs and a number of herbicides, insecticides, and fungicides with current agricultural use. Several instances of these effects occurring due to environmental exposure have been documented for DDT, the marine biocide tributyltin (TBT), and several nonpesticide chemicals. However, evidence for potential disruption of the endocrine system by currently used pesticides has come only from laboratory studies. This issue is discussed in more detail in the next section.

6.2 IMPLICATIONS FOR HEALTH OF AQUATIC ORGANISMS

PESTICIDE CONCENTRATIONS EXCEEDING AQUATIC-LIFE CRITERIA VALUES

The USEPA has established ambient water quality criteria for the protection of aquatic organisms (Nowell and Resek, 1994) for both short-term (acute) and long-term (chronic) exposure to some pesticides. These are nonenforceable guidelines designed to provide a basis for state standards. The derivation of these criteria is described in Nowell and Resek (1994) and will be described here briefly. In general, the acute criteria are concentrations at which 95 percent of the genera of a diverse group of organisms would not be adversely affected on the basis of an exposure time of 1 hour (criteria established before 1985 used an instantaneous exposure time). According to USEPA, if the concentration of a contaminant does not exceed the acute criteria value more than once in 3 years, most aquatic ecosystems can recover. USEPA's assertion is based on the extreme values in the distribution of ambient concentrations, rather than the high concentrations attributed to spills or discharges from point sources. Chronic criteria are based on one of three parameters—the final chronic value (FCV) or, if data were available, the final plant value or the final residue value. The FCV is the highest concentration at which no adverse effects would be expected for 95 percent of the genera of a diverse group of organisms, based on an exposure time of 4 days (criteria established before 1985 used a 24-hour exposure time). According to Nowell and Resek (1994), "...the final plant value is the lowest result from a toxicity test with an important aquatic-plant species in which the endpoint was biologically important, and test-material concentrations were measured." The final residue value incorporates an appropriate bioconcentration factor and is designed to protect both the marketability of fish and the health of wildlife that consume aquatic organisms. The actual chronic criteria is set equal to the lowest of the FCV, final plant value, or final residue value. As with the acute criteria, in the judgement of the USEPA, most aquatic ecosystems should be able to recover if a chronic criterion is not exceeded more than one time in 3 years.

The USEPA has established aquatic life criteria for only 20 of the 118 pesticide compounds targeted in the studies in Tables 2.1 and 2.2. These values are listed in Table 6.1. Of the 96 herbicides, 55 insecticides, and 30 fungicides tabulated by Gianessi and Puffer (1991, 1992a,b)—representing the pesticides with highest agricultural use during 1989 to 1991—only 6 insecticides have USEPA-established aquatic-life criteria. None of the herbicides or fungicides currently used in United States agriculture have USEPA-established criteria. Additional criteria values have been recommended by the National Academy of Sciences and the National Academy of Engineering (NAS/NAE) for some compounds (Nowell and Resek, 1994). These criteria were derived by multiplying acute toxicity values (usually the 96-hour LC_{50} value—the concentration lethal to 50 percent of the population exposed for 96 hours) for the most sensitive species by an appropriate factor, usually 0.01. These values, derived in 1973, are somewhat out-of-date; they are not appropriate for compounds that bioaccumulate because bioaccumulation was not considered in their derivation. Nevertheless, they may be used for compounds that do not have an established USEPA aquatic life criteria. NAS/NAE criteria are listed in Table 6.1 for 23 compounds that do not have a USEPA criteria. Thus, criteria values for 43 of the 121 compounds in Table 6.1 can be compared with the concentrations observed in the reviewed studies.

Pesticides that exceeded an aquatic life criteria in at least one of the reviewed studies (Tables 2.1 and 2.2) are noted in Table 6.1. The same qualifications mentioned for the data on concentrations exceeding drinking water criteria also apply here. Several additional points should be noted for the data on concentrations exceeding aquatic life criteria. First, a reported concentration above a criteria value in a single sample does not necessarily imply that aquatic organisms were adversely affected. USEPA's criteria are based on exposure times of 1 hour (acute) and 4 days (chronic), and the actual exposure time cannot be determined on the basis of the concentration in a single sample. In addition, as mentioned above, the potential for lasting adverse effects on ecosystems is decreased if the criteria are not exceeded more than once in 3 years. In most studies, this information is lacking. Second, analytical detection limits for most of the OCs and several OPs were higher than criteria concentrations in some of the studies reviewed. In these studies, one or more of the compounds may have been present at a concentration above a criteria value without being detected. Third, absence of concentrations above a criteria value does not necessarily mean that no adverse effects occurred, since a limited number of test organisms were used in setting the criteria, which were designed to protect 95 percent of the genera represented by the test organisms.

Despite these limitations, four important points are evident from the data in Table 6.1. First, all of the OCs targeted in the reviewed studies exceeded an aquatic-life criterion, if one has been established. The number of studies in which criteria were exceeded is probably higher than the number shown in the table, as detection limits for the organochlorines were often above the criteria (Tables 2.1 and 2.2). For example, in every study that reported a detection limit for toxaphene, the detection limit was higher than the USEPA chronic criterion. In addition, some studies did not list the maximum concentration observed for each analyte. For the organochlorines, an additional factor must be considered when comparing observed concentrations in ambient waters with aquatic-life criteria. A significant portion of the total amount in the water column may be sorbed to particulates for most of these compounds, as discussed in Section 4.2. Whether this sorbed portion is bioavailable, and should be considered along with the dissolved portion in assessing the potential for adverse effects on aquatic organisms, is not entirely clear. The criteria, for the most part, were developed using whole-water concentrations, and no attempt was made to distinguish between total concentrations and the bioavailable fraction. Thus, it is difficult to compare concentrations of the organochlorines in whole-water samples (which most of the reviewed studies collected) with the criteria, without taking into account the fraction in the bioavailable phase in both the environmental sample and in the toxicity tests used to derive the criteria (Nowell and Resek, 1994). The same can be said for filtered water samples, in which the concentration in the dissolved phase is known. It would be inappropriate to compare these concentrations with the criteria, since the criteria are based on total concentrations, without regard for the fraction in the dissolved phase (Nowell and Resek, 1994).

Second, only four of the OPs—azinphos-methyl, chlorpyrifos, malathion, and parathion—have established USEPA criteria. A number of others have NAS/NAE criteria. Diazinon exceeded the extremely low NAS/NAE criterion of 0.009 µg/L in 18 of 30 studies. In addition, the detection limit for diazinon in every study reviewed was higher than this value (in many of the studies, the detection limit was 0.01 µg/L, slightly above the criterion), so the criterion may have been exceeded in more instances. The NAS/NAE criteria were derived in 1973 and, as mentioned above, some of these values may not reflect current knowledge of the toxicity of particular compounds. More recent estimates of the toxicity of diazinon suggest that concentrations of 0.2 to 0.5 µg/L are toxic to daphnia, with exposure times from 24 hours to 7 days (Kuivila and Foe, 1995). Diazinon concentrations exceeded 0.2 µg/L in seven studies.

Reports of concentrations above criteria values were less frequent for the other OPs. Detection limits in the reviewed studies were higher than the criteria in 4 of 9 studies for chlorpyrifos and in 4 of 22 studies for parathion. Concentrations above criteria levels for these compounds may not have been observed in some studies. Third, none of the herbicides targeted in the studies reviewed have USEPA criteria, and only 12 of 32 have NAS/NAE criteria. Atrazine exceeded the NAS/NAE criterion in 21 studies, although this criterion is established for the protection of marine (saltwater) organisms. Concentrations above criteria values were very rare for any of the other herbicides in the reviewed studies. Fourth, criteria have not been established for any of the high-use agricultural fungicides (Table 3.1). These compounds were analytes in few, or none, of the reviewed studies.

The data shown in Table 6.1 indicate which pesticides have exceeded established criteria, and give some idea of how frequently this has occurred. Because criteria have been established for only a few of the currently used pesticides, and because of limitations in the criteria and reviewed studies, the data in Table 6.1 do not provide much information on whether aquatic ecosystems are adversely affected by pesticides in surface waters. In addition, the aquatic-life criteria for pesticides are based on toxicity tests involving exposure to a single chemical (Nowell and Resek, 1994). Potential synergistic or antagonistic effects of exposure to mixtures of pesticides, and to mixtures of pesticides and pesticide transformation products, are not accounted for in the criteria.

Several studies have raised additional concerns about potential effects of pesticides in surface waters on aquatic (and terrestrial) organisms (Colborn and Clement, 1992). These concerns are based on evidence of adverse effects of certain pesticides and other chemicals on the endocrine systems of a number of different organisms. Effects observed in various species include thyroid dysfunction, decreased fertility, decreased hatching success, birth deformities, compromised immune systems, feminization of males, and defeminization of females (Colborn and Clement, 1992). The effects appear to be related to hormone imbalances and may be due to exposure to chemicals that mimic or block the action of hormones such as estrogen. Effects may be evident in the exposed organism or manifested in the next generation, and, in some cases, not until the progeny reach sexual maturity (Fox, 1992). In addition, the effects of exposure may be different for developing organisms than for adults, so the timing of exposure can be critical in determining the actual effects (Colborn and Clement, 1992). Many of these effects are not accounted for in established criteria, particularly effects on development and reproductive success of future generations of relatively long-lived organisms. Pesticides that have been shown to disrupt endocrine systems of organisms include many of the OCs, such as chlordane, dicofol, dieldrin, DDT and its metabolites, endosulfan, heptachlor, heptachlor epoxide, lindane, methoxychlor, mircx, and toxaphene (Hileman, 1993). Most of these compounds are no longer used in the United States, but some still are detected frequently in surface waters, as discussed in Section 3.3. Several pesticides commonly used in United States agriculture in recent years have also been shown (in laboratory testing) to disrupt the endocrine systems of certain organisms, including the herbicides alachlor, atrazine, 2,4-D, metribuzin, and trifluralin; the insecticides aldicarb, carbaryl, parathion, and synthetic pyrethroids; the fungicides benomyl, mancozeb, maneb, zineb, and ziram; and the marine biocide tributyltin, or TBT (Hileman, 1993).

Much of the evidence for endocrine system disruption caused by pesticides in the environment is based on the effects of DDT and its transformation products (Colborn and Clement, 1992; Fox, 1992; Guillette and others, 1994). However, little information is available on concentrations in the water column in these studies. Effects of DDT and its metabolites on endocrine systems have been observed primarily in terrestrial organisms, such as eagles, gulls, and alligators, exposed to these compounds through their food supply. TBT, a biocide added to marine paints to prevent growth of aquatic organisms on boats, was shown to produce disruption of the endocrine systems of several species of marine snails (Fox, 1992). Observations of adverse effects of TBT in the waters of marinas (Smith, 1981) were supported by laboratory tests, in which extremely low concentrations of TBT (1 to 2 ng/L) caused abnormalities in reproductive organs and sterilization of some female snails (Bryan and others, 1986). Concentrations of 6 to 8 ng/L caused sterilization in all females of one species, and extremely high concentrations led to complete sex reversal in females (Gibbs and others, 1988). Additional evidence of endocrine disruption in the environment can be found in studies on the effects of effluents of pulp and paper plants (Davis and Bortone, 1992; Munkittrick and others, 1992) and sewage treatment plants (Purdom and others, 1994) on fish, although in these studies, the effects apparently were caused by chemicals other than pesticides. In these studies, the specific chemical or chemicals responsible for the observed effects could not be identified. Evidence of endocrine system disruption in the environment caused by herbicides and insecticides currently used in United States agriculture is lacking. No studies were found in which effects on endocrine systems in organisms were related to environmental concentrations of any of the commonly used herbicides, insecticides, or fungicides mentioned above. Some have expressed concern, however, that the existing procedures used to test the toxicity of pesticides are not adequate to assess the potential for effects on the endocrine systems of organisms, including humans (Clement and Colborn, 1992). As described in Section 5.1, a number of high-use pesticides with the potential for effects on endocrine systems, including atrazine, alachlor, and 2,4-D, exhibit a seasonal pattern of elevated concentrations in large parts of the United States. Little data are available on concentrations of several other currently used compounds, such as the fungicides mancozeb, maneb, and benomyl. The issue of whether endocrine systems of organisms, including humans, are being adversely affected from the current use of pesticides, and the resulting concentrations in surface waters, is clearly an area where more study is needed.

A comprehensive search of the literature on the effects of pesticides on aquatic organisms or aquatic ecosystems is beyond the scope of this book. Selected studies were reviewed, however, and studies dealing with two topics—evidence of fish kills from pesticide use and the effects of atrazine on aquatic ecosystems—are discussed briefly in the next sections.

FISH KILLS ATTRIBUTED TO PESTICIDES

Fish kills represent one of the most obvious indications of problems in an aquatic environment. A number of pesticide-caused fish kills occurred during the 1950's and 1960's, mostly as a result of the use of OCs. For example, it has been estimated that 10 to 15 million fish were killed between 1960 and 1963 in the Mississippi and Atchafalaya Rivers and associated bayous in Louisiana. The insecticide endrin was singled out as the major cause of the mortality (Madhun and Freed, 1990). More recent fish-kill data have been evaluated for coastal areas of the United States. A 1991 National Oceanic and Atmospheric Administration (NOAA) report (Lowe and others, 1991) of fish kills in coastal waters (rivers, streams, and estuaries in 22 states) indicates that pesticides caused a relatively small percentage of the reported fish kills between 1980 and 1989. Of the 3,654 reported fish-kill events, 145 (4 percent) were attributed to pesticides. The total number of fish killed in all reported events was estimated as over 407 million. Approximately 0.5 percent of this total (2.2 million fish) were killed in events attributed to pesticides. The NOAA report cautions that there are large sources of potential error in the fish-kill data, mainly related to differences in the type of data collected and the completeness of coverage in the individual states. In approximately 10 percent of the reported fish-kill events (3 percent of the total fish killed), the direct cause could not be determined from available data. In 41 percent of the fish-kill events, the direct cause was low dissolved-oxygen levels caused by both natural phenomena and human activities.

Trim and Marcus (1990) evaluated fish-kill data from coastal areas of South Carolina. Fish kills in this report were defined as mortality in tidal saltwater species, including crustaceans, finfish, gastropods, and bivalves. Of the 259 fish kills that occurred in the tidal saltwaters (estuaries and tidally influenced rivers, lagoons, and harbors) of South Carolina from 1978 to 1988, 91 (35 percent) were attributed to anthropogenic causes. Pesticides were identified as the cause of 49 fish kills (19 percent of the 259 total kills). Of the fish kills caused by pesticides, the authors attributed 18 to agricultural use, 19 to herbicides used for control of aquatic and terrestrial weeds, and 12 to insecticides used for vector control (control of insects for public health purposes). A seasonal pattern was evident, with fish kills caused by natural phenomena occurring primarily from June through October, and anthropogenic causes predominant in early to mid-spring and early to mid-autumn. The authors state that the peaks in anthropogenically caused fish kills correspond to the periods of heaviest application of agricultural- and vector-control pesticides in this area. The authors also mention that ambient water quality monitoring conducted by state agencies during this period detected very few pesticides in South Carolina estuaries, suggesting that ambient monitoring programs cannot be used for early detection of potential fish kills or for identification of pollutant sources of fish kills. However, the ambient monitoring described in this study consisted of annual monitoring of water and biota; the detection limit for OPs in water was 0.1 µg/L. Criteria concentrations established for protection of aquatic organisms are below 0.1 µg/L for many of the OPs, so toxic levels of these compounds could have been present in the estuaries at certain times of the year without ever being detected. It is not clear how the authors were able to attribute specific fish kills to pesticides, and to pesticides used in specific types of applications. No data are given in the study, nor are any referred to, on pesticide concentrations measured in water or biota that were used to determine the cause of specific fish-kill events.

EFFECTS OF ATRAZINE ON AQUATIC ORGANISMS AND ECOSYSTEMS

The USEPA has not established aquatic-life criteria for most of the herbicides, including atrazine. In general, herbicides are substantially less acutely toxic to fish and other aquatic animals than insecticides (Baker and Richards, 1990). Adverse impacts of herbicides on aquatic ecosystems would be expected to occur as a result of their effects on aquatic plants—algae, periphyton, and macrophytes (although several herbicides have been identified as potential endocrine system disrupters, as discussed earlier in this section). Secondary effects on aquatic ecosystems, such as changes in species composition or diversity, also may be observed (Hurlbert, 1975). Numerous studies have investigated the effects of atrazine on aquatic organisms and ecosystems. Laboratory studies have shown that atrazine concentrations as low as 1 to 10 µg/L can suppress the growth rates of certain species of algae (deNoyelles and others, 1982; Shehata and others, 1993). The effects of exposure to atrazine have been observed to vary among different species of algae (deNoyelles and others, 1982; Hersh and Crumpton, 1989). This suggests that long-term exposure to atrazine could result in changes in species composition, with species susceptible to atrazine being replaced by more resistant ones. This was observed by deNoyelles and others (1982) who stated, "Atrazine concentrations of 20 µg/L were shown in the laboratory and field [experimental ponds] to affect both photosynthesis and succession, including the establishment of resistant species within the phytoplankton community." Slight reductions in nutrient uptake by *aufwuchs* (biota attached to submerged surfaces) were observed in a model stream at an atrazine concentration of 24 µg/L, although recovery to normal uptake rates was rapid (Krieger and others, 1988). In another model-stream study, an atrazine concentration of 25 µg/L appeared to have no significant effect on community-level variables, standing biomass, or rates of primary production and respiration (Lynch and others, 1985). In this study, however, the actual atrazine concentration to which the biota were exposed was not confirmed, and "....may have been only a fraction of the nominal concentration" (Kreiger and others, 1988). Results from another study, however, also contradict some of the findings discussed above. In a microcosm study simulating a prairie wetland, atrazine had virtually no effect on algal biomass, primary productivity, or macrophytic biomass at concentrations of 10 and 100 µg/L, although all were affected at a concentration of 1,000 µg/L (Johnson, 1986).

All the above studies were conducted either in a laboratory or in artificial streams or microcosms. The complexity of natural aquatic ecosystems, and the difficulty of finding suitable controls, have prevented, for the most part, an assessment of the effects of atrazine in actual surface waters. The concentrations used in the studies described above occur in streams draining agricultural regions, however, for at least part of the year. Although supported by little direct experimental evidence, it is likely that atrazine does affect aquatic plants and that these effects may be reflected in the aquatic ecosystem as a whole. Agricultural activities result in numerous stresses on aquatic ecosystems, including increased suspended sediment concentrations, decreased dissolved-oxygen concentrations related to eutrophication, increased temperatures from reductions in streamside vegetation, and greater extremes in discharge. Whether the effects of atrazine contamination are significant in comparison to these other anthropogenic influences remains largely unknown at this time.

6.3 ENVIRONMENTAL SIGNIFICANCE OF PESTICIDE TRANSFORMATION PRODUCTS IN SURFACE WATERS

While little is known about the presence and concentrations of pesticide transformation products in surface waters (Section 5.5), even less is known about their impacts on ecosystem and human health. Day (1991) reviewed the literature on pesticide transformation products in surface waters and their effects on aquatic biota. Studies have been done on selected transformation products of a few insecticides and herbicides, and the organometallic biocide tributyltin (Table 6.2). Studies examining the toxicological effects of transformation products indicate that they are less toxic, equally toxic, or more toxic than their parent compounds. Differences in toxicity between parent compound and transformation product also are organism dependent. For example, *p,p'*-DDD is more toxic to some species of fish and less toxic to other species of fish than the parent compound *p,p'*-DDT. Day (1991) suggests that the "...factors which must be considered when evaluating the hazards of transformation products of pesticides in aquatic ecosystems include a) the rate at which the compounds appear and disappear, b) concentrations of residues in the field, c) time of exposure for aquatic biota, and d) compartmentalization of transformation products in the ecosystem."

Table 6.2. Relative toxicity of pesticides and their transformation products to aquatic organisms

[Modified from Day, 1991. Parent Pesticide: Transformation product is indented under the parent pesticide. Toxicity Abbreviations: A, algae; F, fish; I, invertebrate insects]

Parent Pesticide	Toxicity Relative to Parent Compound		
	Less	Equal	More
INSECTICIDES			
DDT			
DDD	F, I		F, I
DDE	A	A, F	F
Endosulfan			
Endosulfan sulfate			F
Endosulfan diol	F	I	
Aldrin			
Photoaldrin	A		
Endrin			
Ketoendrin	A		
Fenitrothion			
Fenitrooxon			I
Aminofenitrothion	F		
Carboxyfenitrothion	F		
Demethylfenitrothion	F		
3-Methyl-4-nitrophenol	F		
3-Methyl-4-aminophenol			F
Malathion			
Diethyl fumarate			F
Dimethyl phosphorodithioic acid	F		
Aldicarb			
Aldicarb sulfoxide		F, I	
Aldicarb sulfone	F, I		
Mexacarbate			
4-Amino-3,5-xylenol			F
4-Dimethylamino-3,5-xylenol	F		
Carbaryl			
1-Naphthol			F, I
HERBICIDES			
Atrazine			
Deethylatrazine	A		
Deisopropylatrazine	A		
Diamino atrazine	A		
Hydroxyatrazine	A		
Propanil			
3,4-Dichloroaniline	A		

Table 6.2. Relative toxicity of pesticides and their transformation products to aquatic organisms— Continued

Parent Pesticide	Toxicity Relative to Parent Compound		
	Less	Equal	More
HERBICIDES--*Continued*			
Chlorpropham			
3-Chloroaniline	A		
Triclopyr			
Pyridinol		F	
Pyridine	F		
BIOCIDES			
Tributyltin			
Dibutyltin	A, I		
Monobutyltin	A, I		

Another area of concern, but where little is known, is the interactive, cumulative, and synergistic effects of combinations of parent compounds and transformation products that may be present simultaneously in the water. In most laboratory toxicity studies, one compound-organism pair is evaluated at a time, but in real surface water systems, mixtures of many organic and inorganic chemicals are present most of the time. Stratton (1984) showed that atrazine, deethylatrazine, and deisopropylatrazine, when mixed together in certain ratios, demonstrated synergistic, antagonistic, and additive effects in the toxicity response of a blue-green alga. Day (1991) generalizes this observation and suggests "...the interactive effects of pesticides, their transformation products, and any other chemicals (toxic and nontoxic) could lead to situations in the natural environment where degradation products of low individual toxicity still pose a serious threat to non-target organisms when in combination."

Another area of concern regarding pesticide transformation products is the potential for bioaccumulation. Although transformation processes generally tend to make the transformation product more polar than the parent, the difference may not be significant. In other words, a hydrophobic pesticide (such as DDT) can have transformation products (such as DDE and DDD) that are still very hydrophobic and have similar tendencies to bioaccumulate in aquatic biota. In the case of DDT and its transformation products in Hexagenia larvae, the bioaccumulation factor of DDE has been reported to be about the same as DDT, whereas that of DDD is about a factor of four less than that of DDT (Capel and Eisenreich, 1990). Bioaccumulation of transformation products of other hydrophobic OCs also has been observed. Nowell (1996), in the review of pesticides in sediments and biota, discusses bioaccumulation of pesticides and some transformation products at length.

The effects of pesticide transformation products on human health are largely unknown. Some pesticide transformation products are more toxic than the parent compound to target organisms (Felsot and Pederson, 1991) and also may be more toxic to humans. In the absence of experimental data, one approach is to assume that the transformation products have the same toxicity as the parent compound. In Wisconsin, for example, the Department of Health and Social Services has proposed that their water quality standard for atrazine be "...applied to total atrazine residues. Thus, detection of 2 µg/L of atrazine and 1 µg/L of each of these [two] dealkylated metabolites would be interpreted as a total atrazine concentration of 4 µg/L, and would exceed the current enforcement standard [3.5 µg/L]" (Belluck and others, 1991). Although this method draws attention to the problem of pesticide metabolites, it does not address the question of which, if any, of the transformation products are a threat to human health.

The general lack of knowledge of the toxicity of pesticide transformation products is matched by a lack of data on the occurrence of most pesticide transformation products in surface waters (Section 5.5). This is due to primarily a lack of studies in which transformation products have been targeted. The observations of Belluck and others (1991), in discussing pesticides and their transformation products in ground water, also are applicable to surface waters as noted below.

> If it is not monitored, it will not be detected. Nondetection of a substance is likely to result in non-regulation. As the Wisconsin experience shows, ground water sampling programs, as indispensable as they are, can grossly mislead us into believing that ignorance about substances in our water is bliss. Unless national and state drinking, surface, and ground water monitoring programs recognize the metabolite issue, and begin looking for parents plus metabolites in samples, the extent of contamination will remain unknown and will be under-reported. Monitoring programs that do not look for the full spectrum of breakdown products produced from parent compounds run the real risk of underestimating the extent of potential harm to the environment and human health.

Unfortunately, looking for the full spectrum of breakdown products is almost impossible from analytical and cost perspectives. Because of these two limitations, scientists usually have been compelled to create a target list of analytes and focus their analytical and interpretive efforts around that list, especially in large-scale monitoring programs. This will most likely remain true in the future, and it must be realized that a biased and incomplete picture of the occurrence of pesticides in surface waters is being generated.

CHAPTER 7

Summary and Conclusions

The information currently available on pesticides in surface waters is synthesized from a combination of monitoring and research studies conducted over the last 35 years. Each study had its unique objectives, sampling schedule, sampling and analytical methods, target analytes, detection limits, and presentation and interpretation of results. Very few of the complete data sets from these studies are available in the open literature. Only a few long-term, systematic monitoring studies have been done. The summed total of all these studies yields a general picture of the occurrence of pesticides in surface waters, as discussed in this book, but there remain many areas in which our understanding is incomplete or nonexistent. The main points discussed in the preceding chapters are summarized below.

Monitoring of pesticides in surface waters of the United States began in the late 1950's, and steadily increased through the following 3 decades. Several large-scale monitoring programs were conducted during the 1960's and 1970's, but in recent years the emphasis has shifted to smaller-scale studies. No national-scale study has been conducted since 1980. Many states have initiated monitoring programs in the 1990's, although data from these studies generally are not available at this time. A shift also has occurred in the types of pesticides targeted in surface waters. Before 1970, attention was focused almost exclusively on the organochlorine insecticides (OCs) used at that time. Use of most OCs was severely restricted in the 1970's, and agricultural use of herbicides increased dramatically during the same period. By the 1980's, the amount of effort spent on monitoring for herbicides was about equal to the amount spent on monitoring for insecticides.

The reviewed studies show that pesticides have been found in surface waters throughout the (conterminous) United States. Of the 118 pesticides and pesticide transformation products included as analytes in the studies reviewed, 76 have been detected in at least one surface water body. Some currently used compounds have been detected very rarely, however, despite heavy agricultural use. Organochlorine insecticides continue to be detected at low levels in surface waters 20 years after their use was banned or severely restricted. Transformation products of several high-use herbicides, including atrazine and alachlor, are detected frequently in surface waters of the central United States. Little data are available on the occurrence of pesticide transformation products in surface waters overall, however, because transformation products of only 15 pesticides were targeted in the reviewed studies.

Agriculture accounts for 75 percent of total pesticide use in the United States and is the primary source of pesticides to surface waters in most areas. Little data are available on the relative importance of urban areas as a source of pesticides to surface waters. Current use of pesticides in forestry probably is not contributing significantly to pesticide contamination of surface waters. Current agricultural use results in distinct seasonal patterns in the occurrence of a number of compounds, particularly herbicides, in surface waters. The timing of this seasonal occurrence, which varies across the United States, is determined by the timing of pesticide application, the size of the drainage basin, and the weather. In the central United States, elevated concentrations of several high-use herbicides, including alachlor, atrazine, cyanazine, and metolachlor, are observed in spring and early summer in streams and rivers draining agricultural areas for weeks to months following the first significant rainfall after application. Peak concentrations generally are highest in streams with small drainage basins. In general, less than 2 percent of the amount of pesticide applied in the drainage basin is transported to surface waters. In contrast to the pattern observed in the central United States, elevated concentrations of several insecticides, including diazinon and methidathion, occur in January and February in streams draining the Central Valley of California following application to orchards during the dormant season.

Annual mean concentrations of pesticides in surface waters used as sources of drinking water have seldom exceeded maximum contaminant levels (MCLs) or lifetime health-advisory levels (HALs) established by the U.S. Environmental Protection Agency. However, peak concentrations of several currently used herbicides commonly exceed the MCL or lifetime HALs for days to weeks each year following the pesticide application period. Human health criteria have not been established for many of the pesticides currently used. Similarly, criteria for protection of aquatic life have been established for very few pesticides in current use. The existing criteria do not account for certain subacute effects, including potential effects on endocrine systems. In addition, potential effects of pesticide transformation products are not accounted for in the criteria, nor are potential effects of mixtures of pesticides or mixtures of pesticides and pesticide transformation products. For the organochlorine compounds, evidence is available on how organisms, primarily terrestrial organisms that rely on aquatic organisms for food, are adversely affected by environmental concentrations in surface waters. However, little data are available on the effects of observed concentrations of currently used pesticides on aquatic or terrestrial biota.

To obtain a complete national perspective on pesticides in surface waters, long-term, systematic monitoring studies that target a wide range of pesticides and transformation products must be done. Such monitoring needs to be done at a variety of sites throughout the nation, covering all seasons. In the United States, federal agencies are in a unique position to conduct such studies. These studies, at the national level, were last conducted in the 1970's, and then only for a limited number of analytes. The conclusions of Gilliom and others (1985), when summarizing the results from the 1975 to 1980 national survey of pesticides in surface water and bed sediments, still hold true and are noted below.

> The low and variable frequency of detection of the pesticides, regional patterns of use, and the constantly changing array of available pesticides, make national-scale monitoring of pesticides a very difficult undertaking. Pesticide use tends to be strongly regional, with most use of each chemical occurring in only one or two regions of the country; for example, most DDT and toxaphene were applied in cotton-growing areas, and most atrazine was applied in corn-growing areas. The types of pesticides used are changing constantly; new chemicals are being introduced each year, and others are being discontinued. Each different type of chemical presents unique sampling and analysis problems. Future pesticide-monitoring efforts will need to respond to changes in the types of pesticides, methods of application, chemical characteristics, and geographic patterns of use. Analytical

methods will need to be developed and improved, and different types of monitoring approaches will need to be applied. As our knowledge about pesticide chemicals and their behavior in the environment increases, efforts to monitor the levels, trends, and geographic distribution of pesticides gradually will become more sophisticated and effective.

It has been a decade since these observations were made. In that time, analytical methods have certainly improved, as has our knowledge of use and environmental behavior of pesticides. The authors are hopeful these tools can be combined with well-conceived research and monitoring programs that are thorough yet flexible, to yield a better understanding of pesticides in surface waters from both local and national perspectives.

APPENDIX

Glossary of Common and Chemical Names of Pesticides and Related Compounds Given in Text

[Chemical Class: miscellaneous N, miscellaneous nitrogen–containing compound; VOC, volatile organic compound. Use class abbreviations: Ac, acaricide; Ad, adjuvant; An, antibiotic; Df, defoliant; Di, disinfectant; Fn, fungicide; Fm, fumigant; H, herbicide; I, insecticide; IGR, insect growth regulator; IR, insect repellant; IS, insecticide synergist; Mi, miticide; Mo, molluscicide; N, nematocide; Ni, nitrification inhibitor; PGR, plant growth regulator; P, piscicide; R, rodenticide; U, unspecified. CAS, Chemical Abstracts Service. In cases where more than one CAS number was available for a given pesticide, the number pertaining to the more structurally specific chemical nomenclature is given. Principal sources: Windholz (1976), U.S. Environmental Protection Agency (1987; 1992), Worthing and Walker (1987), Howard and Neal (1992), Meister Publishing Company (1995), and Milne (1995). Blank cell, information not available or given elsewhere in appendix]

Common Name	Chemical Class	Use	CAS No.	Chemical Nomenclature
Abamectin	miscellaneous	I,Mi	71751-41-2	mixture of avermectin B,a and avermectin B,b
Acephate	organophosphorus	I	30560-19-1	O,S-dimethyl acetylphosphoramidothioate
Acetochlor	amide	H	34256-82-1	2-chloro-N-ethoxymethyl-6′-ethylact-O-tuluidide
Acifluorfen	benzoic acid derivative	H	62476-59-9	sodium 5-[2-chloro-4-(trifluoromethyl)phenoxy]-2-nitrobenzoate
Acrolein	VOC	H,R	107-02-8	2-propenal
Alachlor	acetanilide	H	15972-60-8	2-chloro-N-(2,6-diethylphenyl)-N-(methoxymethyl) acetamide
Alachlor ethanesulfonic acid	alachlor degradate		594-45-6	2-[(2,6-diethylphenyl)(methoxymethyl)amino]-2-oxoethanesulfonic acid
Alanap	see Naptalam			
Aldicarb	carbamate	I,Ac,N	116-06-3	2-methyl-2-(methylthio)propionaldehyde O-(methyl-carbamoyl)oxime
Aldicarb sulfone	carbamate, aldicarb degradate	I,N	1646-88-4	2-methyl-2-(methylsulfonyl)propionaldehyde O-(methylcarbamoyl)oxime
Aldicarb sulfoxide	aldicarb degradate		1646-87-3	2-methyl-2-(methylsulfinyl)propionaldehyde O-(methylcarbamoyl)oxime

Appendix. Glossary of Common and Chemical Names of Pesticides and Related Compounds Given in Text—Continued

Common Name	Chemical Class	Use	CAS No.	Chemical Nomenclature
Aldrin	organochlorine	I	309-00-2	(1α,4α,4aβ,5α,8α,8aβ)-1,2,3,4,10,10-hexachloro-1,4,4a,5, 8,8a-hexahydro-1,4:5,8-dimethanonapthylene
Allethrin	pyrethroid	I	584-79-2	(RS)-3-allyl-2-methyl-4-oxocyclopent-2-enyl (1RS)-cis, trans-chrysanthemate
Ametryn	triazine	H	834-12-8	2-(ethylamino)-4-isopropylamino-6-methylthio-s-triazine
Amiben	see Chloramben			
Aminocarb	carbamate	I	2032-59-9	4-(dimethylamino)-3-methylphenyl methylcarbamate
4-Amino-3,5-xylenol	mexacarbate degradate			
Aminofenitrothion	fenitrothion degradate			
Amitraz	amide	I,Ac,IS	33089-61-1	N'-(2,4-dimethylphenyl)-N-[[(2,4-dimethylphenyl)imino]methyl]-N-methylmethanimidamide
Amitrole	triazole	H	61-82-5	3-amino-1,2,4-triazole
Anilazine	triazine	Fn	101-05-3	4,6-dichloro-N-(2-chlorophenyl)-1,3,5-triazin-2-amine
Antimycin	miscellaneous	P	1397-94-0	
Asulam	carbamate	H	3337-71-1	methyl [(4-aminophenyl)sulfonyl]carbamate
Atratone	triazine	H	1610-17-9	2-(ethylamino)-4-(isopropylamino)-6-methoxy-s-triazine
Atrazine	triazine	H	1912-24-9	6-chloro-N-ethyl-N'-(1-methylethyl)-1,3,5-triazine-2,4-diamine
Azinphos-ethyl	organophosphorus	I	2642-71-9	O,O-diethyl S-[(4-oxo-1,2,3-benzotriazin-3(4H)-yl) methyl] phosphorodithioate

Appendix. Glossary of Common and Chemical Names of Pesticides and Related Compounds Given in Text—Continued

Common Name	Chemical Class	Use	CAS No.	Chemical Nomenclature
Azinphos-methyl	organophosphorus	I	86-50-0	O,O-dimethyl S-[(4-oxo-1,2,3-benzotriazin-3(4H)-yl) methyl] phosphorodithioate
Azinphos-methyl oxon	azinphos-methyl degradate		961-22-8	O,O-dimethyl S-[(4-oxo-1,2,3-benzotriazin-3(4H)-yl) methyl] phosphorothioate
Azodrin	see Monocrotophos			
Bacillus thuringiensis var *kurstaki*	bacterial agent	I	68038-71-1	
Bacillus thuringiensis var *israelensis*	bacterial agent	I		
Baygon	carbamate	I	204-043-8	2-(1-methylethoxy)phenyl methylcarbamate
Bendiocarb	carbamate	I	22781-23-3	2,2-dimethyl-1,3-benzodioxol-4-yl methylcarbamate
Benefin	see Benfluralin			
Benfluralin	dinitroaniline	H	1861-40-1	N-butyl-N-ethyl-α, α,α-trifluoro-2,6-dinitro-p-toluidine
Benomyl	carbamate	Fn	17804-35-2	methyl 1-(butylcarbamoyl)-2-benzimidazol-2-yl carbamate
Bensulfuron-methyl	sulfonylurea	H	83055-99-6	methyl 2-[[[[(4,6-dimethoxypyrimidin-2-yl)amino]-carbonyl]-amino]sulfonyl]methyl]benzoate
Bensulide	organophosphorus	H	741-58-2	S-2-benzenesulfonamidoethyl O,O-di-isopropyl phosphorodithioate
Bentazon	miscellaneous	H	25057-89-0	3-isopropyl-1H-2,1,3-benzothiadiazin-4(3H)-one 2,2-dioxide
Bifenthrin	pyrethroid	I,Mi	82657-04-3	[1α,3α-(Z)]-($\pm$)-(2 methyl[1,1'-biphenyl]-3-yl) methyl 3-(2-chloro-3,3,3-trifluoro-1-propenyl)-2,2-dimethylcyclopropanecarboxylate
Borax	inorganic	H	1303-96-4	sodium tetraborate decahydrate

Appendix. Glossary of Common and Chemical Names of Pesticides and Related Compounds Given in Text—Continued

Common Name	Chemical Class	Use	CAS No.	Chemical Nomenclature
Brodifacoum	miscellaneous	R	56073-10-0	3-[3-(4'-bromo[1,1'-biphenyl]-4-yl)-1,2,3,4-tetrahydro-1-naphthalenyl]-4-hydroxy-2H-1-benzopyran-2-one
Bromacil	uracil	H	314-40-9	5-bromo-6-methyl-3-(1-methylpropyl)-2,4-(1H,3H)-pyrimidinedione
Bromadiolone	miscellaneous	R	28772-56-7	3-[3-(4'-bromo[1,1'-biphenyl]-4-yl)-3-hydroxy-1-phenylpropyl]-4-hydroxy-2H-1-benzopyran-2-one
Bromoxynil	miscellaneous N	H	1689-84-5	3,5-dibromo-4-hydroxybenzonitrile
BSM	see Bensulfuron-methyl			
Bt	see *Bacillus thuringiensis* var. *kurstaki*			
Bti	*Bacillus thuringiensis* var. *israelensis*			
Butylate	thiocarbamate	H	2008-41-5	S-ethyl bis(2-methylpropyl)thiocarbamate
Captan	imide	Fn	133-06-2	N-(trichloromethylthio)-4-cyclohexene-1,2-dicarboximide
Carbaryl	carbamate	I	63-25-2	1-naphthalenyl-N-methylcarbamate
Carbofuran	carbamate	LN	1563-66-2	2,3-dihydro-2,2-dimethyl-7-benzofuranyl methylcarbamate
Carbofuran phenol	carbofuran degradate		1563-38-8	2,3-dihydro-2,2-dimethyl-7-hydroxy-benzofuran
Carbophenothion	see Trithion			
Carboxin	amide	Fm	5234-68-4	5,6-dihydro-2-methyl-N-phenyl-1,4-oxathiin-3-carbox-amide
Carboxyfenitrothion	fenitrothion degradate			
Chloramben	chlorobenzoic acid	H	133-90-4	3-amino-2,5-dichlorobenzoic acid

Appendix. Glossary of Common and Chemical Names of Pesticides and Related Compounds Given in Text—Continued

Common Name	Chemical Class	Use	CAS No.	Chemical Nomenclature
Chlordane	organochlorine	I	57-74-9	1,2,4,5,6,7,8,8-octachloro-3a,4,7,7a-tetrahydro-4,7-methanoindan
cis-Chlordane (α-Chlordane)	organochlorine	I	5103-71-9	1α,2α,4β,5,6,7β,8,8-octachloro-3aα,4,7,7aα-tetrahydro-4,7-methanoindan
trans-Chlordane (γ-Chlordane)	organochlorine	I	5103-74-2	1β,2α,4α,5,6,7α,8,8-octachloro-3aβ,4,7,7aβ-tetrahydro-4,7-methanoindan
2-Chloro-2′,6′-diethylacetanilide	alachlor degradate		6967-29-9	2-chloro-2′,6′-diethylacetanilide
Chlorflurenol, methyl ester	miscellaneous	H	2464-37-1	methyl 2-chloro-9-hydroxyfluorene-9-carboxylate
3-Chloroaniline	chlorpropham degradate		108-42-9	4-chloropheynlurea
4-Chlorophenyl-urea	diflubenzuron degradate		140-38-5	4-chlorophenylurea
Chloropicrin	VOC	Fm	76-06-2	trichloronitromethane
Chlorothalonil	organochlorine	Fn	1897-45-6	2,4,5,6-tetrachloro-1,3-benzenedicarbonitrile
Chlorpropham	carbamate	H	101-21-3	isopropyl 3-chlorophenylcarbamate
Chlorpyrifos	organophosphorus	I	2921-88-2	O,O-diethyl O-(3,5,6-trichloro-2-pyridinyl) phosphorothioate
Chlorthiamid	amide	H	1918-13-4	2,6-dichlorothiobenzamide
Chlorthion	organophosphorus	I	500-28-7	O-(3-chloro-4-nitrophenyl) O,O-dimethyl phosphorothioate
Clomazone	miscellaneous N	H	81777-89-1	2-[(2-chlorophenyl)methyl]-4,4-dimethyl-3-isoxazolidinone
Copper sulfate	inorganic	Fn,H	7758-99-8	copper sulfate
Crufomate	organophosphorus	I	299-86-5	4-tert-butyl-2-chlorophenyl N-methyl O-methylphosphor-amidate
Cryolite	inorganic	I	15096-52-3	sodium fluoaluminate

Appendix. Glossary of Common and Chemical Names of Pesticides and Related Compounds Given in Text—Continued

Common Name	Chemical Class	Use	CAS No.	Chemical Nomenclature
Cyanazine	triazine	H	21725-46-2	2-[[4-chloro-6-(ethylamino)-1,3,5-triazin-2-yl]amino]-2-methylpropionitrile
Cyanazine amide	cyanazine degradate			
Cycloate	thiocarbamate	H	1134-23-2	S-ethyl cyclohexeyl(ethyl)thiocarbamate
Cyfluthrin	pyrethroid	I	68359-37-5	cyano(4-fluoro-3-phenoxyphenyl)methyl 3-(2,2-dichloroethenyl)-2,2-dimethylcyclopropanecarboxylate
λ-Cyhalothrin	pyrethroid	I	91465-08-6	(RS)-a-cyano-3-phenoxybenzyl (Z)-(1R,3R)-3-(2-chloro-3,3,3-trifluoroprop-1-enyl)-2,2-dimethylcyclopropanecarboxylate
Cypermethrin	pyrethroid	I	52315-07-8	(±)-α-cyano-3-phenoxybenzyl-(±)-cis,trans-3-(2,2-dichlorovinyl)-2,2-dimethylcyclopropanecarboxylate
Cyprazine	triazine	H	22936-86-3	2-chloro-4-(cyclopropylamino)-6-(isopropylamino)-s-triazine
Cyromazine	triazine	I	66215-27-8	N-cyclopropyl-1,3,5-triazine-2,4,6-triamine
2,4-D	chlorophenoxy acid	H	94-75-7	(2,4-dichlorophenoxy)acetic acid
2,4-D, BEE	see 2,4-D, butoxyethyl ester			
2,4-D, butoxyethyl ester	chlorophenoxy acid ester	H	1929-73-3	butoxyethyl (2,4-dichlorophenoxy)acetate
2,4-D, dimethylamine salt	chlorophenoxy acid salt	H	2008-39-1	dimethylamine (2,4-dichlorophenoxy)acetate
2,4-D DMA	see 2,4-D, dimethylamine salt			
2,4-D, dodecyltetradecyl-amine salt	chlorophenoxy acid salt	H		dodecyltetradeeyl amine salt of 2,4-dichlorophenoxyacetic acid
2,4-D, methyl ester	chlorophenoxy acid ester	H	1928-38-7	methyl (2,4-dichlorophenoxy)acetate

Appendix. Glossary of Common and Chemical Names of Pesticides and Related Compounds Given in Text—Continued

Common Name	Chemical Class	Use	CAS No.	Chemical Nomenclature
Dacthal	chlorobenzoic acid	H	1861-32-1	dimethyl 2,3,5,6-tetrachloro-1,4-benzenedicarboxylate
Dalapon	organochlorine	H	75-99-0	2,2-dichloropropanoic acid
Dazomet	miscellaneous N	Fn,H,N	533-74-4	tetrahydro-3,5-dimethyl-2H-1,3,5-thiadiazine-2-thione
2,4-DB	chlorophenoxy acid	H	94-82-6	4-(2,4-dichlorophenoxy)butanoic acid
DBCP	see Dibromochloro-propane			
DBT	see Dibutyltin			
DCNA	nitroaniline	Fn	99-30-9	2,6-dichloro-4-nitroaniline
DCP	see 1,2-Dichloropropane			
DCPA	see Dacthal			
o,p'-DDD	o,p'-DDT degradate		53-19-0	1-chloro-2-[2,2-dichloro-1-(4-chlorophenyl)ethyl]benzene
p,p'-DDD	p,p'-DDT degradate	I	72-54-8	1,1-dichloro-2,2-bis(p-chlorophenyl)ethane
o,p'-DDE	o,p'-DDT degradate		3424-82-6	1-chloro-2-[(2,2-dichloro-1-(4-chlorophenyl)ethenyl)] benzene
p,p'-DDE	p,p'-DDT degradate	I	72-55-9	1,1-dichloro-2,2-bis(p-chlorophenyl)ethene
o,p'-DDT	organochlorine	I	789-02-6	1,1,1-trichloro-2-(p-chlorophenyl)-2-(o-chlorophenyl) ethane
p,p'-DDT	organochlorine	I	50-29-3	1,1,1-trichloro-2,2-bis(p-chlorophenyl)ethane
DDVP	see Dichlorvos			
Deet	amide	IR	134-62-3	N,N-diethyl-m-toluamide
Deethylatrazine	atrazine degradate		6190-65-4	2-chloro-4-amino-6-isopropylamino-s-triazine

Appendix. Glossary of Common and Chemical Names of Pesticides and Related Compounds Given in Text—Continued

Common Name	Chemical Class	Use	CAS No.	Chemical Nomenclature
DEF	organophosphorus	Df	78-48-8	*S,S,S*-tributyl phosphorotrithioate
Deispropylatrazine	atrazine degradate		1007-28-9	2-chloro-4-amino-6-ethylamino-*s*-triazine
Deltamethrin	pyrethroid	I	52918-63-5	(*S*)-α-cyano-(3-phenoxyphenyl) (1R,3R)-3-(2,2-dibromo-ethenyl)-2,2-dimethylcyclopropane carboxylate
Demethylfenitrothion	fenitrothion degradate			
Desethylatrazine	see Deethylatrazine			
Desisopropylatrazine	see Deisopropylatrazine			
Desrethyl norflurazon	nonflurazon degradate		112748-69-3	4-chloro-5-amino-2-(α,α,α-trifluoro-m-tolyl)pyridazin-3(2*H*)-one
Diamino atrazine	atrazine degradate			2-chloro-4,6-diamino-*s*-triazine
Diazinon	organophosphorus	I,N	333-41-5	*O,O*-diethyl *O*-[6-methyl-2-(1-methylethyl)-4-pyrimidinyl] phosphorothioate
Dibrom	see Naled			
Dibromochloropropane	VOC	Fm	96-12-8	1,2-dibromo-3-chloropropane
Dibutyltin	organotin	Di, Fn		dibutyltin
Dicarba	chlorobenzoic acid Vel-4207 degradate	H	1918-00-9	3,6-dichloro-2-methoxybenzoic acid
Dichlobenil	organochlorine	H	1194-65-6	2,6-dichlorobenzonitrile
2,4-Dichlorophenol	2,4-D degradate		120-83-2	2,4-dichlorophenol
1,2-Dichloropropane	VOC	Fm,Ad	78-87-5	1,2-dichloropropane
Dichlorprop	see 2,4-DP			

Appendix. Glossary of Common and Chemical Names of Pesticides and Related Compounds Given in Text—Continued

Common Name	Chemical Class	Use	CAS No.	Chemical Nomenclature
Dichlorvos	organophosphorus	Fm,I, Ad	62-73-7	O,O-dimethyl O-(2,2-dichlorovinyl) phosphate
Diclofop	chlorophenoxy acid	H	40843-25-2	(RS)-2-[4-(2,4-dichlorophenoxy)phenoxy]propionic acid
Diclofop-methyl	chlorophenoxy acid ester	H	51338-27-3	methyl (RS)-2-[4-(2,4-dichlorophenoxy)phenoxy] propionate
Dicofol	organochlorine	I	115-32-2	4-chloro-α-(4-chlorophenyl)-α-(trichloromethyl) benzenemethanol
Dicrotophos	organophosphorus	I	141-66-2	(E)-2-dimethylcarbamayl-1-methylvinyl dimethyl
2,6-Diethylaniline	alachlor degradate		579-66-8	
Dieathatyl ethyl	miscellaneous	H	38727-55-8	ethyl, N-(chloroacetyl)-N-(2,6-diethylphenylglycinate)
Dieldrin	organochlorine	I	60-57-1	1,2,3,4,10,10-hexachloro-6,7-epoxy-1,4,4a,5,6,7,8,8a-octahydro (endo,exo) 1,4:5,8-dimethanonaphthalene
Dienochlor	organochlorine	Mi	2227-17-0	decachlorobis(2,4-cyclopentadien-1-yl)
Diethyl fumarate	malathion degradate		623-91-6	
Diethylacetanilide	amide			2′,6′-diethylacetanilide
Diethyltoluamide	see Deet			
Diflubenzuron	urea	I	35367-38-5	1-(4-chlorophenyl)-3-(2,6-difluorobenzoyl)urea
Dimethoate	organophosphorus	I	60-51-5	O,O-dimethyl S-methylcarbamoylmethyl phosphoro-dithioate
Dimethyl phosphorodithioic acid	malathion degradate			dimethyl phosphorodithioic acid
4-Dimethylamino-3,5-xylenol	mexacarbate degradate		6120-10-1	
Dimethylnitros-amine	2,4-D degradate		62-75-9	dimethylnitrosamine

Appendix. Glossary of Common and Chemical Names of Pesticides and Related Compounds Given in Text—Continued

Common Name	Chemical Class	Use	CAS No.	Chemical Nomenclature
Dimilin	see Diflubenzuron			
Dinitramine	miscellaneous N	H	29091-05-2	N^4,N^4-diethyl-α,α,α-trifluoro-3,5-dinitrotoluene-2,4-diamine
Dinocap	miscellaneous N	Ac,Fn	39300-45-3	2,4-dinitro-6-octylphenylcrotonate
Dinoseb	nitrophenol	H	88-85-7	2-sec-butyl-4,6-dinitrophenol
Diphenamid	amide	H	957-51-7	N,N-dimethyl-2,2-diphenylacetamide
Diquat	see Diquat dibromide			
Diquat dibromide	miscellaneous N	H	85-00-7	1,1'-ethylene-2,2'-bipyridylium dibromide, monohydrate
Disulfoton	organophosphorus	I	298-04-4	O,O-diethyl S-[2-(ethylthio)ethyl]phosphorodithioate
Disyston	see Disulfoton			
Diuron	urea	H	330-54-1	3-(3,4-dichlorophenyl)-1,1-dimethylurea
DNOC	miscellaneous N	H,I,Ad	534-52-1	4,6-dinitro-o-cresol
Dodine	amine	Fn	2439-10-3	1-dodecylguanidine acetate
2,4-DP	chlorophenoxy acid derivative	H	120-36-5	($\pm$)-2-(2,4-dichlorophenoxy)propanoic acid
DSMA	organic arsenical	H	144-21-8	disodium methanearsonate
EDB	VOC	Fm,LN	106-93-4	1,2-dibromoethane
Endosulfan	organochlorine	I	115-29-7	6,7,8,9,10,10-hexachloro-1,5,5a,6,9,9a-hexahydro-6,9-methano-2,4,3-benzodioxathiepin-3-oxide

Appendix. Glossary of Common and Chemical Names of Pesticides and Related Compounds Given in Text—Continued

Common Name	Chemical Class	Use	CAS No.	Chemical Nomenclature
Endosulfan I (α-Endosulfan)	organochlorine	I	959-98-8	3α,5aβ,6α,9α,9aβ-6,7,8,9,10,10-hexachloro-1,5,5a,6,9,9a-hexahydro-6,9-methano-2,4,3-benzodioxathiepin-3-oxide
Endosulfan II (β-Endosulfan)	organochlorine	I	33213-65-9	3α,5aα,6β,9β,9aβ-6,7,8,9,10,10-hexachloro-1,5,5a,6,9,9a-hexahydro-6,9-methano-2,4,3-benzodioxathiepin-3-oxide
Endosulfan diol	endosulfan degradate			
Endosulfan sulfate	endosulfan degradate		1031-07-8	3α,5aα,6β,9β,9aβ-6,7,8,9,10,10-hexachloro-1,5,5a,6,9,9a-hexahydro-6,9-methano-2,4,3-benzodioxathiepin-3,3-dioxide
Endothall	miscellaneous	H	145-73-3	7-oxabicyclo[2,2,1]heptane-2,3-dicarboxylic acid
Endothall, di-Na salt	miscellaneous	H	129-67-0	7-oxabicyclo[2,2,1]heptane-2,3-dicarboxylic acid, sodium salt
Endrin	organochlorine	I	72-20-8	1,2,3,4,10,10-hexachloro-6,7-epoxy-1,4,4a,5,6,7,8,8a-octahydro-(endo,endo)-1,4:5,8-dimethanonaphthalene
Endrin aldehyde	endrin degradate		7421-93-4	2,2a,3,3,7-hexachlorodecahydro-1,2,4-methenocyclo-penta-(c,d)-pentalene-r-carboxaldehyde
Eptam	thiocarbamate	H	759-94-7	S-ethyl dipropylthiocarbamate
EPTC	see Eptam			
ESA	see Alachlor ethanesulfonic acid			
Esfenvalerate	pyrethroid	I	66230-04-4	(S)-α-cyano-3-phenoxybenzyl (S)-2-(4-chlorophenyl)-3-methylbutyrate
Ethalfluralin	dinitroaniline	H	55283-68-6	N-ethyl-N-(2-methyl-2-propenyl)-2,6-dinitro-4-(trifluoromethyl)benzenamine
Ethanesulfonic acid	see Alachlor ethanesulfonic acid			
Ethion	organophosphorus	I,Ac	563-12-2	S,S'-methylene bis(O,O-diethyl phosphorodithioate)

Appendix. Glossary of Common and Chemical Names of Pesticides and Related Compounds Given in Text—Continued

Common Name	Chemical Class	Use	CAS No.	Chemical Nomenclature
Ethoprop	organophosphorus	I,N	13194-48-4	O-ethyl S,S-dipropyl phosphorodithioate
Ethylan	see Perthane			
Ethylene dibromide	see EDB			
Ethylene thiourea	urea, ethylene bis-dithiocarbamate (EBDC) degradate	Ad	96-45-7	1,3-ethylene-2-thiourea
Etridiazole	miscellaneous N	Fn	2593-15-9	5-ethoxy-3-trichloromethyl-1,2,4-thiadiazole
ETU	see Ethylene thiourea			
Fenamiphos	organophosphorus	N	2224-92-6	O-ethyl[3-methyl-4-(methylthio)phenyl] (1-methylethyl phosphoramidate)
Fenarimol	miscellaneous N	Fn	60168-88-9	a-(2-chlorophenyl)-a-(4-chlorophenyl)-5-pyrimidine methanol
Fenbutatin-oxide	organotin	I,Ac	13356-08-6	bis[tris (2-methyl-2-phenylpropyl)tin] oxide
Fenitrooxon	fenitrothion degradate			
Fenitrothion	organophosphorus	I, Ac	122-14-5	O,O-dimethyl O-(3-methyl-4-nitrophenyl) phosphorothioate
Fensulfothion	organophosphorus	I	115-90-2	O,O-diethyl O-[4-(methylsulfinyl)phenyl] phosphorothioate
Fenthion	organophosphorus	I	55-38-9	O,O-dimethyl O-[3-methyl-4-(methylthio)phenyl] phospho-rothioate
Fenvalerate	pyrethroid	I	51630-58-1	cyano-(3-phenoxyphenyl)methyl-4-chloro-(1-methylethyl)-benzeneacetate
Ferbam	thiocarbamate	Fn	14484-64-1	ferric dimethyldithiocarbamate
Fluazifop-butyl	miscellaneous N	H	69806-50-4	butyl (RS)-2-[4-[[5-(trifluoromethyl)-2-pyridinyl]oxy] phenoxy]propanoate

Appendix. Glossary of Common and Chemical Names of Pesticides and Related Compounds Given in Text—Continued

Common Name	Chemical Class	Use	CAS No.	Chemical Nomenclature
Fluometuron	urea	H	2164-17-2	1,1-dimethyl-3-(α,α,α-trifluoro-*m*-tolyl)urea
Fluorodifen	miscellaneous N	U	15457-05-3	4-nitrophenyl α,α,α-trifluoro-2-nitro-p-tolyl ether
Fluridone	miscellaneous N	H	59756-60-4	1-methyl-3-phenyl-5-[3-(trifluoromethyl)phenyl]-4(1*H*)-pyridinone
Folex	organophosphorus	Df	150-50-5	*S,S,S*-tributyl phosphorotrithioite
Folpet	imide	Fn	133-07-3	*N*-[(trichloromethyl)thio]phthalimide
Fonofos	organophosphorus	I	944-22-9	*O*-ethyl *S*-phenyl ethylphosphonodithioate
Formetanate HCl	carbamate	I, Ac	23422-53-9	3-dimethylaminomethyleneaminophenyl methylcarbamate hydrochloride
Fosamine	see Fosamine ammonium			
Fosamine ammonium	organophosphorus	PGR	25954-13-6	ammonium ethyl carbamoylphosphonate
Fosetyl aluminum	organophosphorus	Fn	39148-24-8	aluminum tris(*O*-ethyl phosphonate)
Fosetyl-Al	see Fosetyl aluminum			
Furadan	see Carbofuran			
Glyphosate	amino acid derivative	H	1071-83-6	*N*-(phosphonomethyl)glycine, isopropylamine salt
HCB	organochlorine	Fn	118-74-1	hexachlorobenzene
α-HCH	organochlorine	I	319-84-6	1α,2α,3β,4α,5β,6β-hexachlorocyclohexane
β-HCH	organochlorine	I	319-85-7	1α,2β,3α,4β,5α,6β-hexachlorocyclohexane
γ-HCH	organochlorine	I	58-89-9	1α,2α,3β,4α,5α,6β-hexachlorocyclohexane
δ-HCH	organochlorine	I	319-86-8	1α,2α,3α,4β,5α,6-hexachlorocyclohexane

Appendix. Glossary of Common and Chemical Names of Pesticides and Related Compounds Given in Text—Continued

Common Name	Chemical Class	Use	CAS No.	Chemical Nomenclature
HCH, technical	organochlorine	I	608-73-1	1,2,3,4,5,6-hexachlorocyclohexane (isomer mixture)
Heptachlor	organochlorine	I	76-44-8	1,4,5,6,7,8,8-heptachloro-3a,4,7,7a-tetrahydro-4,7-methano-1H-indene
Heptachlor epoxide	heptachlor degradate		1024-57-3	2,3,4,5,6,7,8-heptachloro-1a,1b,5,5a,6,6a,-hexahydro-2,5-methano-2H-indeno(1,2b)oxirene
Hexachlorobenzene	see HCB			
Hexazinone	triazine	H	51235-04-2	3-cyclohexyl-6-(dimethylamino)-1-methyl-1,3,5-triazine-2,4(1H,3H)-dione
Hydramethylnon	miscellaneous N	I	67485-29-4	tetrahydro-5,5-dimethyl-2(1H)-pyrimidinone [3-[4-(trifluoromethyl)phenyl]-1-[2-[4-(trifluoromethyl) phenyl]ethenyl]-2-propenylidene]hydrazone
Hydramethylon	see Hydramethylnon			
2-Hydroxy-2'6'diethylacetanilide	alachlor degradate		52559-52-1	2-hydroxy-2'6'diethylacetanilide
Hydroxyatrazine	atrazine degradate		2163-68-0	6-hydroxy-N-ethyl-N'-(1-methylethyl)-1,3,5-triazine-2,4-diamine
3-Hydroxycarbofuran	carbofuran degradate		16655-82-6	2,3-dihydro-2,2-dimethyl-7-methylcarbamate, 3,7-benzofurandiol
1-Hycroxychlordene	chlordane degradate		24009-05-0	4,5,6,7,8,8-hexachloro-3a,4,7,7a-tetrahydro-(endo,exo)-4,7-methanoinden-1-ol
Imazapyr	imidazolinone	H	81334-34-1	2-(4-isopropyl-4-methyl-5-oxo-2-imidazolin-2-yl)-nicotinic acid
Imazaquin	imidazolinone	H	81335-37-7	(±)-2-[4,5-dihydro-4-methyl-4-(1-methylethyl)-5-oxo-1H-imidazol-2-yl]-3-quinolinecarboxylic acid
Imidan	see Phosmet			
Iprodione	amide	Fn	36734-19-7	3-(3,5-dichlorophenyl)-N-(1-methylethyl)-2,4-dioxo-1-imidazolidinecarboxamide

Appendix. Glossary of Common and Chemical Names of Pesticides and Related Compounds Given in Text—Continued

Common Name	Chemical Class	Use	CAS No.	Chemical Nomenclature
Isazofos	organophosphate	I,N	42509-80-8	O-[5-chloro-1-(methylethyl)-1H-1,2,4-triazol-3-yl] O,O-diethyl phosphorothioate
Isofenphos	organophosphorus	I	25311-71-1	1-methylethyl 2-[[ethoxy[(1-methylethyl)amino] phos-phinothioyl]oxy]benzoate
Isophenphos	see Isofenphos			
Kelthane	see Dicofol			
Kepone	organochlorine	I	143-50-0	1,1a,3,3a,4,5,5,5a,5b,6-decachlorooctahydro-1,3,4-metheno-2H-cyclobuta-[c,d]-pentalen-2-one
Ketoendrin	endrin degradate		53494-70-5	
2-Ketomolinate	molinate degradate			S-ethyl hexahydro-1H-azepine-2-one-1-carbothioate
4-Ketomolinate	molinate degradate			S-ethyl hexahydro-1H-azepine-4-one-1-carbothioate
Leptophos	organophosphorus	I	21609-90-5	O-(4-bromo-2,5-dichlorophenyl) O-methylphenyl phospho-nothioate
Lindane	see γ-HCH			
Linuron	urea	H	330-55-2	N'-(3,4-dichlorophenyl)-N-methoxy-N-methylurea
Malathion	organophosphorus	I	121-75-5	O,O-dimethyl S-[1,2-bis(ethoxycarbonyl)ethyl] dithiophosphate
Malathion oxon	malathion degradate		1634-78-2	O,O-dimethyl S-[1,2-bis(ethoxycarbonyl)ethyl] phosphorothioate
Maloxon	see Malathion oxon			
Mancozeb	ethylene bisdithiocarbamate	Fn	8018-01-7	coordination product of zinc ion and manganese ethylene bisdithiocarbamate
Maneb	ethylene bisdithiocarbamate	Fn	12427-38-2	manganese ethylenebisdithiocarbamate

Appendix. Glossary of Common and Chemical Names of Pesticides and Related Compounds Given in Text—Continued

Common Name	Chemical Class	Use	CAS No.	Chemical Nomenclature
MCFA	chlorophenoxy acid	H	94-74-6	(4-chloro-2-methyl)phenoxyacetic acid
MCPB	chlorophenoxy acid	H	94-81-5	4-(4-chloro-2-methylphenoxy)butanoic acid
MCPP	chlorophenoxy acid salt	H	1929-86-8	potassium (RS)-2-(4-chloro-2-methylphenoxy)propanoate
Mecoprop	chlorophenoxy acid	H	7085-19-0	(RS)-2-(4-chloro-2-methylphenoxy)propanoic acid
Metalaxyl	amino acid derivative	Fn	57837-19-1	N-(2,6-dimethylphenyl)-N-(methoxyacetyl)-DL-alanine methyl ester
Metaldehyde	miscellaneous	Mo	108-62-3	2,4,6,8-tetramethyl-1,3,5,7-tetroxocane
Metam-sodium	thiocarbamate	Fm,Fn, H,I,N	137-42-8	sodium N-methyldithiocarbamate
Methamidophos	organophosphorus	I	10265-92-6	O,S-dimethyl phosphoramidothioate
Methidathion	organophosphorus	I,Ac	950-37-8	[(5-methoxy-2-oxo-1,3,4-thiadiazol-3(2H)-yl)methyl] O,O-dimethylphosphorodithioate
Methiocarb	carbamate	I,Ac, Mo	2032-65-7	3,5-dimethyl-4-(methylthio)phenyl methylcarbamate
Methomyl	carbamate, thiodicarb degradate	I	16752-77-5	methyl N-[[(methylamino)carbonyl]oxy] ethanimidothioate
Methoprene	miscellaneous	IGR	40596-69-8	isopropyl (2E,4E,7S)-11-methoxy-3,7,11-trimethyldodeca-2,4-dienoate
Methoxychlor	organochlorine	I	72-43-5	2,2-bis(4-methoxyphenyl)-1,1,1-trichloroethane
3-Methyl-4-aminophenol	fenitrothion degradate			
Methyl bromide	VOC	Fm,Ad	74-83-9	bromomethane
3-Methyl-4-nitrophenol	fenitrothion degradate		2581-34-2	

Appendix. Glossary of Common and Chemical Names of Pesticides and Related Compounds Given in Text—Continued

Common Name	Chemical Class	Use	CAS No.	Chemical Nomenclature
Methyl parathion	organophosphorus	I	298-00-0	*O,O*-dimethyl *O*-(4-nitrophenyl) phosphorothioate
Methyl trithion	organophosphorus	I,Ac	953-17-3	*S*-[[(4-chlorophenyl)thio]methyl] *O,O*-dimethyl phosphorodithioate
Methylarsonic acid, mono-sodium salt	organic arsenical	H	2163-80-6	monosodium methanearsonate
Metiram	thiocarbamate	Fn	9006-42-2	tris[ammine-[ethylen bis(dithiocarbamato)]zinc(II)] [tetrahydro-1,2,4,7-dithiadiazocine-3,8-dithione] polymer
Metolachlor	acetanilide	H	51218-45-2	2-chloro-*N*-(2-ethyl-6-methylphenyl)-*N*-(2-methoxy-1-methylethyl)acetamide
Metoxuron	urea	H	19937-59-8	*N'*-(3-chloro-4-methoxyphenyl)-*N,N*-dimethylurea
Metribuzin	triazine	H	21087-64-9	4-amino-6-(1,1-dimethylethyl)-3-(methylthio)-1,2,4-triazin-5(4*H*)-one
Mevinphos	organophosphorus	I, Ac	7786-34-7	methyl 3-[(dimethoxyphosphinyl)oxy]-2-butenoate
Mexacarbate	carbamate	I,Mo, Ac	315-18-4	4-dimethylamino-3,5-xylyl methylcarbamate
MGK 264	imide	IS	113-48-4	*N*-(2-ethylhexyl)bicyclo(2,2,1)-hept-5-ene-2,3-dicarboximide
Mirex	organochlorine	I	2385-85-5	1,1a,2,2,3,3a,4,5,5,5a,5b,6-dodecachlorooctahydro-1,3,4-metheno-1*H*-cyclobuta[cd]pentalene
Molinate	thiocarbamate	H	2212-67-1	*S*-ethyl hexahydro-1*H*-azepine-1-carbothioate
Monobutyltin	tributyltin degradate		78763-54-9	monobutyl tin
Monocrotophos	organophosphorus	I, Ac	6923-22-4	dimethyl (E)-[1-methyl-3-(methylamino)-3-oxo-1-pro-penyl] phosphate
MSMA	see Methylarsonic acid, monosodium salt			

Appendix. Glossary of Common and Chemical Names of Pesticides and Related Compounds Given in Text—Continued

Common Name	Chemical Class	Use	CAS No.	Chemical Nomenclature
Myclobutanil	triazole	Fn	88671-89-0	a-butyl-a-(4-chlorophenyl)-1H-1,2,4-triazole-1-propanenitrile
N-methylformamide	fluridone degradate		123-39-7	N-methylformamide
Naled	organophosphorus	I	300-76-5	1,2-dibromo-2,2-dichloroethyl dimethyl phosphate
1-Naphthol	carbaryl degradate		90-15-3	1-naphthalenol
Napropamide	amide	H	15299-99-7	(RS)-N,N-diethyl-2-(1-naphthyloxy)propionamide
Naptalam	amine	H	132-66-1	sodium 2-[(1-naphthalenylamino)carbonyl]benzoate
Nitralin	dinitroaniline	H	4726-14-1	4-methylsulfonyl-2,6-dinitro-N,N-dipropylaniline
Nitrapyrin	miscellaneous N	Ni	1929-82-4	2-chloro-6-(trichloromethyl)pyridine
cis-Nonachlor	organochlorine	I	5103-73-1	1,2,3,4,5,6,7,8,8-nonachloro-2,3,3a,4,7,7a-hexahydro-4,7-methano-1H-indene (combined nomenclature for cis and trans isomers)
trans-Nonachlor	organochlorine	I	39765-80-5	see cis-Nonachlor
Norea	urea	H	18530-56-8	3-(hexahydro-4,7-methanoindan-5-yl)-1,1-dimethylurea
Norflurazon	amine	H	27314-13-2	4-chloro-5-methylamino-2-(α,α,α-trifluoro-m-tolyl) pyridazin-3(2H)-one
Oryzalin	dinitroaniline	H	19044-88-3	3,5-dinitro-N₄,N₄-dipropylsulfanilamide
Oxamyl	carbamate	I,N,Ac	23135-22-0	S-methyl N',N'-dimethyl-N-(methylcarbamoyloxy)-1-thio-oxaminidate
Oxychlordane	chlordane degradate		27304-13-8	2,3,4,5,6,6a,7,7-octachloro-1a,1b,5,5a,6,6-hexahydro-2,5-methano-2H-indeno(1,2b)oxirene

Appendix. Glossary of Common and Chemical Names of Pesticides and Related Compounds Given in Text—Continued

Common Name	Chemical Class	Use	CAS No.	Chemical Nomenclature
Oxydemeton-methyl	organophosphorus	I	301-12-2	S-[2-(ethylsulfinyl)ethyl] O,O-dimethyl phosphorothioate
Oxyfluorfen	diphenyl ether	H	42874-03-3	2-chloro-1-(3-ethoxy-4-nitrophenoxy)-4-(trifluoromethyl benzene)
Oxytetracycline	amide	An	79-57-2	4-(dimethylamino)-1,4,4a,5,5a,6,11,12a-octahydro-3,5,6,10,12,12a-hexahydroxy-6-methyl-1,11-dioxo-2-naphthacenecarboxamide
Oxythioquinox	dithiocarbonate	Ac,Fm, Fn	2439-01-2	6-methyl-1,3-dithiolo[4,5-b]quinoxalin-2-one
Paranitrophenol	Methyl parathion degradate		100-02-7	4-nitrophenol
Paraquat	miscellaneous N	H	1910-42-5	1,1'-dimethyl-4,4'-bipyridinium ion, dichloride salt
Parathion	organophosphorus	I	56-38-2	O,O-diethyl O-(4-nitrophenyl) phosphorothioate
para-Dichlorobenzene	miscellaneous	I	106-46-7	1,4-dichlorobenzene
PCNB	organochlorine	Fn	82-68-8	pentachloronitrobenzene
PCP	organochlorine	Fn,Mo, Ad	87-86-5	pentachlorophenol
Pebulate	thiocarbamate	H	1114-71-2	S-propyl butyl(ethyl)thiocarbamate
Pendimethalin	dinitroaniline	H	40487-42-1	N-(1-ethylpropyl)-3,4-dimethyl-2,6-dinitrobenzeneamine
Pentachlorophenol	see PCP			
Permethrin	pyrethroid	I	52645-53-1	(3-phenoxyphenyl)methyl (±)-cis,trans-3-(2,2-dichloro- ethenyl)-2,2-dimethylcyclopropanecarboxylate
cis-Permethrin	pyrethroid	I	61949-76-6	(3-phenoxyphenyl)methyl cis-3-(2,2-dichloroethenyl)-2,2-dimethylcyclopropanecarboxylate

Appendix. Glossary of Common and Chemical Names of Pesticides and Related Compounds Given in Text—Continued

Common Name	Chemical Class	Use	CAS No.	Chemical Nomenclature
trans-Permethrin	pyrethroid	I		(3-phenoxyphenyl)methyl trans-3-(2,2-dichloroethenyl)-2,2-dimethylcyclopropanecarboxylate
Perthane	organochlorine	I	72-56-0	1,1-dichloro-2,2-bis(4-ethylphenyl)ethane
Phorate	organophosphorus	I	298-02-2	O,O-diethyl S-ethylthiomethyl phosphorodithioate
Phosalone	organophosphorus	Ac,I	2310-17-0	S-[(6-chloro-2-oxo-3(2H)-benzoxazolyl)methyl] O,O-diethyl phosphorodithioate
Phosdrin	see Mevinphos			
Phosmet	organophosphorus	I	732-11-6	N-(mercaptomethyl)phthalimide-S-(O,O-dimethylphos-phorodithioate)
Phosphamidon	organophosphorus	I	13171-21-6	O,O-dimethyl O-(2-chloro-2-diethylcarbamoyl-1-methyl-vinyl)-phosphate
Photoaldrin	aldrin degradate		13350-71-5	8-hydroxy-1,1a,2,2,3,3a,4,5,5,5a,5b,6-dodecachlorooctahydro-1,3,4-metheno-1H-cyclobuta[cd]pentalene
Photomirex	mirex degradate		39801-14-4	
Picloram	amine	H	1918-02-1	4-amino-3,5,6-trichloropicolinic acid
Profenfos	organophosphorus	Ac,I	41198-08-7	O-4-bromo-2-chlorophenyl O-ethyl S-propyl phosphorothioate
Profluralin	dinitroaniline	H	26399-36-0	2,6-dinitro-N-cyclopropylmethyl-N-propyl-4-(trifluoromethyl)benzenamine
Prometon	triazine	H	1610-18-0	6-methoxy-N,N'-bis(1-methylethyl)-1,3,5-triazine-2,4-diamine
Prometone	see Prometon			
Prometryn	triazine	H	7287-19-6	N,N'-bis(1-methylethyl)-6-(methylthio)-1,3,5-triazine-2,4-diamine
Pronamide	amide	H	23950-58-5	3,5-dichloro-N-(1,1-dimethyl-2-propynyl)benzamide

Appendix. Glossary of Common and Chemical Names of Pesticides and Related Compounds Given in Text—Continued

Common Name	Chemical Class	Use	CAS No.	Chemical Nomenclature
Propachlor	acetanilide	H	1918-16-7	2-chloro-N-(1-methylethyl)-N-phenylacetanilide
Propanil	amide	H	709-98-8	N-(3,4-dichlorophenyl)propanamide
Propargite	sulfite ester	Ac	2312-35-8	2-[4-(1,1-dimethylethyl)phenoxy]cyclohexyl-2-propynyl sulfite
Propazine	triazine	H	139-40-2	6-chloro-N,N'-bis(1-methylethyl)-1,3,5-triazine-2,4-diamine
Propham	carbamate	H,PGR	122-42-9	1-methylethylphenyl carbamate
Propiconazole	triazole	Fn	60207-90-1	1-[[2-(2,4-dichlorophenyl)-4-propyl-1,3-dioxolan-2-yl] methyl]-1H-1,2,4-triazole
Propoxur	see Baygon	I	10453-86-8	(5-benzyl-3-furyl)methyl cis,trans-2,2-dimethyl-3-(2-methylpropenyl)-cyclopropanecarboxylate
Pyridine	tributyltin degradate		110-86-1	pyridine
Pyridinol	triclopyr degradate			pyridinol
Resmethrin	pyrethroid	I	10453-86-8	(5-benzyl-3-furyl)methyl cis,trans-2,2-dimethyl-3-(2-methylpropenyl)-cyclopropanecarboxylate
Ronilan	organochlorine	Fn	50471-44-8	3-(3,5-dichlorophenyl)-5-methyl-5-vinyl-1,3-oxazolidine-2,4-dione
Ronnel	organophosphorus	I	299-84-3	O,O-dimethyl O-(2,4,5-trichlorophenyl)phosphorothioate
Rotenone	miscellaneous	I	83-79-4	1,2,12,12a-tetrahydro-8,9-dimethoxy-2-(1-methylethenyl)-[1]benzopyrano[3,4-b]furo[2,3-h][1]-benzopyran-6(6H)-one
Sethoxydim	miscellaneous N	H	74051-80-2	2-[1-(ethoxyimino)butyl]-5-[2-(ethylthio)propyl]-3-hydroxy-2-cyclohexen-1-one
Sevin	see Carbaryl			

Appendix. Glossary of Common and Chemical Names of Pesticides and Related Compounds Given in Text—Continued

Common Name	Chemical Class	Use	CAS No.	Chemical Nomenclature
Silvex	see 2,4,5-TP			
Sinazine	triazine	H	122-34-9	2-chloro-4,6-bis(ethylamino)-s-triazine
Simetone	triazine	H	673-04-1	2,4-bis(ethylamino)-6-methoxy-s-triazine
Simetryn	triazine	H	1014-70-6	2,4-bis(ethylamino)-6-methylmercapto-s-triazine
Streptomycin	amine	An	57-92-1	O-2-deoxy-2-(methylamino)-a-L-glucopyranosyl-(1Æ2)-O-5-deoxy-3-C-formyl-a-L-lyxofuranosyl-(1Æ4)-N,N-bis(aminoiminomethyl)-D-streptamine
Strobane	organochlorine	I	8001-50-1	mixture of polychlorinated camphene, pinene and related terpenes
Sulfometuron-methyl	sulfonylurea	H	74222-97-2	methyl 2-[[[[(4,6-dimethyl-2-pyrimidinyl)amino]-carbonyl]amino]sulfonyl]benzoate
Sulprofos	organophosphorus	I	35400-43-2	O-ethyl O-[(4-methylthio)phenyl] S-propyl phosphoro-dithioate
Sumithrin	pyrethroid	I	26046-85-5	3-phenoxybenzyl (1R)-cis,trans-chrysanthemate
2,4,5-T	chlorophenoxy acid	H	93-76-5	2,4,5-trichlorophenoxyacetic acid
TBT	see Tributyltin			
TCDD	impurity in 2,4,5-T and other organochlorines		1746-01-6	2,3,7,8-tetrachlorodibenzo-p-dioxin
p,p'-DE	see p,p'-DDD			
Tebuthiuron	urea	H	34014-18-1	N-[5-(1,1-dimethylethyl)-1,3,4-thiadiazol-2-yl]-N,N'-dimethylurea
Tefluthrin	pyrethroid	I	79538-32-2	1α,3α(Z)-(±)-(2,3,5,6-tetrafluoro-4-methylphenyl)methyl-3-(2-chloro-3,3,3-trifluoro-1-propenyl)-2,2-dimethylcyclopropanecarboxylate

Appendix. Glossary of Common and Chemical Names of Pesticides and Related Compounds Given in Text—Continued

Common Name	Chemical Class	Use	CAS No.	Chemical Nomenclature
Temephos	organophosphorus	I	3383-96-8	O,O'-thiodi-4,1-phenylene O,O,O',O'-dimethyl phosphorothioate
Terbufos	organophosphorus	I,N	13071-79-9	S-[[(1,1-dimethylethyl)thio]methyl] O,O diethyl phosphorodithioate
Terbufos sulfone	terbufos degradate		56070-16-7	S-[[(1,1-dimethylethyl)sulfonyl]methyl] O,O diethylphosphorodithioate
Terbufos sulfoxide	terbufos degradate		10548-10-4	S-[[(1,1-dimethylethyl)sulfinyl]methyl] O,O diethylphosphorodithioate
Terbuthylazine	triazine	H	5915-41-3	2-(tert-butylamino)-4-chloro-6-(ethylamino)-s-triazine
Terbutryn	triazine	H	886-50-0	2-(tert-butylamino)-4-(ethylamino)-6-(methylthio)-s-triazine
Tetradifon	organochlorine	Ac	116-29-0	1,2,4-trichloro-5-[(4-chlorophenyl)sulfonyl]benzene
Tetramethrin	pyrethroid	I	7696-12-0	3,4,5,6-tetrahydrophthalimidomethyl (1RS)-cis,trans-chrysanthemate
TFM	see 3-Trifluoromethyl-4-nitrophenyl			
Thiabendazole	imidazole	Fn	148-79-8	2-(4'-thiazolyl)-benzimidazole
Thiobencarb	thiocarbamate	H	28249-77-6	S-4-chlorobenzyl diethylthiocarbamate
Thiodicarb	carbamate	I	59669-26-0	dimethyl N,N'-[thiobis[(methylimino)carbonyloxy]] bis(ethanimidothioate)
Thiophanate-methyl	carbamate	Fn	23564-05-8	dimethyl 4,4'-o-phenylenebis(3-thioallophanate)
Thiram	thiocarbamate	Fn	137-26-8	bis(dimethylthiocarbamyl) disulfide
Thymol	miscellaneous	Fn	89-83-8	5-methyl-2-(1-methylethyl)phenol
Toxaphene	organochlorine	I	8001-35-2	polychlorinated camphene
2,4,5-TP	chlorophenoxy acid	H	93-72-1	(±)-2-(2,4,5-trichlorophenoxy)propanoic acid

Appendix. Glossary of Common and Chemical Names of Pesticides and Related Compounds Given in Text—Continued

Common Name	Chemical Class	Use	CAS No.	Chemical Nomenclature
Tralomethrin	pyrethroid	I	66841-25-6	(S)-α-cyano-3-phenoxybenzyl(1R,3S)-3-[(1'RS)(1',2',2',2'-tetrabromoethyl)]-2,2-dimethylcyclopropanecarboxylate
Triadimefon	triazole	Fn	43121-43-3	1-(4-chlorophenoxy)-3,3-dimethyl-1-(1H-1,2,4-triazol-1-yl)-2-butanone
Triallate	thiocarbamate	H	2303-17-5	S-(2,3,3-trichloro-2-propenyl) bis(1-methylethyl) thiocarbamate
Tributyltin	organotin (marine biocide)	Di, Fn	56573-85-4	tributyltin
Trichlorfon	organophosphorus	I	52-68-6	dimethyl (2,2,2-trichloro-1-hydroxyethyl)-phosphonate
Triclopyr	organochlorine	H	55335-06-3	(3,5,6-trichloro-2-pyridinyloxy)acetic acid
Tricosene	hydrocarbon	I	27519-02-4	cis-tricos-9-ene
3-Trifluoromethyl-4-nitrophenol	miscellaneous	P	88-30-2	3-trifluoromethyl-4-nitrophenyl
Trifluralin	dinitroaniline	H	1582-09-8	2,6-dinitro-N,N-dipropyl-4-(trifluoromethyl) benzenamine
Triforine	amide	Fn	26644-46-2	N,N'-[1,4-piperazinediylbis(2,2,2-trichloroethylidene)] bis(formamide)
Trimethacarb	carbamate	I	12407-86-2	3,4,5- (or 2,3,5-)trimethylphenyl methylcarbamate
Triphenyltin hydroxide			76-87-9	fentin hydroxide
Trithion	organophosphorus	I,Ac	786-19-6	S-[[(4-chlorophenyl)thio]methyl] O,O-diethyl phosphoro-dithioate
Vernolate	thiocarbamate	H	1929-77-7	S-propyl dipropylthiocarbamate
Vinclozolin	see Ronilan			
Warfarin	miscellaneous	Ro	81-81-2	3-(a-acetonylbenzyl)-4-hydroxycoumarin

Appendix. Glossary of Common and Chemical Names of Pesticides and Related Compounds Given in Text—Continued

Common Name	Chemical Class	Use	CAS No.	Chemical Nomenclature
Xylene	miscellaneous	H	68920-06-9	dimethyl benzene
Zineb	thiocarbamate	Fn	12122-67-7	[[1,2-ethanediylbis[carbamodithioato]](-2-)]zinc complex
Ziram	thiocarbamate	Fn	137-30-4	zinc bis(dimethyldithiocarbamate)

REFERENCES

Adams, C.D., and Randtke, S.J., 1992, Ozonation byproducts of atrazine in synthetic and natural waters: *Environ. Sci. Technol.*, v. 26, no.11, pp. 2218-2227.

Alexander, M., 1981, Biodegradation of chemicals of environmental concern: *Science*, v. 211, no. 4478, pp. 132-138.

Ambrose, R.B., Jr., HIll, S.I., and Mulkey, L.A., 1983, Users manual for the chemical transport and fate model (TOXIWASP), version I: U.S. Environmental Protection Agency Report EPA/600/3-83/005, 95 p.

Andrews, F.L., and Schertz, T.L., 1986, Statistical summary and evaluation of the quality of surface water in the Colorado River basin, Texas, 1973-82 water years: U.S. Geological Survey Water-Resources Investigations Report 85-4181, 97 p.

Andrilenas, P., 1974, Farmers' use of pesticides in 1971—Quantities: U.S. Department of Agriculture, Economic Research Service, Agricultural Economic Report 252, 56 p.

Armstrong, D.E., and Chesters, G.W., 1968, Adsorption catalyzed chemical hydrolysis of atrazine: *Environ. Sci. Technol.*, v. 2, no. 9, pp. 683-689.

Arruda, J.A., Cringan, M.S., Layher, W.G., Kersh, G., and Bever, C., 1988, Pesticides in fish tissue and water from Tuttle Creek Lake, Kansas: *Bull. Environ. Contam. Toxicol.*, v. 41, no. 4, pp. 617-624.

Artman, J., 1994, personal communication, telephone conference, March, Virginia Department of Forestry, Charlottesville, Va.

Aspelin, A.L., 1994, Pesticides industry sales and usage, 1992 and 1993 market estimates: U.S. Environmental Protection Agency, Office of Pesticide Programs, Biological and Economic Analysis Division, Economic Analysis Branch Report 733-K-92-001, 33 p.

Aspelin, A.L., Grube, A.H., and Torla, R., 1992, Pesticide industry sales and usage; 1990 and 1991 market estimates: U.S. Environmental Protection Agency, Office of Pesticide Programs, Biological and Economic Analysis Division, Economic Analysis Branch Report 733K-92-001, 37 p.

Baker, D.B., 1985, Regional water quality impacts of intensive row-crop agriculture: A Lake Erie basin case study: *J. Soil Water Conserv.*, v. 40, no. 1, pp. 125-132.

——1988a, Overview of rural nonpoint pollution in the Lake Erie basin, *in* Logan, T.L., Davidson, J.M., Baker, J.L., and Overcash, M.R., eds., *Effects of conservation tillage on groundwater quality: Nitrates and pesticides*: Lewis Publishers, Chelsea, Mich., pp. 65-91.

——1988b, Sediment, nutrient and pesticide transport in selected lower Great Lakes tributaries: U.S. Environmental Protection Agency, Great Lakes National Program Office Report 1, EPA-905/4-88-001, 225 p.

Baker, D.B., and Richards, R.P., 1990, Transport of soluble pesticides through drainage networks in large agricultural river basins, *in* Kurtz, D.A., ed., *Long range transport of pesticides*: Lewis Publishers, Chelsea, Mich., pp. 241-270.

Baker, J.E., Eisenreich, S.J., Johnson, T.C., and Halfman, B.M., 1985, Chlorinated hydrocarbon cycling in the benthic nepheloid layer of Lake Superior: *Environ. Sci. Technol.*, v. 19, no. 9, pp. 854-861.

Baker, J.L., Johnson, H.P., Borcherding, M.A., and Payne, W., 1979, Nutrient and pesticide movement from field to stream: A field study, *in* Loehr, R.C., Haith, D.A., Walter, M.F., and Martin, C.S., eds., *Best management practices for agriculture and silviculture: Proceedings of the 1978 Cornell Agricultural Waste Management Conference*: Ann Arbor Science, Ann Arbor, Mich., pp. 213-245.

Barbash, J.E., and Resek, E.A., 1996, *Pesticides in ground water: Distribution, trends, and governing factors*: Ann Arbor Press, Chelsea, Mich., Pesticides in the Hydrologic System series, v. 2, 590 p.

Baughman, G.L., and Lassiter, R.R., 1978, Prediction of environmental pollutant concentration, *in* Cairns, J. Jr., Dickson, K.L., and Maki, A.W., eds., Estimating the hazard of chemical substances to aquatic life: American Society for Testing and Materials Special Technical Publication 657, p. 35.

Becker, R.L, Herzfeld, D., Ostlie, K.R., and Stamm-Katovich. E.J., 1989, Pesticides: Surface runoff, leaching, and exposure concerns: University of Minnesota, Minnesota Extension Service Publication AG-BU-3911, 32 p.

Bedding, N.D., Mcintyre, A.E., Perry, R., and Lester, J.N., 1983, Organic contaminants in the aquatic environment: II. behavior and fate in the hydrological cycle: *Sci. Total Environ.*, v. 26, pp. 255-312.

Belluck, D.A., Benjamin, S.L., and Dawson, T., 1991, Groundwater contamination by atrazine and its metabolites: Risk assessment, policy, and legal implications, *in* Somasundaram, L., and Coats, J.R., eds., *Pesticide transformation products: Fate and significance in the environment*: American Chemical Society Symposium series, v. 459, pp. 254-273.

Bengtson, R.L., Southwick, L.M., Willis, G.H., and Carter, C.E., 1990, The influence of subsurface drainage practices on herbicide losses: *Trans. Am. Soc. Agric. Eng.*, v. 33, no. 2, pp. 415-418.

Bennett, D., 1990, Evaluation of the fate of pesticides in water and sediment, *in* Hutson, D.H., and Roberts, T.R., eds., *Environmental fate of pesticides*: John Wiley, Chichester, England, Progress in Pesticide Biochemistry and Toxicology series, v. 7, pp. 149-173.

Bero, A.S., and Gibbs, R.J., 1990, Mechanisms of pollutant transport in the Hudson estuary: *Sci. Total Environ.*, v. 97/98, pp. 9-22.

Berryhill, W.S., Jr., Lanier, A.L., and Smolen, M.D., 1989, The impact of conservation tillage and pesticide use on water quality: research needs, *in* Weigmann, D.L., ed., *Pesticides in terrestrial and aquatic* environments: Proceedings of a national research conference: Virginia Polytechnic Institute and State University, Virginia Water Resources Center, Blacksburg, Va., pp. 397-404.

Bevenue, A., Hylin, J.W., Kawano, Y., and Kelley, T.W., 1972, Organochlorine pesticide residues in water, sediment, algae, and fish, Hawaii—1970-71: *Pest. Monit. J.*, v. 6, no. 1, pp. 56-64.

Biberhofer, J., and Stevens, R., 1987, Organochlorine contaminants in ambient waters of Lake Ontario: Environment Canada, Burlington, Ont., Canada. Inland Waters Directorate scientific series, v. 159, 11 p.

Bicknell, B.R., Donigian, A.S., Barnwell, T.A., 1985, Modeling water quality and the effects of agricultural best management practices in the Iowa River basin: *Water Sci. Technol.*, v. 17, pp. 1141-1153.

Biggar, J.W., and Seiber, J.N., eds., 1987, Fate of pesticides in the environment: Proceedings of a technical seminar: University of California, Division of Agriculture and Natural Resources, Agricultural Experiment Station Publication 3320, 157 p.

Blevins, R.L., Frye, W.W., Baldwin, P.L., and Robertson, S.D., 1990, Tillage effects on sediment and soluble nutrient losses from a Maury silt loam soil: *J. Environ. Qual.*, v. 19, no. 4, pp. 683-686.

Bollag, J.M., and Liu, S.Y., 1990, Biological transformation processes of pesticides, *in* Cheng, H.H., ed., *Pesticides in the soil environment: Processes, impacts, and modeling*: Soil Science Society of America, Madison, Wis., pp. 169-212.

Borthwick, P.W., Duke, T.W., Wilson, A.J., Jr., Lowe, J.I., Patrick J. M., Jr., and Oberheu, J.C., 1973, Accumulation and movement of mirex in selected estuaries of South Carolina, 1969-71: *Pest. Monit. J.*, v. 7, no. 1, pp. 6-26.

Bowmer, K.H., 1987, Herbicides in surface water, *in* Hutson, D.H., and Roberts, T.R., eds., *Herbicides*: Wiley, Chichester, England, Progress in Pesticide Biochemistry and Toxicology series, v. 6, pp. 271-355.

Bowmer, K.H. and Saintly, G.R., 1977, Management of aquatic plants with acrolein: *J. Aquatic Plant Manag.*, v. 15, pp. 40-46.

Bradshaw, J.S., Loveridge, E.L., Rippee, K.P., Peterson, J.L., White, D.A., Barton, J.R., and Fuhriman, D.K., 1972, Seasonal variations in residues of chlorinated hydrocarbon pesticides in the water of the Utah lake drainage system—1970 and 1971: *Pest. Monit. J.*, v. 6, no. 3, pp.166-170.

Breidenbach, A.W., Gunnerson, C.G., Kawahara, F.K., Lichtenberg, J.J., and Green, R.S., 1967, Chlorinated hydrocarbon pesticides in major river basins, 1957-65: *Publ. Hea. Rev.*, v. 82, no. 2, pp. 139-156.

Briggs, J.C., and Feiffer, J.S., 1986, Water quality of Rhode Island streams: U.S. Geological Survey Water-Resources Investigations Report 84-4367, 51 p.

Brodtmann, N.V., Jr., 1976, Continuous analysis of chlorinated hydrocarbon pesticides in the lower Mississippi River: *Bull. Environ Contam. Toxicol.*, v. 15, no. 1, pp. 33-39.

Brown, E., and Nishioka, Y.A., 1967, Pesticides in western streams—A contribution to the national program: *Pest. Monit. J.*, v. 1, no. 2, pp. 38-46.

Bruce, R.R., Harper, L.A., Leonard, R.A., Snyder, W.M., and Thomas, A.W., 1975, A model for runoff of pesticide from small upland watersheds: *J. Environ. Qual.*, v. 4, no. 4, pp. 541-548.

Bryan, G.W., Gibbs, P.E., Hummerstone, L.G., and Burt, G.R., 1986, The decline of the gastropod *Nucella lapillus* around south-west England: Evidence for the effect of tributyltin from antifouling paints: *J. Mar. Biol. Assoc. U.K.*, v 66, pp. 611-640

Buchman, M.F., 1989, A review and summary of trace contaminant data for coastal and estuarine Oregon: U.S. Department of Commerce, National Oceanic and Atmospheric Administration, National Ocean Service, NOAA Technical Memorandum NOS OMA 42, 115 p.

Burkhard, N., and Guth, J.A., 1981, Chemical hydrolysis of 2-chloro-4,6-bis-(alkylamino)-1,3,5-triazine herbicides and their breakdown in soil under the influence of adsorption: *Pest. Sci.*, v. 12, no. 1, pp. 45-52.

Burns, L.A., 1983, Fate of chemcials in aquatic systems: Process models and computer codes, *in* Swann, R.L., and Eschenroeder, A., eds., *Fate of chemicals in the environment: Compartmental and multimedia models for predictions*: American Chemical Society Symposium series, v. 225, pp. 25-40.

Burns, L.A., and Cline, D.M., 1985, Exposure analysis modeling system: Reference manual for EXAMS II: U.S. Environmental Protection Agency, Office of Research and Development, Environmental Research Laboratory Report EPA/600/3-85/038, 84 p.

Burridge, L.E., and Haya, K., 1989, The use of a fugacity model to assess the risk of pesticides to the aquatic environment on Prince Edward Island: *Advan. Environ. Sci. Technol.*, v. 22, pp. 193-203.

Buser, H.-R., 1990, Atrazine and other *s*-triazine herbicides in lakes and in rain in Switzerland: *Environ. Sci. Technol.*, v. 24, no. 7, pp. 1049-1058.

Bushway, R.J., Litten, W., Porter, K., and Wertam, J., 1982, A survey of azinphos methyl and azinphos methyl oxon in water and blueberry samples from Hancock and Washington Counties of Maine: *Bull. Environ. Contam. Toxicol.*, v. 28, no. 3, pp. 341-347.

Butler, D.L., 1987, Pesticide data for selected Wyoming streams, 1976-78: U.S. Geological Survey Water-Resources Investigations Report 83-4127, 41 p.

Butler, D.L., Krueger, R.P., Osmundson, B.C., Thompson, A.L., and McCall, S.K., 1991, Reconnaissance investigation of water quality, bottom sediment, and biota associated with irrigation drainage in the Gunnison and Uncompahgre River basins and at Sweitzer Lake, west-central Colorado, 1988-89: U.S. Geological Survey Water-Resources Investigations Report 91-4103, 99 p.

Butler, M.K., and Arruda, J.A., 1985, Pesticide monitoring in Kansas surface waters: 1973–1984, in Perspectives on nonpoint source pollution: Proceedings of a national conference, Kansas City, Missouri: U.S. Environmental Protection Agency Report EPA 440/5-85-001, pp. 196-200.

Butler, P.A., 1966, Problem of pesticides in estuaries, in A symposium on estuarine fisheries, American Fisheries Society Special Publication 3: Trans. Amer. Fish. Soc., v. 95, no. 4 supp., pp. 110-115.

Buttle, J.M., 1990, Metolachlor transport in surface runoff: J. Environ. Qual., v. 19, no. 3, pp. 531-538.

Bysshe, S.E., 1990, Bioconcentration factor in aquatic organisms, in Lyman, W.J., Reehl, W.F., and Rosenblatt, D.H., 1990, eds., Handbook of chemical property estimation methods: Environmental behavior of organic compounds: American Chemical Society, Washington, D.C., chap. 5.

California Department of Food and Agriculture, 1991, Summary of pesticide use report data annual 1991: California Environmental Protection Agency, Department of Pesticide Regulation, Information Services Branch, Sacramento, California, p. 87-88.

Callahan, M.A., Slimak, M.W., Gabel, N.W., May, I.P., Fowler, C.F., Freed, J.R., Jennings, P., Durfee, R.L., Whitmore, F.C., Maestri, B., Mabey, W.R., Holt, B.R., and Gould, C., 1979 [1980], Water-related environmental fate of 129 priority pollutants, Volume 1: U.S. Environmental Protection Agency, Office of Water and Waste Management, Office of Water Planning and Standards Report EPA-440/4-79-029a, 542 p.

Capel, P.D., 1991, Wet deposition of herbicides in Minnesota, in Mallard, G.E., and Aronson, D.A., eds., U.S. Geological Survey Toxic Substances Hydrology Program—Proceedings of the technical meeting, Monterey, California, March 11-15, 1991: U.S. Geological Survey Water-Resources Investigations Report 91-4034, pp. 334-337.

Capel, P.D. and Eisenreich, S.J., 1990, Relationship between chlorinated hydrocarbons and organic carbon in sediment and porewater: J. Great Lakes Res., v. 16, pp. 245-257.

Capel, P.D., Giger, W., Reichert, P., Wanner, O., 1988, Accidental input of pesticides into the Rhine River: Environ. Sci. Technol., v. 22, pp. 992-997.

Carey, J.H. and Fox, M.E., 1981, Photodegradation of the lampricide 3-trifluoromethyl-4-nitrophenol (TFM): 1. Pathway of the direct photolysis in solution: J. Great Lakes Res., v. 7, pp. 234-241.

Carroll, B.R., Willis, G.H., and Graves, J.B., 1981, Permethrin concentration on cotton plants, persistence in soil, and loss in runoff: J. Environ. Qual., v. 10, no. 4, pp. 497-500.

Carsel, R.F., Mulkey, L.A., Lorber, M.N., and Baskin, L.B., 1985, The pesticide root zone model (PRZM): A procedure for evaluating pesticide leaching threats: Ecolog. Model., v., 30, pp. 49-69.

Casper, V.L., 1967, Galveston Bay Pesticide Study—Water and oyster samples analyzed for pesticide residues following mosquito control program: Pest. Monit. J., v. 1, no. 3, pp. 13-15.

Chapra, S.C., and Boyer, J.M., 1992, Fate of environmental pollutants: Water Environ. Res., v. 64, no. 4, pp. 581-593.

Chiou, C.T., 1990, Roles of organic matter, minerals and moisture in sorption of nonionic compounds and pesticides by soil, *in* MacCarthy, P., Clapp, C.E., Malcolm, R.L., and Bloom, P.R., eds., *Humic substances in soil and crop sciences*: American Society of Agronomy and Soil Science Society of America, Madison, Wis., pp. 111-159.

Ciba-Geigy, 1992a, Historical surface water monitoring for atrazine in the Mississippi River near Baton-Rouge—St. Gabriel, Louisiana: Ciba-Geigy Corporation, Agricultural Division, Environmental and Public Affairs Department, Technical Report 1-92, 17 p.

——1992b, Historical surface water monitoring for atrazine in Chesapeake Bay: Ciba-Geigy Corporation, Agricultural Division, Environmental and Public Affairs Department, Technical Report 3-92, 54 p.

——1992c, A review of historical surface water monitoring for atrazine in eleven states in the central United States: Ciba-Geigy Corporation, Agricultural Division, Environmental and Public Affairs Department, Technical Report 11-92, 84 p.

——1992d, A review of historical surface water monitoring for atrazine in the Mississippi, Missouri, and Ohio Rivers: Ciba-Geigy Corporation, Agricultural Division, Environmental and Public Affairs Department, Technical Report 6-92, 69 p.

——1992e, Best management practices to reduce runoff of pesticides into surface water: A review and analysis of supporting research: Ciba-Geigy Corporation, Agricultural Division, Environmental and Public Affairs Department, Technical Report 9-92, 57 p.

——1992f, A review of historical surface water monitoring for atrazine in Illinois: Ciba-Geigy Corporation, Agricultural Division, Environmental and Public Affairs Department, Technical Report 5-92, 67 p.

——1992g, Herbicides in drinking water sources: A treatment technology overview: Overview: Ciba-Geigy Corporation, Agricultural Division, Environmental and Public Affairs Department, Technical Report 2-92, 4 p.

——1993a, Atrazine and drinking water sources: An exposure assessment for populations using the greater Mississippi River system: Ciba-Geigy Corporation, Agricultural Division, Environmental and Public Affairs Department, Technical Report 2-93, 20 p.

——1993b, Atrazine and drinking water sources: A preliminary exposure assessment for Illinois: Ciba-Geigy Corporation, Agricultural Division, Environmental and Public Affairs Department, Technical Report 3-93, 23 p.

——1994a, A review of historical surface water monitoring for atrazine in Iowa: Ciba-Geigy Corporation, Agricultural Division, Environmental and Public Affairs Department, Technical Report 2-94, 185 p.

——1994b, Atrazine and drinking water sources: A preliminary exposure assessment for Iowa: Ciba-Geigy Corporation, Agricultural Division, Environmental and Public Affairs Department, Technical Report 1-94, 22 p.

Clement, C.R., and Colborn, T., 1992, Herbicides and fungicides: A perspective on potential human exposure, *in* Colborn, T., and Clement, C.R., eds., *Chemically-induced alterations in sexual and functional development: The wildlife/human connection*: Princeton Publishing, Princeton, N.J., Advances in Modern Environmental Toxicology series, v. 21, pp. 347-364.

Colborn, T., and Clement, C.R., eds., 1992, *Chemically-induced alterations in sexual and functional development: The wildlife/human connection*: Princeton Publishing, Princeton, N.J., Advances in Modern Environmental Toxicology series, v. 21, 403 p.

Cole, H., Jr., Barry, D., Frear, D.E.H., and Bradford, A., 1967, DDT levels in fish, streams, stream sediments, and soil before and after DDT aerial spray application for fall cankerworm in northern Pennsylvania: *Bull. Environ. Contam. Toxocol.*, v. 2, no. 3, pp. 127-146.

Cole, R.H., Frederick, R.E., Healy, R.P., and Rolan, R.G., 1983, *NURP (Nationwide Urban Runoff Program) Priority Pollutant Monitoring Project: Summary of findings*: Dalton-Dalton-Newport, Cleveland, Ohio, 149 p.

——1984, Preliminary findings of the Priority Pollutant Monitoring Project of the Nationwide Urban Runoff Program: *J. Water Poll. Control Fed.*, v. 56, no. 7, pp. 898-908.

Cooper, C.M., 1991, Persistent organochlorine and current use insecticide concentrations in major watershed components of Moon Lake, Mississippi, USA: *Arch. Hydrob.*, v. 121, no. 1, pp. 103-113.

Cooper, C.M., Dendy, F.E., McHenry, J.R., and Ritchie, J.C., 1987, Residual pesticide concentrations in Bear Creek, Mississippi, 1976 to 1979: *J. Environ. Qual.*, v. 16, no. 1, pp. 69-72.

Coupe, R.H., Goolsby, D.A., Iverson, J.L., Markovchick, D.J., and Zaugg, S.D., 1995, Pesticide, nutrient, water-discharge and physical-property data for the Mississippi River and some of its tributaries, April 1991-September 1992: U.S. Geological Survey Open-File Report 93-657, 116 p.

Crawford, N.H. and Donigian, A.S., 1973, Pesticide transport and runoff model for agricultural lands: U.S. Environmental Protection Agency, Office of Research and Development Report EPA-660/2-74/013, 211 p.

Crepeau, K.L., Kuivila, K.M., and Domagalski, J.L., 1996, Concentrations of dissolved rice pesticides in the Colusa Basin Drain and Sacramento River, California, 1990-92, *in* Morganwalp, D.W., and Aronson, D.A., eds., U.S. Geological Survey Toxic Substances Hydrology Program—Proceedings of the technical meeting, Colorado Springs, Colorado, September 20-24, 1993: U.S.Geological Survey Water-Resources Investigations Report 94-4015, p. 711-718.

Crossland, N.O., Bennett, D., Wolff, C.J.M., and Swannell, P.R.J., 1986, Evaluation of models used to assess the fate of chemicals in aquatic systems: *Pest. Sci.*, v. 17, pp. 297-304.

Cunningham, P.A., and Myers, L.E., 1986, Dynamics of diflubenzuron (dimilin) concentrations in water and sediment of a supratidal saltmarsh site following repetitive aerial applications for mosquito control: *Environ. Poll. Ser. A*, v. 41, no. 1, pp. 63-88.

Dagley, S., 1983, Biodegradation and biotransformation of pesticides in the earth's carbon cycle, *in* Gunther, F.A., and Gunther, J.D., eds., *Residue reviews: Residues of pesticides and other contaminants in the total environment*: Springer-Verlag, N.Y., pp. 127-137.

Davis, W.P. and Bortone, S.A., 1992, Effects of kraft mill effluent on the sexuality of fishes—An environmental early warning?, *in* Colborn, T. and Clement, C.R., eds., 1992, *Chemically-induced alterations in sexual and functional development: The wildlife/human connection*: Princeton Scientific Publishing, Princeton, N.J., Advances in Modern Environmental Toxicology series, v. 21, pp. 113-127.

Day, K.E., 1991, Pesticide transformation products in surface waters, effects on aquatic biota, *in* Somasundaram, L., and Coats, J.R., eds., *Pesticide transformation products: Fate and significance in the environment*: American Chemical Society Symposium series, v. 459, pp. 217-241.

Decoursey, D.G., 1992, Developing models with more detail: Do more algorithms give more truth?: *Weed Technol.*, v. 6, no. 3, pp. 709-715.

Del Re, A.A.M., Cova, D., Ragozza, L., Rondelli, E., Rossini, L., and Trevisan, M., 1989, Pesticide residues in the Po River watershed—Application of a mathematical model: *Agri. Ecosys. Envir.*, v. 27, pp. 539-553.

DeLeon, I.R., Byrne, C.J., Peuler, E.A., Antoine, S.R., Schaeffer, J., and Murphy, R,C., 1986, Trace organic and heavy metal pollutants in the Mississippi River: *Chemosphere*, v. 15, no. 6, pp. 795-805.

DeMarco, J., Symons, J.M., and Robeck, G.G., 1967, Behavior of synthetic organics in stratified impoundments: *J. Am. Water Works Assoc.*, v. 59, pp. 965-976.

Demoute, J.P., 1989, A brief review of the environmental fate and metabolism of pyrethroids: *Pest. Sci.*, v. 27, no. 4, pp. 375-385.

deNoyelles, F., Kettle, W.D., and Sinn, D.E., 1982, The responses of plankton communities in experimental ponds to atrazine, the most heavily used pesticide in the United States: *Ecology*, v. 63, no. 5, pp. 1285-1293.

Desideri, P., Lepri, L., Santianni, D., and Checchini, L., 1991, Chlorinated pesticides in sea water and pack ice in Terra Nova Bay (Antarctica): *Ann. Chim.*, v. 81, nos. 9-10, pp. 533-540.

Deubert, K.H., and Kaczmarek, G.Z., 1989, Quantitation of nonpoint source pollution associated with cranberry production in Massachusetts, *in* Weigmann, D.L., ed., *Pesticides in terrestrial and aquatic environments: Proceedings of a national research conference*: Virginia Polytechnic Institute and State University, Virginia Water Resources Center, Blacksburg, Va., pp. 214-219.

Dietrich, A.M., Millington, D.S., and Seo, Y.H., 1988, Specific identification of synthetic organic chemicals in river water using liquid-liquid extraction and resin adsorption coupled with electron impact, chemical ionization and accurate mass measurement gas chromatography-mass spectrometry analyses: *J. Chromat.*, v. 436, no. 2, pp. 229-241.

DiToro, D.M., and Horzempa, L.M., 1982, Reversible and reversible components of PCB adsorption-desorption: Isotherms: *Environ. Sci. Technol.*, v. 16, no. 10, pp. 594-602.

DiToro, D.M., O'Conner, D.J., Thomann, R.V., and St. John, J.P., 1982, Modeling hydrophobic organic chemicals in lakes and streams, *in* Dickson, K.L., Cairne, J., Jr., and Maki, A.W., eds., *Modeling the fate of chemicals in the aquatic environment*: Ann Arbor Science Publishers, Ann Arbor, Mich., pp. 165-190.

Domagalski, J.L., and Kuivila, K.M., 1991, Transport and transformation of dissolved rice pesticides in the Sacramento River Delta, California, *in* Mallard, G.E., and Aronson, D.A., eds., U.S. Geological Survey Toxic Substances Hydrology Program—Proceedings of the technical meeting, Monterey, California, March 11-15, 1991: U.S. Geological Survey, Water-Resources Investigations Report 91-4034, pp. 664-666.

Donigian, A.S., Jr., and Carsel, R.J., 1992, Developing computer simulation models for estimating risks of pesticide use: Research vs. user needs: *Weed Technol.*, v. 6, pp. 677- 682.

Donigian, A.S., Jr., and Crawford, N.H., 1976, Modeling pesticides and nutrients on agricultural lands: U.S. Environmental Protection Agency Report EPA-600/2-7-76/043, 318 p.

Donigian, A.S., Jr., Imhoff, J.C., and Bicknell, B.R., 1983, Predicting water quality resulting from agricultural nonpoint source pollution via simulation—HSPF Hydrologic Simulation Program—FORTRAN, *in* Schaller, F.W., and Bailey, G.W., eds., *Agricultural management and water*: Iowa State University Press, Ames, Iowa, pp. 200-249.

Donigian, A.S., Jr., Meier, D.W., and Jowsie, P.P., 1986, Stream transport and agricultural runoff of pesticides for exposure assessmnet—A methodology: U.S. Environmental Protection Agency Report, EPA/600/3-86/011a, variously paged.

Dowd, J.F., Bush, P.B., Neary, D.G., Taylor, J.W., Berisford, Y.C., 1993, Modeling pesticide movement in forested watersheds—Use of PRZM for evaluating pesticide options in loblolly pine stand management: *Environ. Toxicol. Chem.*, v. 12, pp. 429-439.

Duce, R.A., Liss, P.S., Merrill, J.T., Atlas, E.L., Buat-Menard, P., Hicks, B.B., Miller, J.M., Prospero, J.M., Arimoto, R., Church, T.M., Ellis, W., Galloway, J.N., Hansen, L., Jickells, J.D., Knap, A.H., Reinhart, K.H., Schneider, B., Soudine, A., Tokos, J.J., Tsunogai, S., Wollast, R., and Zhou, M., 1991, The atmospheric input of trace species to the world ocean: *Global Biogeochem. Cycles*, v. 5, no. 3, pp. 193-259.

Dudley, D.R., and Karr, J.R., 1980, Pesticides and PCB residues in the Black Creek watershed, Allen County, Indiana—1977–78: *Pest. Monit. J.*, v. 13, no. 4, pp. 155-157.

Dupuy, A.J., Manigold, D.B., and Schulze, J.A., 1970, Biochemical oxygen demand, dissolved oxygen, selected nutrients, and pesticide records of Texas surface waters, 1968: Texas Water Development Board Report 108, 37 p.

Eadie, B.J., and Robbins, J.A., 1987, The role of particulate matter in the movement of contaminants in the Great Lakes, *in* Eisenreich, S.J., and Hites, R.A., eds., *Sources and fates of aquatic pollutants*: American Chemical Society Advances in Chemistry series, v. 219, pp. 319-355.

Eichers, T.R., Andrilenas, P.A., and Anderson, T.W., 1978, Farmers' use of pesticides in 1976: U.S. Department of Agriculture, Economics, Statistics, and Cooperatives Service, Agricultural Economics Report 418, 58 p.

Eichers, T.R., Andrilenas, P., Blake, H., Jenkins, R., and Fox, A., 1970, Quantities of pesticides used by farmers in 1966: U.S. Department of Agriculture, Economic Research Service, Economic Report 179, 61 p.

Eidt, D.C., 1985, Toxicity of *Bacillus thuringiensis* var. *kurstaki* to aquatic insects: *Can. Entom.*, v. 117, pp. 829-837.

Eidt, D.C., Hollebone, J.E., Lockhart, W.L., Kingsbury, P.D., Gadsby, M.C., and Ernst, W.R., 1989, Pesticides in forestry and agriculture—Effects on aquatic habitats: *Advan. Environ. Sci. Technol.*, v. 22, pp. 245-284.

Eisenreich, S.J., Elzerman, A.W., and Armstrong, D.E., 1978, Enrichment of micronutrients, heavy metals, and chlorinated hydrocarbons in wind-generated lake foam: *Environ. Sci. Technol.*, v. 12, no. 4, pp. 413-417.

Eisenreich, S.J., Looney, B.B., and Thornton, J.D., 1981, Airborne organic contaminants in the Great lakes ecosystem: *Environ. Sci. Technol.*, v. 15, no. 1, pp. 30-38.

El-Shaarawi, A.H., Esterby, S.R., Warry, N.D., and Kuntz, K.W., 1985, Evidence of contaminant loading to Lake Ontario from the Niagara River: *Can. J. Fish. Aqua. Sci.*, v. 42, pp. 1278-1289.

Ellis, S.R., 1978, Hydrologic data for urban storm runoff from three localities in the Denver metropolitan area, Colorado: U.S. Geological Survey Open-File Report 78-410, 162 p.

Elzerman, A.W., and Coates, J.T., 1987, Hydrophobic organic compounds on sediments: Equilibria and kinetics of sorption, *in* Eisenreich, S.J., and Hites, R.A., eds., *Sources and fates of aquatic pollutants*: American Chemical Society Advances in Chemistry series, v. 219, pp. 263-317.

Erickson, R.E., and Essig, M.G., 1981, Pesticides in Lower Flathead Valley drainage: Montana State University, Montana Water Resources Research Center Report 123, 18 p.

Evans, M.S., Noguchi, G.E., and Rice, C.P., 1991, The biomagnification of polychlorinated biphenyls, toxaphene, and DDT compounds in a Lake Michigan offshore food web: *Arch. Environ. Contam. Toxicol.*, v. 20, no. 1, pp. 87-93.

Fahey, J.E., Butcher, J.W., and Turner, M.E., 1968, Monitoring the effects of the 1963–64 Japanese Beetle control program on soil, water, and silt in the Battle Creek area of Michigan: *Pest. Monit. J.*, v. 1, no. 4, pp. 30-33.

Faust, S.D., 1977, Chemical mechanisms affecting the fate of organic pollutants in natural aquatic environments, *in* Suffet, I.H., ed., *Fate of Pollutants in the Air and Water Environments*: Wiley, N.Y., Advances in Environmental Sciences and Technology series, v. 8, pp. 317-364.

Felsot, A.S., Mitchell, J.K., and Kenimer, A.L., 1990, Assessment of management practices for reducing pesticide from sloping cropland in Illinois: *J. Environ. Qual.*, v. 19, no. 3, pp. 539-545.

Felsot, A.S., and Pederson, W.L., 1991, Pesticidal activity of degradation products, *in* Somasundaram, L., and Coats, J.R., eds., *Pesticide transformation products: Fate and significance in the environment*: American Chemical Society Symposium series, v. 459, pp. 172-187.

Fischer, S.A., and Hall, L.W., 1992, Environmental concentrations and aquatic toxicity data on diflubenzuron (Dimilin): *Crit. Rev. Toxicol.*, v. 22, no. 1, pp. 45-79.

Fishel, D.K., 1984, Water-quality and chemical loads of the Susquehanna River at Harrisburg, Pennsylvania, April 1980 to March 1981: U.S. Geological Survey Water-Resources Investigations Report 83-4164, 90 p.

Fleck, J., Ross, L., Tran, D., Melvin, J., and Fong, B., 1991, Off-target movement of endosulfan from artichoke fields in Monterey County: California Department of Food and Agriculture, Environmental Hazards Assessment Program Report EH 91-5, 30 p.

Fong, W.G., and Brooks, G.M., 1989, Regulation of chemicals for aquaculture use: *Food Technol.*, v. 23, pp. 88-93.

Forbes, R.B., 1968, Water quality studies, Zellwood drainage and water control district: *Soil Crop Sci. Soc. Florida Proc.*, v. 28, p. 42-48.

Fox, G.A., 1992, Epidemiological and pathobiological evidence of contaminant-induced alterations in sexual development in free-living wildlife, *in* Colborn, T., and Clement, C.R., eds., *Chemically-induced alterations in sexual and functional development: The wildlife/ human connection*: Princeton Publishing, Princeton, N.J., Advances in Modern Environmental Toxicology series, v. 21, pp. 147-158.

Frere, M.H., Onstad, C.A., and Holtan, H.N., 1975, ACTMO: An agricultural chemical transport model: U.S. Department of Agriculture, Agricultural Research Service Report ARS-H-3, 54 p.

Fuhrer, G.J., 1984, Chemical analyses of elutriates, native water, and bottom material from the Chetco, Rogue, and Columbia Rivers in western Oregon: U.S. Geological Survey Open-File Report 84-133, 80 p.

Fuhrer, G.J., and Rinella, F.A., 1983, Analyses of elutriates, native water, and bottom material in selected rivers and estuaries in western Oregon and Washington: U.S. Geological Survey Open File Report 82-922, 157 p.

Fujii, R., 1988, Water-quality and sediment-chemistry data of drain water and evaporation ponds from Tulare Lake Drainage District, Kings County, California, March 1985 to March 1986: U.S. Geological Survey Open-File Report 87-700, 19 p.

Fukuto, T.R., 1987, Organophosphorus and carbamate esters—The anticholinesterase insecticides, *in* Biggar, J.W., and Seiber, J.N., eds., Fate of pesticides in the environment: Proceedings of a technical seminar: University of California, Division of Agriculture and Natural Resources, Agricultural Experiment Station Publication 3320, pp. 5-18.

Funk, W.H., Hindin, E., Moore, B.C., and Wasem, C.R., 1987, Water quality benchmarks in the North Cascades: State of Washington Water Research Center Report 68, 83 p.

Gangstad, E.O., 1982, Dissipation of 2,4-D residues in large reservoirs: *J. Aquatic Plant Manag.*, v. 20, pp. 13-16.

Gentile, I.A., Ferraris, L., and Crespi, S., 1989, The degradation of methyl bromide in some natural fresh waters: Influence of temperature, pH, and light: *Pest. Sci.*, v. 25, pp. 261-272.

Gianessi, L.P., and Puffer, C., 1991, *Herbicide use in the United States*: Resources for the Future, Quality of the Environment Division, Washington, D.C., 128 p.

———1992a, *Insecticide use in U.S. crop production*: Resources for the Future, Quality of the Environment Division, Washington, D.C., 180 p.

——1992b, *Fungicide use in U.S. crop production*: Resources for the Future, Quality of the Environment Division, Washington, D.C., 170 p.

Gibbs, P.E., Pascoe, P.L., and Burt, Gr.R., 1988, Sex change in the female dogwhelk, *Nucella lapillus*, induced by triorganotin from antifouling paints: *J. Mar. Biol. Assoc. U.K.*, v. 68, pp. 715-731.

Gilliom, R.J., 1985, Pesticides in rivers of the United States, *in* National water summary 1984: Hydrologic events, selected water-quality trends, and ground-water resources: U.S. Geological Survey Water-Supply Paper 2275, pp. 85-92.

Gilliom, R.J., Alexander, R.B., and Smith, R.A., 1985, Pesticides in the Nation's rivers, 1975-1980, and implications for future monitoring: U.S. Geological Survey Water-Supply Paper 2271, 26 p.

Gilliom, R.J., Alley, W.M., and Gurtz, M.E., 1995, Design of the National Water-Quality Assessment Program: Occurrence and distribution of water-quality conditions: U.S. Geological Survey Circular 1112, 33 p.

Gilliom, R.J., and Clifton, D.G., 1990, Organochlorine pesticide residues in bed sediments of the San Joaquin River, California: *Water Resour. Bull.*, v. 26, no. 1, pp. 11-24.

Ginn, T.M., and Fisher, F.M., Jr., 1974, Studies on the distribution and flux of pesticides in waterways associated with a ricefield—marshland ecosystem: *Pest. Monit. J.*, v. 8, no. 1, pp. 23-32.

Glooschenko, W.A., Strachan, W.M.J., and Sampson, R.C.J., 1976, Distribution of pesticides and polychlorinated biphenyls in water, sediments, and seston of the upper Great Lakes—1974: *Pest. Monit. J.*, v. 10, no. 2, pp. 61-67.

Glotfelty, D.E., Taylor, A.W., Isensee, A.R., Jersey, J., and Glenn, S., 1984, Atrazine and simazine movement to Wye River estuary: *J. Environ. Qual.*, v. 13, no. 1, pp. 115-121.

Glotfelty, D.E., Williams, G.H., Freeman, H.P., and Leech, M.M., 1990, Regional atmospheric transport and deposition of pesticides in Maryland, *in* Kurtz, D.A., ed., *Long range transport of pesticides*: Lewis Publishers, Chelsea, Mich., pp. 199-221.

Gold, A.J., and Loudon, T.L., 1982, Nutrient, sediment, and herbicide losses in tile drainage under conservation and conventional tillage, in *1982 winter meeting, American Society of Agricultural Engineers, Chicago, Illinois*: American Society of Agricultural Engineers, St. Joseph, Mich., pp. 1-14.

Gonzalez, D., Bisbiglia, M., and Denton, D., 1989, Monitoring study for tributyltin contamination in California lakes, 1987: California Department of Food and Agriculture, Environmental Hazards Assessment Program Report EH 89-7, variously paged.

Goolsby, D.A., and Battaglin, W.A., 1993, Occurrence, distribution and transport of agricultural chemicals in surface waters of the midwestern United States, *in* Goolsby, D.A., Boyer, L.L., and Mallard, G.E., compilers, Selected papers on agricultural chemicals in water resources of the midcontinental United States: U.S. Geological Survey Open-File Report 93-418, pp. 1-25.

Goolsby, D.A., Battaglin, W.A., Fallon, J.D., Aga, D.S., Kolpin, D.W., and Thurman, E.M., 1993, Persistence of herbicides in selected reservoirs in the midwestern United States: Some preliminary results, *in* Goolsby, D.A., Boyer, L.L., and Mallard, G.E., compilers, Selected papers on agricultural chemicals in water resources of the midcontinental United States: U.S. Geological Survey Open-File Report 93-418, pp. 51-63.

Goolsby, D.A., Boyer, L.L., and Battaglin, W.A., 1994, Plan of study to determine the effect of changes in herbicide use on herbicide concentrations in midwestern streams, 1989-94: U.S. Geological Survey Open-File Report 94-347, 14 p.

Goolsby, D.A., Thurman, E.M., and Kolpin, D.W., 1991a, Herbicides in streams—Midwestern United States, in *Irrigation and drainage: Proceedings of the 1991 national conference, Honolula,, Hawaii*, July 22-26, 1991: American Society of Civil Engineers, pp. 17-23.

——1991b, Geographic and temporal distribution of herbicides in surface waters of the upper midwestern United States, 1989-90, *in* Mallard, G.E., and Aronson, D.A., eds., U.S. Geological Survey Toxic Substances Hydrology Program—Proceedings of the technical meeting, Monterey, California, March 11-15, 1991: U.S. Geological Survey Water-Resources Investigations Report 91-4043, pp. 183-188.

Goss, D.W., 1992, Screening procedure for soil and pesticides relative to potential water quality impacts: *Weed Technol.*, v. 6, pp. 701-708.

Graczyk, D.J., 1986, Water quality in the St. Croix National Scenic Riverway, Wisconsin: U.S. Geological Survey Water-Resources Investigations Report 85-4319, 48 p.

Granstrom, M.L., Ahlert, R.C., and Wiesenfeld, J., 1984, The relationships between the pollutants in the sediments and in the water of the Delaware and Raritan Canal: *Water Sci. Technol.*, v. 16, nos. 5-7, pp. 375-380.

Green, R.E., and Karickhoff, S.W., 1990, Sorption estimates for modeling, *in* Cheng, H.H., ed., *Pesticides in the soil environment: Processes, impacts, and modeling*: Soil Science Society of America, Madison, Wis., pp. 79-102.

Green, R.S., Gunnerson, C.G., and Lichtenberg, J.J., 1967, *Pesticides in our National waters, agriculture and the quality of our environment*: American Association for the Advancement of Science, Washington, D.C., pp. 137-156.

Gregor, D.J., 1990, Deposition and accumulation of selected agricultural pesticides in Canadian Arctic snow, *in* Kurtz, D.A., ed., *Long range transport of pesticides*: Lewis Publishers, Chelsea, Mich., pp. 373-386.

Greve, P.A., and Wit, S.L., 1971, Endosulfan in the Rhine River: *J. Water Poll. Control Fed.*, v. 43, no. 12, pp. 2338-2348.

Guillette, L.J.Jr., Gross, T.S., Masson, G.R., Matter, J.M., Percival, H.F., and Woodward, A.R., 1994, Developmental abnormalities of the gonad and abnormal sex hormone concentrations in juvenile alligators from contaminated and control lakes in Florida: *Environ. Health Pers.*, v. 102, no. 8, pp. 680-688.

Gustafson, D.I., 1990, Field calibration of SURFACE: A model of agricultural surface waters: J. *Environ. Sci. Hea. B*, v. 25, no. 5, pp. 665-687

Halfon, E., 1986, Modelling the fate of mirex and lindane in Lake Ontario, off the Niagara River mouth: *Ecolog. Model.*, v. 33, pp. 13-33.

——1987, Modeling of mirex loadings to the bottom sediments of Lake Ontario within the Niagara River plume: *J. Great Lakes Res.*, v. 13, pp. 18-23.

Hall, J.K., Pawlus, M., and Higgins, E.R., 1972, Losses of atrazine in runoff water and soil sediment: *J. Environ. Qual.*, v. 1, no. 2, pp. 172-176.

Hannon, M.R., Greichus, Y.A., Applegate, R., and Fox, A.C., 1970, Ecological distribution of pesticides in Lake Poinsett, South Dakota: *Trans. Amer. Fish. Soc.*, v. 3, pp. 496-500.

Hansen, G.W., Oliver, F.E., and Otto, N.E., 1983, *Herbicide manual: A guide to supervise pest management and to train O&M personnel*: U.S. Department of the Interior, Bureau of Reclamation, Denver, Colo., 346 p.

Hardy, J.T., Crecelius, E.A., Antrim, L.D., Broadhurst, V.L., Apts, C.W., Gurtisen, J.M., and Fortman, T.J., 1987, The sea-surface microlayer of Puget Sound—Part II. Concentrations of contaminants and relation to toxicity: *Mar. Environ. Res.*, v. 23, pp. 251-271.

Hargrave, B.T., Harding, G.C., Vass, W.P., Erickson, P.E., Fowler, B.R., and Scott, V., 1992, Organochlorine pesticides and polychlorinated biphenyls in the Arctic Ocean food web: *Arch. Environ. Contam. Toxicol.*, v. 22, pp. 41-54.

Hargrave, B.T., Vass, W.P., Erickson, P.E., and Fowler, B.R., 1988, Atmospheric transport of organochlorines to the Arctic Ocean: *Tellus*, v. 40B, no. 5, pp. 480-493.

Harrison, S.A., Watschke, T.L., Mumma, R.O., Jarrett, A.R., and Hamilton, G.W., Jr., 1993, Nutrient and pesticide concentrations in water from chemically treated turfgrass, *in* Racke, K.D., and Leslie, A.R., eds., *Pesticides in urban environments: Fate and significance*: American Chemical Society Symposium Series 522, pp. 191-207.

Hawxby, K.W., and Mehta, R., 1979, The fate of aquazine in a small pond: *Proc. Okla. Acad. Sci.*, v. 59, pp. 16-19.

Heinis, L.J., and Knuth, M.L., 1992, The mixing, distribution and persistence of esfenvalerate within littoral enclosures: *Environ. Toxicol. Chem.*, v. 11, no. 1, pp. 11-25.

Hellawell, J.M., 1988, Toxic substances in rivers and streams: *Environ. Poll.*, v. 50, pp. 61-85.

Hendrick, R.D., Bonner, F.L., Everett, T.R., and Fahey, J.E., 1966, Residue studies on aldrin and dieldrin in soils, water, and crawfish from rice fields having insecticide contamination: *J. Econ. Ent.*, v. 59, no. 6, pp. 1388-1391.

Hersh, C.M., and Crumpton, W.G., 1989, Atrazine tolerance of algae isolated from two agricultural streams: *Environ. Toxicol. Chem.*, v. 8, pp. 327-332.

Hester, P.G., Rathburn, C.B., Jr., and Boike, A.H., Jr., 1980, Effects of methoprene on non-target organisms when applied as a mosquito larvicide: *Proc. Fla. Anti-Mosquito Assoc.*, v. 5, pp. 16-20.

Hickman, J.S., Harward, M.E., and Montgomery, M.L., 1983, Herbicides in runoff from agricultural watersheds in a high-winter-rainfall zone: Oregon State University, Water Resources Research Institute Report WRRI 86, 107 p.

Hileman, B., 1993, Concerns broaden over chlorine and chlorinated hydrocarbons: *Chem. Eng. News*, April 19, pp. 11-20.

Hodge, J.E., 1993, Pesticide trends in the professional and consumer markets, *in* Racke, K.D., and Leslie, A.R., eds., *Pesticides in urban environments: Fate and significance*: American Chemical Society Symposium series, v. 522, pp. 11-17.

Hoeppel, R.E., and Westerdahl, H.E., 1983, Dissipation of 2,4-D DMA and BEE from water, mud, and fish at Lake Seminole, Georgia: *Water Resour. Bull.*, v. 19, no. 2, pp. 197-204.

Hoigné, J., Faust, B.C., Haag, W.R., Scully, F.E., Jr., and Zepp, R.G., 1989, Aquatic humic substances as sources and sinks of photochemically produced transient reactants, *in* Suffet, I.H., and MacCarthy, P., eds., *Aquatic humic substances: Influence on fate and treatment of pollutants*: American Chemical Society Advances in Chemistry series, v. 219, pp. 363-381.

Howard, P.H., ed., 1991, *Pesticides*, v. 3 of *Handbook of environmental fate and exposure data for organic chemicals*: Lewis Publishers, Chelsea, Mich., 684 p.

Howard, P.H., Boethling, R.S., Jarvis, W.F., Meylan, W.M., and Michalenko, E.M., 1991, *Handbook of environmental degradation rates*: Lewis Publishers, Chelsea, Mich., 725 p.

Howard, P.H., and Neal, M., 1992, *Dictionary of chemical names and synonyms*: Lewis Publishers, Chelsea, Mich., variously paged.

Huggett, R.J., and Bender, M.E., 1980, Kepone in the James River: *Environ. Sci. Technol.*, v. 14, no. 8, pp. 918-923.

Huggett, R.J., Unger, M.A., Seligman, P.F., and Valkirs, A.O., 1992, The marine biocide tributyltin: Assessing and managing the environmental risks: *Environ. Sci. Technol.*, v. 26, no. 2, pp. 233-237.

Hurlbert, S.H., 1975, Secondary effects of pesticides on aquatic ecosystems: *Residue Rev.*, v. 57, pp. 81-148.

Immerman, F.W., and Drummond, D.J., 1984, National urban pesticide applicator survey: Final report—overview and results: U.S. Environmental Protection Agency, Office of Pesticides Programs, Economic Analysis Branch, variously paged.

Iwata, H., Tanabe, S., Sakai, N., Tatsukawa, R., 1993, Distribution of persistent organochlorines in the oceanic air and surface seawater and the role of ocean on their global transport and fate: *Environ. Sci. Technol.*, v. 27, no. 6, pp. 1080-1098.

Jaffe, P.R., Parker, F.L., and Wilson, D.J., 1982, Distribution of toxic substances in rivers: *J. Environ. Eng., Proc. Am. Soc. Civil Eng.*, v. 108, no. EE4, pp. 639-649.

Johanson, R.C., Imhoff, J.C., and Davis, H.H., Jr., 1980, Users manual for hydrological simulation program—FORTRAN (HSPF): U.S. Environmental Protection Agency, Office of Research and Development, Environmental Research Laboratory Report EPA-600/9-80-015, 678 p.

Johnsen, T.N., Jr., and Warskow, W.L., 1980, Picloram dissipation in a small southwestern stream: *Weed Sci.*, v. 28, no. 5, pp. 612-615.

Johnson, A., Norton, D., and Yake, B., 1988, Persistence of DDT in the Yakima River drainage, Washington: *Arch. Environ. Contam. Toxicol.*, v. 17, pp. 289-297.

Johnson, B.T., 1986, Potential impact of selected agricultural chemical contaminants on a northern prairie wetland: A Microcosm Evaluation: *Environ. Toxicol. Chem.*, v. 5, pp. 473-485.

Johnson, H.E., and Ball, R.C., 1972, Organic pesticide pollution in an aquatic environment, *in* Faust, S.D., ed., *Fate of organic pesticides in the aquatic environment*: American Chemical Society Advances in Chemistry series, v. 111, pp. 1-10.

Johnson, L.G., and Morris, R.L., 1971, Chlorinated hydrocarbon pesticides in Iowa rivers: *Pest. Monit. J.*, v. 4, no. 4, pp. 216-219.

Johnson, S., 1988, 1988 Forestry chemical update, *in* Tomascheski, J.H., ed., *Proceedings, tenth annual Forest Vegetation Management Conference, Eureka, California*: Forest Vegetation Management Conference, pp. 116-119.

Johnson, W.D., Lee, G.F., and Spyridakis, D., 1966, Persistence of toxaphene in treated lakes: *Air and Water Poll.*, v. 10, no. 8, p. 555-560.

Joyce, J.C., and Sikka, H.C., 1977, Residual 2,4-D levels in the St. Johns River, Florida: *J. Aquatic Plant Manag.*, v. 15, pp. 76-82.

Junk, G.A., Richard, J.J., Svec, H.J., and Fritz, J.S., 1976, Simplified resin sorption for measuring selected contaminants: *J. Am. Water Works Assoc.*, v. 68, no. 4, pp. 218-222.

Kaiser, K.L.E., 1974, Mirex—An unrecognized contaminant of fishes from Lake Ontario: *Science*, v. 185, pp. 523-525.

Kaiser, K.L.E., Comba, M.E., Hunter, H., Maguire, R.J., Tkacz, R.J., and Platford, R.F., 1985, Trace organic contaminants in the Detroit River: *J. Great Lakes Res.*, v. 11, no. 3, pp. 386-399.

Kard, B.M. and McDaniel, C.A., 1993, Field evaluation of the persistence and efficacy of pesticides used for termite control, *in* Racke, K.D., and Leslie, A.R., eds., *Pesticides in urban environments: Fate and significance*: American Chemical Society Symposium series, v. 522, pp. 46-61.

Karickhoff, S.W., 1984, Organic pollutant sorption in aquatic systems: *J. Hydraul. Eng.*, v. 110, no. 6, pp. 707-735.

Karickhoff, S.W., and Morris, K.R., 1985, Sorption dynamics of hydrophobic pollutants in sediment suspensions: *Environ. Toxicol. Chem.*, v. 4, no. 3, pp. 469-479.

Kauss, P.B., 1983, Studies of trace contaminants, nutrients, and bacteria levels in the Niagara River: *J. Great Lakes Res.*, v. 9, no. 2, pp. 249-273.

Kellogg, R.L., and Bulkley, R.V., 1976, Seasonal concentrations of dieldrin in water, channel catfish, and catfish-food organisms, Des Moines River, Iowa—1971-73: *Pest. Monit. J.*, v. 9, no. 4, pp. 186-194.

Kelly, T.J., Czuczwa, J.M., Sticksel, P.R., Sverdrup, G.M., Koval, P.J., and Hodanbosi, R.F., 1991, Atmospheric and tributary inputs of toxic substances to Lake Erie: *J. Great Lakes Res.*, v. 17, no. 4, pp. 504-516.

Kenaga, E.E., and Goring, C.A.I., 1980, Relationship between water solubility, soil-sorption, octanol-water partioning, and bioconcentration of chemicals in biota, *in* Eaton, J.G., Parrish, P.R., and Hendricks, A.C., eds., *Aquatic toxicology: Proceedings of the Third Annual Symposium on Aquatic Toxicology*: American Society for Testing and Materials Special Technical Publication 707, pp. 78-115.

Kent, J.C., and Johnson, D.W., 1979, Organochlorine residues in fish, water, and sediment of American Falls Reservoir, Idaho, 1974: *Pest. Monit. J.*, v. 13, no. 1, pp. 28-34.

Kent, R.A., 1991, Canadian water quality guidelines for metolachlor: Environment Canada, Inland Waters Directorate, Water Quality Branch Scientific Series 184, 34 p.

Kent, R.A., and Pauli, B.D., 1991, Canadian water quality guidelines for captan: Environment Canada, Inland Waters Directorate, Water Quality Branch Scientific Series 188, 23 p.

Kent, R.A., Pauli, B.D., and Caux, P.-Y., 1991, Canadian water quality guidelines for dinoseb: Environment Canada, Inland Waters Directorate, Water Quality Branch Scientific Series 189, 35 p.

Kent, R.A., Trotter, D.M., and Gareau, J., 1992, Canadian water quality guidelines for triallate: Environment Canada, Ecosystem Sciences and Evaluation Directorate, Eco-Health Branch Scientific Series 195, 47 p.

Klaassen, H.E., and Kadoum, A.M., 1975, Insecticide residues in the Tuttle Creek Reservoir ecosystem, Kansas—1970-71: *Pest. Monit. J.*, v. 9, no. 2, pp. 89-93.

Klaseus, T.G., Buzicky, G.C., and Schneider, E.C., 1988, Pesticides and groundwater: surveys of selected Minnesota wells: Minnesota Department of Health and Minnesota Department of Agriculture report prepared for the Legislative Commission on Minnesota Resources, 95 p.

Knapton, J.R., Jones, W.E., and Sutphin, J.W., 1988, Reconnaissance investigation of water quality, bottom sediment, and biota associated with irrigation drainage in the Sun River area, west-central Montana, 1986-87: U.S. Geological Survey Water-Resources Investigations Report 87-4244, 78 p.

Knisel, W.G., 1980, CREAMS: A field-scale model for chemcials, runoff, and erosion from agricultural management systems: U.S. Department of Agriculture, USDA-SEA Conservation Research Report 26, 653 p.

Knutson, H., Kadoum, A.M., Hopkins, T.L., Swoyer, G.F., and Harvey, T.L., 1971, Insecticide usage and residues in a newly developed Great Plains irrigation district: *Pest. Monit. J.*, v. 5, no. 1, pp. 17-27.

Kolpin, D.W., and Kalkhoff, S.J., 1993, Atrazine degradation in a small stream in Iowa: *Environ. Sci. Technol.*, v. 27, no. 1, pp. 134-139.

Krämer, W., and Ballschmiter, K., 1988, Content and pattern of polychloro-cyclohexanes (HCH) and -biphenyls (PCB), and content of hexachlorobenzene in the water column of the Atlantic Ocean: *Fres. Zeit. fur Anal. Chem.*, v. 330, pp. 524-526.

Kreutzweiser, D.P., Holmes, S.B., Capell, S.S., and Eichenberg, D.C., 1992, Lethal and sublethal effects of *Bacillus thuringiensis* var. *kurstaki* on aquatic insects in laboratory bioassays and outdoor stream channels: *Bull. Environ. Contam. Toxicol.*, v. 49, pp. 252-258.

Kreutzweiser, D.P., Kingsbury, P.D., and Feng, J.C., 1989, Drift response of stream invertebrates to aerial applications of glyphosate: *Bull. Environ. Contam. Toxicol.*, v. 42, pp. 331-338.

Krieger, K.A., Baker, D.B., and Kramer, J.W., 1988, Effects of herbicides on stream aufwuchs productivity and nutrient uptake: *Arch. Environ. Contam. Toxicol.*, v. 17, pp. 299-306.

Kuivila, K.M., and Foe, C.G., 1995, Concentrations, transport, and biological effects of dormant spray pesticides in the San Francisco Estuary, California: *J. Environ. Toxicol. Chem.*, v. 14, no. 7, pp. 1141-1150.

Kuntz, K.W., and Warry, N.D., 1983, Chlorinated organic contaminants in water and suspended sediments of the lower Niagara River: *J. Great Lakes Res.*, v. 9, no. 2, pp. 241-248.

Kurtz, D.A., 1978, Residues of polychlorinated biphenyls, DDT, and DDT metabolites in Pennsylvania streams, community water supplies, and reservoirs, 1974-76: *Pest. Monit. J.*, v. 11, no. 4, pp. 190-198.

Kutz, F.W., and Carey, A.E., 1986, Pesticides and toxic substances in the environment: *J. Arboricul.*, v. 12, no. 4, pp. 92-95.

Lambing, J.H., Jones, W.E., and Sutphin, J.W., 1988, Reconnaissance investigation of water quality, bottom sediment, and biota assocated with irrigation drainage in Bowdoin National Wildlife Refuge and adjacent areas of the Milk River basin, northeastern Montana, 1986-87: U.S. Geological Survey Water-Resources Investigations Report 87-4243, 71 p.

Larson, S.J., Capel, P.D., Goolsby, D.A., Zaugg, S.D., and Sandstrom, M.W., 1995, Relations between pesticide use and riverine flux in the Mississippi River basin: *Chemosphere*, v. 31, no. 5, pp. 3305-3321.

Lau, Y.L., Oliver, B.G., and Krishnappan, B.G., 1989, Transport of some chlorinated contaminants by the water, suspended sediments, and bed sediments in the St. Clair and Detroit Rivers: *Environ. Toxicol. Chem.*, v. 8, no. 4, pp. 293-301.

Lavy, T.L., Mattice, J.D., and Kochenderfer, J.N., 1989, Hexazinone persistence and mobility of a steep forested watershed: *J. Environ. Qual.*, v. 18, no. 4, pp. 507-514.

LeBel, G.L., Williams, D.T., and Benoit, F.M., 1987, Use of large-volume resin cartridges for the determination of organic contaminants in drinking water derived from the Great Lakes, *in* Suffet, I.H., and Malaiyandi, M., eds. *Organic pollutants in water: Sampling, analysis, and toxicity testing*: American Chemical Society Advances in Chemistry series, v.. 214, pp. 309-325.

Leonard, R.A., 1988, Herbicides in surface waters, *in* Grover, R., ed., *Environmental chemistry of herbicides*: CRC Press, Boca Raton, Fla., v. 1, pp. 45-88.

——1990, Movement of pesticide into surface waters, *in* Cheng, H.H., ed., *Pesticides in the soil environment: Processes, impacts, and modeling*: Soil Science Society of America, Madison, Wis., pp. 303-349.

Leonard, R.A., Knisel, W.G., and Still, D.A., 1987, GLEAMS: Groundwater loading effects of agricultural management systems: *Trans. Am. Soc. Agric. Eng.*, v. 30, pp. 1403-1418.

Leonard, R.A., Langdale, G.W., and Fleming, W.G., 1979, Herbicide runoff from upland Piedmont watersheds—Data and implications for modeling pesticide transport: *J. Environ. Qual.*, v. 8, no. 2, pp. 223-229.

Leung, S.-Y.T., Bulkley, R.V., and Richard, J.J., 1981, Persistence of dieldrin in water and channel catfish from the Des Moines River, Iowa, [USA] 1971-73 and 1978: *Pest. Monit. J.*, v. 15, no. 2, pp. 98-102.

——1982, Pesticide accumulation in a new impoundment in Iowa: *Water Resour. Bull.*, v. 18, no. 3, pp. 485-93.

Lewis, D.L., Freeman, L.F., III, and Watwood, M.E., 1986, Seasonal effects on microbial transformation rates of an herbicide in a freshwater stream—Application of laboratory data to a field site: *Environ. Toxicol. Chem.*, v. 5, pp. 791-796.

Lewis, M.A., 1986, Impact of a municipal wastewater effluent on water quality, periphyton, and invertebrates in the Little Miami River near Xenia, Ohio: *Ohio J. Sci.*, v. 86, no. 1, pp. 2-8.

Lewis, M.E., Garrett, J.W., and Hoos, A.B., 1992, Nonpoint-source pollutant discharges of the three major tributaries to Reelfoot Lake, west Tennessee, October 1987 through September 1989: U.S. Geological Survey Water-Resources Investigations Report 91-4031, 24 p.

Lichtenberg, J.J., Eichelberger, J.W., Dressman, R.C., and Longbottom, J.E., 1970, Pesticides in surface waters of the United States—A 5-year summary, 1964-68: *Pest. Monit. J.*, v. 4, no. 2, pp. 71-86.

Lick, W., 1982, The transport of contaminants in the Great Lakes: *Ann. Rev. Earth and Planet. Sci.*, v. 10, pp. 327-353.

Lietman, P.L., Ward, J.R., and Behrendt, T.E., 1983, Effects of specific land uses on nonpoint sources of suspended sediment, nutrients, and herbicides—Pequea Creek basin, Pennsylvania, 1979-80: U.S. Geological Survey Water-Resources Investigations Report 83-4113, 96 p.

Lin, J.C., and Graney, R.L., 1992, Combining computer simulation with physical simulation: An attempt to validate turf runoff models: *Weed Technol.*, v. 6, pp. 688-695.

Lockerbie, D.M., and Clair, T.A., 1988, Organic contaminants in isolated lakes of southern Labrador, Canada: *Bull. Environ. Contam. Toxicol.*, v. 41, no. 4, pp. 625-32.

Logan, T.J., 1987, Diffuse (non-point) source loading of chemicals to Lake Erie: *J. Great Lakes Res.*, v. 13, no. 4, pp. 649-658.

Lowe, J.A., Farrow, D.R.G., Pait, A.S., Arenstam, S.J., and Lavan, E.F., 1991, *Fish kills in coastal waters 1980–1989*: National Oceanic and Atmospheric Administration, National Ocean Service, Office of Ocean Resources Conservation and Assessment, Strategic Environmental Assessments Division, [Washington, D.C.], 69 p.

Lum, K.R., Kaiser, K.L.E., and Comba, M.E., 1987, Export of mirex from Lake Ontario to the St. Lawrence estuary: *Sci. Total Environ.*, v. 67, no. 1, pp. 41-51.

Lunsford, C.A., 1981, Kepone distribution in the water column of the James River estuary—1976-78: *Pest. Monit. J.*, v. 14, no. 4, pp. 119-124.

Lurry, D.L., 1983, Analyses of native water, bottom material, elutriate samples, and dredged material from selected southern Louisiana waterways and selected areas in the Gulf of Mexico, 1979-81: U.S. Geological Survey Open-File Report 82-690, 113 p.

Lykins, B.W., Koffskey, W., and Miller, R.G., 1987, Organic contaminant control: Pilot scale studies at Jefferson Parish, Louisiana, *in* Huck, P.M., and Toft, P., eds., *Treatment of drinking water for organic contaminants: proceedings of the Second National Conference on Drinking Water, Edmonton, Canada, April 7 and 8, 1986*: Pergamon Press, N.Y., pp. 263-282.

Lym, R.G., and Messersmith, C.G., 1987, Survey for picloram in North Dakota water: North Dakota Water Resources Research Institute, Research Project Technical Completion Report, Project 02-01, 36 p.

Lyman, W.F., 1990, Adsorption coefficient for soils and sediments, *in* Lyman, W.F., Reehl, W.F., and Rosenblatt, D.H., eds., *Handbook of chemical property estimation methods: Environmental behavior of organic compounds*: American Chemical Society, Washington, D.C., chap. 4.

Lyman, W.F., Reehl, W.F., and Rosenblatt, D.H., eds., 1990, *Handbook of chemical property estimation methods: Environmental behavior of organic compounds*: American Chemical Society, Washington, D.C., variously paged.

Lynch, T.R., Johnson, H.E., and Adams, W.J., 1985, Impact of atrazine and hexachlorobiphenyl on the structure and function of model stream ecosystems: *Environ. Toxicol. Chem.*, v. 4, pp. 399-413.

Mabey, M., and Mill, T., 1978, Critical review of hydrolysis of organic compounds in water under environmental conditions: *J. Phys. Chem. Ref. Data*, v. 7, no. 2, pp. 383-425.

Macalady, D.L., Tratnyek, P.G., and Grundl, T.J., 1986, Abiotic reduction reactions of anthropogenic organic chemicals in anaerobic systems: A critical review: *J. Contam. Hydrol.*, v. 1, no. 1, pp. 1-28.

Mack, G.L., Corcoran, S.M., Gibbs, S.D., Gutenmann, W.H., Reckahn, J.A., and Lisk, D.J., 1964, The DDT content of some fishes and surface waters of New York State: *N.Y. Fish Game J.*, v. 11, no. 2, pp. 148-153.

Mackay, D., 1991, *Multimedia environmental models: The fugacity approach*: Lewis Publishers, Chelsea, Mich., Toxicology and Environmental Health series, 257 p.

Mackay, D., Paterson, S., and Shiu, W.Y., 1992, Generic models for evaluating the regional fate of chemicals: *Chemosphere*, v. 24, pp. 695-717.

Madhun, Y.A., and Freed, V.H., 1990, Impact of pesticides on the environment, *in* Cheng, H.H., ed., *Pesticides in the soil environment: Processes, impacts, and modeling*: Soil Science Society of America, Madison, Wis., pp. 429-466.

Majewski, M.S., and Capel, P.D., 1995, *Pesticides in the atmosphere: Distribution, trends, and governing factors*: Ann Arbor Press, Chelsea, Mich., Pesticides in the Hydrologic System series, v. 1, 214 p.

Manigold, D.B., and Schulze, J.A., 1969, Pesticides in water: Pesticides in selected western streams—a progress report: *Pest. Monit. J.*, v. 3, no. 2, pp. 124-135.

Marston, R.B., Schults, D.W., Shiroyama, T., and Snyder, L.V., 1968, Pesticides in water: Amitrole concentrations in creek waters downstream from an aerially sprayed watershed sub-basin: *Pest. Monit. J.*, v. 2, no. 3, pp. 123-128.

Marston, R.B., Tyo, R.M., and Middendorff, S.C., 1969, Endrin in water from treated Douglas Fir seed: *Pest. Monit. J.*, v. 2, no. 4, pp. 167-171.

Mattraw, H.C., Jr., 1975, Occurrence of chlorinated hydrocarbon insecticides, Southern Florida—1968-72: *Pest. Monit. J.*, v. 9, no. 2, pp. 106-114.

Mauck, W.L., Mayer, F.L., Jr., and Holz, D.D., 1976, Simazine residue dynamics in small ponds: *Bull. Environ. Contam. Toxicol.*, v. 16, no. 1, pp. 1-8.

Mayack, D.T., Bush, P.B., Neary, D.G., and Douglass, J.E., 1982, Impact of hexazinone on invertebrates after application to forested watersheds: *Arch. Environ. Contam. Toxicol.*, v. 11, no. 2, pp. 209-217.

Mayer, J.R., and Elkins, N.R., 1990, Potential for agricultural pesticide runoff to a Puget Sound estuary: Padilla Bay, Washington: *Bull. Environ. Contam. Toxicol.*, v. 45, no. 2, pp. 215-222.

Mayeux, H.S., Jr., Richardson, C.W., Bovey, R.W., Burnett, E., Merkle, M.G., and Meyer, R.E., 1984, Dissipation of picloram in storm runoff: *J. Environ. Qual.*, v. 13, no. 1, pp. 44-49.

McDowell, L.L., Willis, G.H., Murphree, C.E., and Southwick, S.S., 1981, Toxaphene and sediment yields in runoff from a Mississippi watershed: *J. Environ. Qual.*, v. 10, no. 1, p. 120-125.

McFall, J.A., Antoine, S.R., and DeLeon, I.R., 1985, Organics in the water column of Lake Pontchartrain: *Chemosphere*, v. 14, no. 9, pp. 1253-1265.

McKinley, R.S., Arron, G.P., 1987, Distribution of 2,4-D and picloram residues in environmental components adjacent to a treated right-of-way: Canada Department of Energy and Mines Research Report OH/R-87/49/K, 33 p.

Meister Publishing Company, 1995, *Farm chemicals handbook '95*: Meister Publishing, Willoughby, Ohio, variously paged.

Menon, A.S., and De Mestral, J., 1985, Survival of *Bacillus Thuringiensis* var. Kurstaki in waters: *Water, Air, Soil Poll.*, v. 25, pp. 265-274.

Merriman, J.C., 1988, Distribution of organic contaminants in water and suspended solids of the Rainy River: *Water Poll. Res. J. Can.*, v. 23, no. 4, pp. 590-600.

Metcalf, R.L., 1977, Biological fate and transformation of pollutants in water, *in* Suffet, I.H., ed., *Fate of Pollutants in the Air and Water Environments*: Wiley, N.Y., Advances in Environmental Sciences and Technology series, v. 8, pp. 195-221.

Metropolitan Mosquito Control District, 1993, *Annual operations report 1993*: Metropolitan Mosquito Control District, [St. Paul, Minnesota], 91 p.

Michael, J.L., and Neary, D.G., 1993, Herbicide dissipation studies in southern forest ecosystems: *Environ. Toxicol. Chem.*, v. 12, no. 3, pp. 405-410.

Michael, J.L., Neary, D.G., and Wells, M.J.M., 1989, Picloram movement in soil solution and streamflow from a coastal plain forest: *J. Environ. Qual.*, v. 18, no. 1, pp. 89-95.

Miller, J.H., and Bace, A.C., 1980, Streamwater contamination after aerial application of a pelletized herbicide: U.S. Department of Agriculture, Southern Forest Experiment Station (New Orleans, La.), Forest Service Research Note SO-255, 4 p.

Milne, G.W.A., 1995, *CRC handbook of pesticides*: CRC Press, Boca Raton, Fla. 402 p.

Miltner, R.J., Baker, D.B., Speth, T.F., and Fronk, C.A., 1989, Treatment of seasonal pesticides in surface waters: *J. Am. Water Works Assoc.*, v. 81, no. 1, pp. 43-52.

Minnesota Environmental Quality Board, 1992, *Silvicultural systems: A background paper for a generic environmental impact statement on timber harvesting and forest management in Minnesota*: Jaakko Poyry Consulting, St. Paul, Minn., table 4.1.

Mischke, T., Brunetti, K., Acosta, V., Weaver, D., and Brown, M, 1985, Agricultural sources of DDT residues in California's environment: California Department of Food and Agriculture, Environmental Hazards Assessment Program report prepared in response to House Resolution 53 (1984), 42 p.

Miyamoto, J., Mikami, N., and Takimoto, Y., 1990, The fate of pesticides in aquatic ecosystems, *in* Hutson, D.H., and Roberts, T.R., eds., *Environmental fate of pesticides*: John Wiley, Chichester, England, Progress in Pesticide Biochemistry and Toxicology series, v. 7, pp. 123-147.

Moody, J.A., and Goolsby, D.A., 1993, Spatial variability of triazine herbicides in the lower Mississippi River: *Environ. Sci. Technol.*, v. 27, no. 10, pp. 2120-2126.

Moore, D.E., Hall, J.D., and Hug, W.L., 1974, Endrin in forest streams after aerial seeding with endrin-coated Douglas-Fir seed: U.S. Department of Agriculture, Pacific Northwest Forest and Range Experiment Station, Forest Service Research Note PNW-219, 14 p.

Moore, D.R.G., 1992, Canadian water quality guidelines for organotins: Environment Canada, Ecosystem Sciences and Evaluation Directorate, Eco-Health Branch Scientific Series 191, 145 p.

Morris, K., and Jarman, R., 1981, Evaluation of water quality during herbicide applications to Kerr Lake, Oklahoma [USA]: *J. Aquatic Plant Manag.*, v. 19, pp. 15-18.

Morris, R.L., Johnson, L.G., and Ebert, D.W., 1972, Pesticides and heavy metals in the aquatic environment: *Health Lab. Sci.*, v. 9, no. 2, pp. 145-151.

Moyer, L., and Cross, J., 1990, Pesticide monitoring—Illinois EPA's summary of results 1985—1989: Illinois Environmental Protection Agency, Division of Water Pollution Control Report IEPA/WPC/90-297, 143 p.

Mudambi, A.R., Hassett, J.P., McDowell, W.H., and Scrudato, R.J., 1992, Mirex-photomirex relationships in Lake Ontario: *J. Great Lakes Res.*, v. 18, no. 3, pp. 405-14.

Muir, D.C.G.. 1991, Dissipation and transformation in water and sediment, in Grover, R., ed., *Environmental chemistry of herbicides*: CRC Press, Boca Raton, Fla., v. 2, pp. 3-88.

Muir, D.C.G., Grift, N.P., Ford, C.A., Reiger, A.W., Hendzel, M.R., and Lockhart, W.L., 1990, Evidence for long-range transport of toxaphene to remote arctic and subarctic waters from monitoring of fish tissue, *in* Kurtz, D.A., ed., *Long range transport of pesticides*: Lewis Publishers, Chelsea, Mich., pp. 329-346.

Munkittrick, K.R., Van Der Kraak, G.J., McMaster, M.E., and Portt, C.B., 1992, Response of hepatic MFO activity and plasma sex steroids to secondary treatment of bleached kraft mill effluent and mill shutdown: *Environ. Toxicol. Chem.*, v. 11, pp. 1427-1439.

Murphy, T.J., 1984, Atmospheric inputs of chlorinated hydrocarbons to the Great Lakes, *in* Nriagu, J.O., and Simmins, M.S., eds., *Toxic contaminants in the Great Lakes*: Wiley, N.Y., Advances in Environmental Science and Technology series, v. 14, pp. 53-79.

Murphy, K.J., and Barrett, P.R.F., 1990, Chemical control of aquatic weeds, *in* Pieterse, A.H., and Murphy, K.J., eds., *Aquatic weeds: The ecology and management of nuisance aquatic vegetation*: Oxford University Press, Oxford, England, pp. 136-173.

Murray, H.E., Ray, L.E., and Giam, C.S., 1981, Phthalic acid esters, total DDTs, and polychlorinated biphenyls in marine samples from Galveston Bay, Texas: *Bull. Environ. Contam. Toxicol.*, v. 26, no. 6, pp. 769-774.

Neary, D.G., Bush, P.B., and Douglass, J.E., 1983, Off-site movement of hexazinone in stormflow and baseflow from forested watersheds: *Weed Sci.*, v. 31, pp. 543-551.

Neary, D.G., Bush, P.B., Douglass, J.E., and Todd, R.L., 1985, Picloram movement in an Appalachian hardwood forest watershed: *J. Environ. Qual.*, v. 14, no. 4, pp. 585-592.

Neary, D.G., Bush, P.B., and Grant, M.A., 1986, Water quality of ephemeral forest streams after site preparation with the herbicide hexazinone: *Forest Ecol. Manag.*, v. 14, no. 1, pp. 23-40.

Neary, D.G., Bush, P.B., and Michael, J.L., 1993, Fate, dissipation and environmental effects of pesticides in southern forests: A review of a decade of research: *Environ. Toxicol. Chem.*, v. 12, no. 3, pp. 411-428.

Neary, D.G., Douglass, J.E., and Fox, W., 1979, Low picloram concentrations in streamflow resulting from forest application of Tordon-10K, in *Proceedings, thirty-second annual meeting, Southern Weed Science Society*: Southern Weed Science Society, [Raleigh, N.C.], v. 32, pp. 182-197.

Neary, D.G., and Michael, J.L., 1989, Effect of sulfometuron methyl on ground water and stream quality in coastal plain forest watersheds: *Water Resour. Bull.*, v. 25, no. 3, pp. 617-623.

Neely, W.B. and Mackay, D., 1982, Evaluative model for estimating environmental fate, *in* Dickson, K.L., Maki, A.W., and Cairns, J., Jr., eds., *Modeling the fate of chemicals in the aquatic environment*: Ann Arbor Science, Ann Arbor, Mich., pp. 127-144.

Nicholson, H.P., Grzenda, A.R., Lauer, G.J., Cox, W.S., and Teasley, J.I., 1964, Water Pollution by Insecticides in an Agricultural River Basin. I. Occurrence of Insecticides in River and Treated Municipal Water: *Limn. Ocean.*, v. 9, pp. 310-314.

Nicosia, S., Carr, N., and Gonzales, D.A., 1990, Off-field movement of soil-incorporated carbofuran from three commercial rice fields and potential discharge in agricultural runoff water: California Department of Food and Agriculture, Environmental Hazards Assessment Program Report EH 90-4, [80] p.

Nicosia, S., Carr, N., Gonzales, D.A., and Orr, M.K., 1991a, Off-field movement and dissipation of soil-incorporated carbofuran from three commercial rice fields: *J. Environ. Qual.*, v. 20, no. 3, pp. 532-539.

Nicosia, S., Collison, C., and Lee, P., 1991b, Bensulfuron methyl dissipation in California rice fields, and residue levels in agricultural drains and the Sacramento River: *Bull. Environ. Contam. Toxicol.*, v. 47, no. 1, pp. 131-137.

Norris, L.A., 1981, Phenoxy herbicides and TCDD in forests: *Residue Rev.*, v. 80, pp. 65-135.

Norris, L.A., and Montgomery, M.L., 1975, Dicamba residues in streams after forest spraying: *Bull. Environ. Contam. Toxicol.*, v. 13, no. 1, pp. 1-8.

Norris, L.A., Montgomery, M.L., Warren, L.E., and Mosher, W.D., 1982, Brush control with herbicides on hill pasture sites in southern Oregon: *J. Range Manag.*, v. 35, no. 1, pp. 75-80.

Nowell, L.H., 1996, personal communication, manuscript in preparation: U.S. Geological Survey, Sacramento, Calif.

Nowell, L.H., and Resek, E.A., 1994, National standards and guidelines for pesticides in water, sediment, and aquatic organisms: Application to water-quality assessments: *Rev. Environ. Contam. Toxicol.*, v. 140, pp. 1-164.

Nutter, W.L., Knisel, W.G., Jr., Bush, P.B., and Taylor, J.W., 1993, Use of GLEAMS to predict pesticide losses from pine seed orchards: *Environ. Toxicol. Chem.*, v. 12, pp. 441-452.

Nutter, W.L., Tkacs, T., Bush, P.B., and Neary, D.G., 1984, Simulation of herbicide concentrations in stormflow from forested watersheds: *Water Resour. Bull.*, v. 20, no. 6, pp. 851-857.

O'Connor, D.J., 1988, Models of sorptive toxic substances in freshwater systems. II: Lakes and Reservoirs: *J. Environ. Eng.*, v. 114, pp. 533-551.

O'Connor, D.J., Mueller, J.A., and Farley, K.J., 1983, Distribution of Kepone in the James River estuary: *J. Environ. Eng.*, v. 109, no. 2, pp. 396-413.

Oliver, B.G., and Charlton, M.N., 1984, Chlorinated organic contaminants on settling particulates in the Niagara River vicinity of Lake Ontario [Canada]: *Environ. Sci. Technol.*, v. 18, no. 12, pp. 903-908.

Oliver, B.G., and Kaiser, K.L.E., 1986, Chlorinated organics in nearshore waters and tributaries of the St. Clair River: *Water Poll. Res. J. Can.*, v. 21, no. 3, pp. 344-350.

Oliver, B.G., and Nicol, K.D., 1984, Chlorinated contaminants in the Niagara River, 1981–1983: *Sci. Total Environ.*, v. 39, pp. 57-70.

Oltmann, R.N., Guay, J.R., and Shay, J.M., 1985, Rainfall and runoff quantity and quality data collected at four urban land-use catchments in Fresno, California, October 1981-April 1983: U.S. Geological Survey Open-File Report 84-718, 139 p.

Oltmann, R.N., and Schulters, M.V., 1989, Rainfall and runoff quantity and quality characteristics of four urban land-use catchments in Fresno, California, October 1981-April 1983: U.S. Geological Survey Water-Supply Paper 2335, 114 p.

Page, G.W., 1981, Comparison of groundwater and surface water for patterns and levels of contamination by toxic substances: *Environ. Sci. Technol.*, v. 15, no. 12, pp. 1475-1481.

Palmer, H.D., Tzou, K.T.S., and Swain, A., 1975, Transport of chlorinated hydrocarbons in sediments of the upper Chesapeake Bay: U.S. Department of the Interior, Office of Water Resources Research, Final Report, contract 14-31-0001-4204, variously paged.

Pantone, D.J., Young, R.A., Buhler, D.D., Eberlein, C.K.W.C., and Forcella, F., 1992, Water quality impacts associated with pre- and postemergence applications of atrazine in maize: *J. Environ. Qual.*, v. 21, no. 4, pp. 567-573.

Paterson, K.G., and Schnoor, J.L., 1992, Fate of alachlor and atrazine in a riparian zone field site: *Water Environ. Res.*, v. 64, no. 3, pp. 274-283.

Patrick, J.E., 1990, Pesticides in Ontario municipal drinking water from surface sources, 1987: Ontario Ministry of the Environment, Water Resources Branch, Drinking Water Section, 35 p.

Pauli, B.D., Kent, R.A., and Wong, M.P., 1990, Canadian water quality guidelines for metribuzin: Environment Canada, Inland Waters Directorate, Water Quality Branch Scientific Series 179, 44 p.

——1991a, Canadian water quality guidelines for cyanazine: Environment Canada, Inland Waters Directorate, Water Quality Branch Scientific Series 180, 26 p.

——, 1991b, Canadian water quality guidelines for simazine: Environment Canada, Inland Waters Directorate, Water Quality Branch Scientific Series 187, 40 p.

Periera, W.E., and Hostettler, F.D., 1993, Nonpoint source contamination of the Mississippi River and its tributaries by herbicides: *Environ. Sci. Technol.*, v. 27, no. 8, pp. 1542-1552.

Pereira, W.E., and Rostad, C.E., 1990, Occurrence, distributions, and transport of herbicides and their degradation products in the lower Mississippi River and its tributaries: *Environ. Sci. Technol.*, v. 24, no. 9, pp. 1400-1406.

Pereira, W.E., Rostad, C.E., and Leiker, T.J., 1990, Distribution of agrochemicals in the lower Mississippi River and its tributaries: *Sci. Total Environ.*, v. 97/98, pp. 41-53.

——1992, Synthetic organic agrochemicals in the lower Mississippi River and its major tributaries: Distribution, transport and fate: *J. Contam. Hydrol.*, v. 9, nos. 1-2, pp. 175-188.

Peterman, P.H., Delfino, J.J., Dube, D.J., Gibson, T.A., and Priznar, F.J., 1980, Chloro-organic compounds in the lower Fox River, Wisconsin: *Environ. Sci. Res.*, v. 16, pp. 145-160.

Petersen, J.C., 1990, Trends and comparison of water quality and bottom material of northeastern Arkansas streams, 1974-85, and effects of planned diversions: U.S. Geological Survey Water-Resources Investigations Report 90-4017, 215 p.

Pieper, G.R., 1979, Residue analysis of carbaryl on forest foliage and in stream water using HPLC: *Bull. Environ. Contam. Toxicol.*, v. 22, nos. 1-2, pp. 167-171.

Pierce, R.H., Jr., Brent, C.R., Williams, H.P., and Reeves, S.G., 1977, Pentachlorophenol distribution in a fresh water ecosystem: *Bull. Environ. Contam. Toxicol.*, v. 18, no. 2, pp. 251-258.

Pionke, H.B., and Chesters, G., 1973, Pesticide-sediment-water interactions: *J. Environ. Qual.*, v. 2, no. 1, pp. 29-45.

Platford, R.F., Maguire, R.J., Tkacz, R.J., Comba, M.E., and Kaiser, K.L.E., 1985, Distribution of hydrocarbons and chlorinated hydrocarbons in various phases of the Detroit River: *J. Great Lakes Res.*, v. 11, no. 3, pp. 379-385.

Pope, L.M., Arruda, J.A., and Vahsholtz, A.E., 1985, Water-quality reconnaissance of selected water-supply lakes in eastern Kansas: U.S. Geological Survey Water-Resources Investigations Report 85-4058, 51 p.

Poulton, D.J., 1987, Trace contaminant status of Hamilton Harbor: *J. Great Lakes Res.*, v. 13, no. 2, pp. 193-201.

Purdom, E.E., Hardiman, P.A., Bye, V.J., Eno, N.C., Tyler, C.R., and Sumpter, J.P., 1994, Estrogenic effects of effluents from sewage treatment works: *Chem. Ecol.*, v. 14, no. 4, pp. 725-731.

Que Hee, S.S., and Sutherland, R.G., 1981, *Chemistry, analysis, and environmental pollution*, v. 1 of *The phenoxyalkanoic herbicides*: CRC Press, Boca Raton, Fla., CRC Series in Pesticide Chemistry.

Radtke, D.B., Kepner, W.G., and Effertz, R.J., 1988, Reconnaissance investigation of water quality, bottom sediment, and biota associated with irrigation drainage in the Lower Colorado River valley, Arizona, California, and Nevada, 1986-87: U.S. Geological Survey Water-Resources Investigation Report 88-4002, 77 p.

Rapaport, R.A., and Eisenreich, S.J., 1986, Atmospheric deposition of toxaphene to eastern North America derived from peat accumulation: *Atmospheric Environ.*, v. 20, no. 12, pp. 2367-2379.

——1988, Historical atmospheric inputs of high molecular weight chlorinated hydrocarbons to eastern North America: *Environ. Sci. Technol.*, v. 22, no. 8, pp. 931-941.

Rapaport, R.A., Urban, N.R., Capel, P.D., Baker, J.E., Looney, B.B., Eisenreich, S.J., and Gorham, E., 1985, "New" DDT inputs to North America—Atmospheric deposition: *Chemosphere*, v. 14, no. 9, pp. 1167-1173.

Ray, L.E., Murray, H.E., and Giam, C.S., 1983, Analysis of water and sediment from the Nueces estuary/Corpus Christi Bay (Texas) for selected organic pollutants: *Chemosphere*, v. 12, nos. 7/8, pp. 1039-1045.

Rice, C.P., and Evans, M.S., 1984, Toxaphene in the Great Lakes, *in* Nriagu, J.O., and Simmins, M.S., eds., *Toxic contaminants in the Great Lakes*: Wiley, N.Y., Advances in Environmental Science and Technology series, v. 14, pp. 163-194.

Richards, R.P., and Baker, D.B., 1990, Estimates of human exposure to pesticides through drinking water: A preliminary risk assessment, *in* Kurtz, D.A., ed., *Long range transport of pesticides*: Lewis Publishers, Chelsea, Mich., pp. 387-403.

——1993, Pesticide concentration patterns in agricultural drainage networks in the Lake Erie Basin: *Environ. Toxicol. Chem.*, v. 12, no. 1, pp. 13-26.

Richards, R.P., Baker, D.B., Christensen, B.R., and Tierney, D.P., 1995, Atrazine exposures through drinking water: Exposure assessments for Ohio, Illinois, and Iowa: *Environ. Sci. Technol.*, v. 29, no. 2, pp. 406-412.

Rihan, T.I., Mustafa, H.T., Caldwell, G., Jr., and Frazier, L., 1979, Chlorinated pesticides in streams and lakes of northern Mississippi water: *Egypt. J. Chem.*, v. 19, no. 6, pp. 1123-1126.

Rinella, F.A., and Schuler, C.A., 1992, Reconnaissance investigation of water quality, bottom sediment, and biota associated with irrigation drainage in the Malheur National Wildlife Refuge, Harney County, Oregon, 1988-89: U.S. Geological Survey Water-Resources Investigations Report 91-4085, 106 p.

Rinella, J.F., Hamilton, P.A., and McKenzie, S.W., 1993, Persistence of the DDT pesticide in the Yakima River basin, Washington: U.S. Geological Survey Circular 1090, 24 p.

Risebrough, R.W., Huggett, R.J., Griffin, J.J., and Goldberg, E.D., 1968, Pesticides: transatlantic movements in the Northeast trades: *Science*, v. 159, no. 3820, pp. 1233-1236.

Ritter, W.F., 1988, Reducing impacts of nonpoint source pollution from agricultural: A review: *J. Environ. Sci. Hea.*, v. A23, no. 7, pp. 645-667.

Ritter, W.F., Johnson, H.P., Lovely, W.G., and Molnau, M., 1974, Atrazine, propachlor, and diazinon residues on small agricultural watersheds: Runoff losses, persistence, and movement: *Environ. Sci. Technol.*, v. 8, no. 1, pp. 38-42.

Rivera, J., Caixach, J., Ventura, F., and Espadaler, I., 1985, Herbicide and surfactant spill analysis of an industrial dumping at Llobregat River (Spain): *Chemosphere*, v. 14, no. 5, pp. 395-402.

Rogers, R.J., 1984, Chemical quality of the Saw Mill River, Westchester County, New York, 1981-83: U.S. Geological Survey Water-Resources Investigations Report 84-4225, 51 p.

Rohde, W.A., Asmussen, L.E., Hauser, E.W., Hester, M.L., and Allison, H.D., 1981, Atrazine persistence in soil and transport in surface and subsurface runoff from plots in the Coastal Plain of the southern United States: *Agro-Ecosys.*, v. 7, pp. 225-238.

Rohde, W.A., Asmussen, L.E., Hauser, E.W., Wauchope, R.D., and Allison, H.D., 1980, Trifluralin movement in runoff from a small agricultural watershed: *J. Environ. Qual.*, v. 9, no. 1, pp. 37-42.

Rowe, D.R., Canter, L.W., Snyder, P.J., and Mason, J.W., 1971, Dieldrin and endrin concentrations in a Louisiana estuary: *Pest. Monit. J.*, v. 4, no. 4, pp. 177-83.

Ruelle, R., 1991, *A pesticide and toxicity evaluation of wetland waters and sediments on national wildlife refuges in South Dakota*: U.S. Fish and Wildlife Service, Fish and Wildlife Enhancement, South Dakota State Office, Pierre, S. Dak., 26 p.

Russell, M.H., and Layton, R.J., 1992, Models and modeling in a regulatory setting: Considerations, applications, and problems: *Weed Technol.*, v. 6, no. 3, pp. 673-676.

Sato, C. and Schnoor, J.L., 1991, Applications of three completely mixed compartment models to the long-term fate of dieldrin in a reservoir: *Water Res.*, v. 25, pp. 621-631.

Schacht, R.A., 1974, Pesticides in the Illinois waters of Lake Michigan: U.S. Environmental Protection Agency, Office of Research and Development, Ecological Research series, EPA-600/3-74-002, 55 p.

Schacht, R.A., Corley, C., and Rogers, K., 1978, *An intensive water quality survey of the Rock River from Rockford to Byron, Illinois, May-October, 1978*: Illinois Environmental Protection Agency, Division of Water Pollution Control, Springfield, Ill. 44 p.

Schafer, M.L., Peeler, J.T., Gardner, W.S., and Campbell, J.E., 1969, Pesticides in drinking water: Waters from the Mississippi and Missouri Rivers: *Environ. Sci. Technol.*, v. 3, no. 12, pp. 1261-1269.

Schnoor, J.L. and McAvoy, D.C., 1981, Pesticide transport and bioconcentration model: *J. Environ. Eng. Proc. Am. Soc. Civil Eng.*, v. 107, pp. 1229-1246.

Schnoor, J.L., Mossman, D.J., Borzilov, V.A., Novitsky, M.A., Voszhennikov, O.I., and Gerasimneko, A.K., 1992, Mathematical model for chemical spills and distributed source runoff to large rivers, *in* Schnoor, J.L., ed., *Fate of pesticides and chemicals in the environment*: Wiley, N.Y., Environmental Science and Technology series, pp. 347-370.

Schnoor, J.L., Rao, N.B., Cartwright, K.J., and Noll, R.M., 1982, Fate and transport modeling for toxic substances, *in* Dickson, K.L., Cairne, J., Jr., and Maki, A.W., eds., *Modeling the fate of chemicals in the aquatic environment*: Ann Arbor Science Publishers, Ann Arbor, Mich., pp. 145-163.

Schottler, S.P., Eisenreich, S.J., and Capel, P.D., 1994, Atrazine, alachlor, and cyanazine in a large agricultural river system: *Environ. Sci. Technol.*, v. 28, no. 6, pp. 1079-1089.

Schroeder, D.C., 1987, The fate of alachlor during treatment of public water: *Trans. Kans. Acad. Sci.*, v. 90, nos. 1-2, pp. 41-45.

Schroeder, M.H., and Sturges, D.L., 1980, Spraying of big sagebrush with 2,4-D causes negligible stream contamination: *J. Range Manage.*, v. 33, no. 4, pp. 311-312.

Schroeder, R.A., Palawski, D.U., and Skorupa, J.P., 1988, Reconnaisance investigation of water quality, bottom sediment, and biota associated with irrigation drainage in the Tulare Lake Bed area, southern Joaquin Valley, California, 1986-87: U.S. Geological Survey Water-Resources Investigations Report 88-4001, 86 p.

Schultz, D.P., and Gangstad, E.O., 1976, Dissipation and residues of 2,4-D in water, hydrosoil, and fish: *J. Aquatic Plant Manag.*, v. 14, pp. 43-45.

Schultz, D.P., and Whitney, E.W., 1974, Monitoring 2,4-D residues at Loxahatchee National Wildlife Refuge: *Pest. Monit. J.*, v. 7, nos. 3/4, pp. 146-152.

Schulze, J.A., Manigold, D.B., and Andrews, F.L., 1973, Pesticides in selected western streams—1968–71: *Pest. Monit. J.*, v. 7, no. 1, pp. 73-84.

Scow, K.M., 1990, Biodegradation, *in* Lyman, W.F., Reehl, W.F., and Rosenblatt, D.H., eds., *Handbook of chemical property estimation methods: Environmental behavior of organic compounds*: American Chemical Society, Washington, D.C., chap. 9.

Scrudato, R.J., and DelPrete, A., 1982, Lake Ontario sediment-mirex relationships: *J. Great Lakes Res.*, v. 8, no. 4, pp. 695-699.

Segawa, R.T., Sitts, J.A., White, J.H., Marade, S.J., and Powell, S.J., 1991, Environmental monitoring of malathion aerial applications used to eradicate Mediterranean fruit flies in southern California, 1990: California Department of Food and Agriculture, Division of Pest Management, Environmental Monitoring and Pest Management Branch, Environmental Hazards Assessment Program Report EH 91-3.

Seiber, J.N., 1987, Solubility, partition coefficient, and bioconcentration factor, *in* Biggar, J.W., and Seiber, J.N., eds., Fate of pesticides in the environment: Proceedings of a technical seminar: University of California, Division of Agriculture and Natural Resources, Agricultural Experiment Station Publication 3320, pp. 53-59.

Setmire, J.G., and Bradford, W.L., 1980, Quality of urban runoff, Tecolote Creek drainage area, San Diego County, California: U.S. Geological Survey Water-Resources Investigations Report 80-70, 33 p.

Setmire, J.G., Wolfe, J.C., and Stroud, R.K., 1990, Reconnaissance investigation of water quality, bottom sediment, and biota associated with irrigation drainage in the Salton Sea area, California, 1986-87: U.S.Geological Survey Water-Resources Investigations Report 89-4102, 76 p.

Shehata, S.A., El-Dib, M.A., and Abou-Waly, H.F., 1993, Effect of triazine compounds on freshwater algae: *Bull. Environ. Contam. Toxicol.*, v. 50, pp. 369-376.

Smith, B.S., 1981, Male characteristics on female snails caused by antifouling bottom paints: *J. Appl. Toxicol.*, v. 1, pp. 22-25.

Smith, G.E., and Isom, B.G., 1967, Investigation of effects of large-scale applications of 2,4-D on aquatic fauna and water quality: *Pest. Monit. J.*, v. 1, no. 3, pp. 16-21.

Smith, R.L., 1989, A computer assisted, risk-based screening of a mixture of drinking water chemicals: *Trace Subst. Environ. Hea.*, v. 22, pp. 215-232.

Smith, S., Reagan, T.E., Flynn, J.L., and Willis, G.H., 1983, Azinphosmethyl and fenvalerate runoff loss from a IPM system: *J. Environ. Qual.*, v. 12, no. 4, pp. 534-537.

Southwick, L.M., Willis, G.H., Bengtson, R.L., and Lormand, T.J., 1990, Effect of subsurface drainage on runoff losses of atrazine and metolachlor in southern Louisiana: *Bull. Environ. Contam. Toxicol.*, v. 45, no. 1, pp. 113-119.

Spalding, R.F., and Snow, D.D., 1989, Stream levels of agrichemicals during a spring discharge event: *Chemosphere*, v. 19, nos. 8/9, pp. 1129-1140.

Sparr, B.I., Appleby, W.G., DeVries, D.M., Osmun, J.V., McBride, J.M., and Foster, G.L., 1966, Insecticide residues in waterways from agricultural use: *Advan. Chem. Ser.*, v. 60, pp. 146-62.

Spencer, W.F., 1987, Volatilization of pesticide residues, *in* Biggar, J.W., and Seiber, J.N., eds., *Fate of pesticides in the environment: Proceedings of a technical seminar*: University of California, Division of Agriculture and Natural Resources, Agriculture Experiment Station Publication 3320, pp. 53-60.

Spencer, W.F., and Cliath, M.M., 1991, Pesticide losses in surface runoff from irrigated fields: *Environ. Sci. Res.*, v. 42, , pp. 277-289.

Squillace, P.J., Caldwell, J.P., Schulmeyer, P.M., and Harvey, C.A., 1996, Movement of agricultural chemicals between surface water and ground water, lower Cedar River basin, Iowa: U.S. Geological Survey Water-Supply Paper 2448, 59 p.

Squillace, P.J., and Engberg, R.A., 1988, Surface-water quality of the Cedar River basin, Iowa-Minnesota, with emphasis on the occurrence and transport of herbicides, May 1984 through November 1985: U.S. Geological Survey Water Resources Investigations Report 88-4060, 81 p.

Squillace, P.J., and Thurman, E.M., 1992, Herbicide transport in rivers: Importance of hydrology and geochemistry in nonpoint-source contamination: *Environ. Sci. Technol.*, v. 26, no. 3, pp. 538-545.

Squillace, P.J., Thurman, E.M., and Furlong, E.T., 1993, Groundwater as a nonpoint source of atrazine and deethylatrazine in a river during base flow conditions: *Water Resour. Res.*, v. 29, no. 6, pp. 1719-1729.

Stanley, J.G., and Trial, J.G., 1980, Disappearance constants of carbaryl from streams contaminated by forest spraying: *Bull. Environ. Contam. Toxicol.*, v. 25, p. 771-776.

Staples, C.A., Werner, A.F., and Hoogheem, T.J., 1985, Assessment of priority pollutant concentrations in the United States using storet database: *Environ. Toxicol. Chem.*, v. 4, pp. 131-142.

Steenhuis, T.S. and Walter, M.R., 1980, Closed form solution for pesticide loss in runoff water: *Trans. Am. Soc. Agric. Eng.*, v. 23, pp. 615-620.

Stephens, D.W., 1984, Water-quality investigations of the Jordan River, Salt Lake County, Utah, 1980-82: U.S. Geological Survey Water-Resources Investigations Report 84-4298, 45 p.

Stevens, R.J.J., and Neilson, M.A., 1989, Inter- and intralake distributions of trace organic contaminants in surface waters of the Great Lakes: *J. Great Lakes Res.*, v. 15, no. 3, pp. 377-393.

Stoltz, R.L., and Pollock, G.A., 1982, Methoxychlor residues in treated irrigation canal water in southcentral Idaho: *Bull. Environ. Contam. Toxicol.*, v. 28, pp. 473-476.

Strachan, W.M.J., 1985, Organic substances in the rainfall of Lake Superior: 1983: *Environ. Toxicol. Chem.*, v. 4, no. 5, pp. 677-683.

Strachan, W.M.J., and Edwards, C.J., 1984, Organic pollutants in Lake Ontario, *in* Nriagu, J.O., and Simmins, M.S., eds., *Toxic contaminants in the Great Lakes*: Wiley, N.Y., Advances in Environmental Science and Technology series, v. 14, pp. 239-262.

Strachan, W.M.J., and Eisenreich, S.J., 1990, Mass balance accounting of chemicals in the Great lakes, *in* Kurtz, D.A., ed., *Long range transport of pesticides*: Lewis Publishers, Chelsea, Mich., pp. 291-301.

Strachan, W.M.J., and Huneault, H., 1979, Polychlorinated biphenyls and organochlorine pesticides in Great Lakes precipitation: *J. Great Lakes Res.*, v. 5, no. 1, pp. 61-68.

Stratton, G.W., 1984, Effects of the herbicide atrazine and its degration products, alone and in combination, on phototrophic microorganisms: *Arch. Environ. Contam. Toxicol.*, v. 13, no. 1, pp. 35-42.

Sullivan, J.F., and Atchison, G.J., 1977, Impact of an urban methoxychlor spraying program on the Rouge River, Michigan: *Bull. Environ. Contam. Toxicol.*, v. 17, no. 1, pp. 121-126.

Swackhamer, D.L., and Skogland, R.S., 1993, Bioaccumulation of PCBs by algae: Kinetics versus equilibrium: *Environ. Toxicol. Chem.*, v. 12, no. 5, pp. 831-838.

Swain, W.R., 1978, Chlorinated organic residues in fish, water, and precipitation from the vicinity of Isle Royale, Lake Superior: *J. Great Lakes Res.*, v. 4, pp 398-407.

Takita, C.S., 1984, Summary of data from a toxics screening survey of the lower Susquehanna River and major tributaries: Susquehanna River Basin Commission Publication 86, 47 p.

Tanabe, S., Hidaka, H., and Tatsukawa, R., 1983, PCBs and chlorinated hydrocarbon pesticides in Antarctic atmosphere and hydrosphere: *Chemosphere*, v. 12, no. 2, pp. 277-288.

Tanabe, S., Tatsukawa, R., Kawano, M., and Hidaka, H., 1982, Global distribution and atmospheric transport of chlorinated hydrocarbons: HCH (BHC) isomers and DDT compounds in the western Pacific, eastern Indian and Antarctic Oceans: *J. Ocean. Soc. Jap.*, v. 38, pp. 137-148.

Tanita, R., Johnson, J.M., Chun, M., and others., 1976, Organochlorine pesticides in the Hawaii Kai Marina, 1970-74: *Pest. Monit. J.*, v. 10, no. 1, pp. 24-29.

Terry, R.D., and Hughes, G.M., 1976, Pollution effects on surface waters and waters: *J. Water Poll. Control Fed.*, v. 48, no. 6, pp. 1420-1433.

Thomas, R.G., 1990, Volatilization from water, *in* Lyman, W.F., Reehl, W.F., and Rosenblatt, D.H., eds., *Handbook of chemical property estimation methods: Environmental behavior of organic compounds*: American Chemical Society, Washington, D.C., pp. 15-1 to 15-34.

Thompson, K.R., 1984, Reconnaissance of toxic substances in the Jordan River, Salt Lake County, Utah: U.S. Geological Survey Water-Resources Investigations Report 84-4155, 31 p.

Thurman, E.M., Goolsby, D.A., Meyer, M.T., and Kolpin, D.W., 1991, Herbicides in surface waters of the midwestern United States: The effect of spring flush: *Environ. Sci. Technol.*, v. 25, no. 10, pp. 1794-1796.

Thurman, E.M., Goolsby, D.A., Meyer, M.T., Mills, M.S., Pomes, M.L., and Kolpin, D.W., 1992, A reconnaissance study of herbicides and their metabolites in surface water of the midwestern United States using immunoassay and gas chromatography/mass-spectrometry: *Environ. Sci. Technol.*, v. 26, no. 12, pp. 2440-2447.

Toppin, K.W., 1983, Appraisal of water-quality conditions, lower Black River, Windsor County, Vermont: U.S. Geological Survey Water-Resources Investigations Report 82-4116, 61 p.

Trichell, D.W., Morton, H.L., and Merkle, M.G., 1968, Loss of herbicides in runoff water: *Weed Sci.*, v. 16, pp. 447-449.

Trim, A.H., and Marcus, J.M., 1990, Integration of long-term fish kill data with ambient water quality monitoring data and application to water quality management: *Environ. Manag.*, v. 14, no. 3, pp. 389-396.

Troelstrup, N.H., Jr., and Perry, J.H., 1989, Water quality in southeastern Minnesota streams: Observations along a gradient of land use and geology: *J. Minn. Acad. Sci.*, v. 55, no. 1, pp. 6-13.

Trotter, D.M., 1990, Canadian water quality guidelines for atrazine: Environment Canada, Inland Waters Directorate, Water Quality Branch Scientific Series 168, 106 p.

Trotter, D.M., Kent, R.A., and Wong, M.P., 1991, Aquatic fate and effect of carbofuran: *Crit. Rev. Environ. Control*, v. 21, no. 2, pp. 137-176.

Trotter, D.M., Wong, M.P., and Kent, R.A., 1989, Canadian water quality guidelines for carbofuran: Environment Canada, Inland Waters Directorate, Water Quality Branch Scientific Series 169, 34 p.

——1990, Canadian water quality guidelines for glyphosate: Environment Canada, Inland Waters Directorate, Water Quality Branch Scientific Series 170, 27 p.

Truhlar, J.F., and Reed, L.A., 1976, Occurrence of pesticide residues in four streams draining different land-use areas in Pennsylvania, 1969-71: *Pest. Monit. J.*, v. 10, no. 3, pp. 101-110.

Ulrich, M.M., Mudieresisller, S.R., Singer, H.P., Imboden, D.M., and Schwarzenbach, R.P., 1994, Input and dynamic behavior of organic pollutants tetrachloroethane, atrazine, and NTA in a lake: A study combining mathematical modelling and field measurements: *Environ. Sci. Technol.*, v. 28, pp. 1674-1685.

U.S. Department of Commerce, 1990, *Agricultural atlas of the United States*, pt. 1 *of 1987 Census of Agriculture*: U.S. Department of Commerce, Bureau of the Census, AC87-S-1, 199 p.

U.S. Environmental Protection Agency, 1987, FIFRA scientific advisory approval of proposed changes to inert ingredients lists 1 and 2: U.S. Environmental Protection Agency internal memorandum from F.S. Bishop, 5 p.

——1988, *Pesticide fact handbook*: Park Ridge, N.J.Noyes Data Corporation, v. 1, 827 p.

——1990, National survey of pesticides in drinking water wells: Phase I report: U.S. Environmental Protection Agency, Office of Pesticides and Toxic Substances Report EPA 570/9-90-015, 424 p.

——1991, Environmental fate one-liner database: U.S. Environmental Protection Agency, Office of Pesticide Programs, Freedom of Information Office, floppy disk.

——1992, Pesticides in ground water database: A compilation of monitoring studies, 1971–1991: U.S. Environmental Protection Agency Report EPA 734-12-92-001, variously paged.

——1994a, Questions and answers—Conditional registration of acetochlor: U.S. Environmental Protection Agency, Prevention, Pesticides and Toxic Substances, 18 p.

——1994b, Consolidated rule summary for the chemical phases: U.S. Environmental Protection Agency, Office of Water Report EPA 812-S-94-001, 34 p.

——1994c, Consolidated regulations for the chemical phases (draft): U.S. Environmental Protection Agency, Office of Water Report EPA 812-D-94-001, 42 p.

U.S. Forest Service, 1978, *Report of the Forest Service, fiscal year 1977*: U.S. Department of Agriculture - Forest Service, Washington, D.C., variously paged.

——1985, *Report of the Forest Service, fiscal year 1984*: U.S. Department of Agriculture— Forest Service, Washington, D.C., variously paged.

——1989, *Report of the Forest Service, fiscal year 1988*: U.S. Department of Agriculture— Forest Service, Washington, D.C., variously paged.

——1990, *Report of the Forest Service, fiscal year 1989*: U.S. Department of Agriculture— Forest Service, Washington, D.C., variously paged.

——1991, *Report of the Forest Service, fiscal year 1990*: U.S. Department of Agriculture— Forest Service, Washington, D.C., variously paged.

——1992, *Report of the Forest Service, fiscal year 1991*: U.S. Department of Agriculture— Forest Service, Washington, D.C., variously paged.

——1993, *Report of the Forest Service, fiscal year 1992*: U.S. Department of Agriculture— Forest Service, Washington, D.C., variously paged.

——1994a, *Report of the Forest Service, fiscal year 1993*: U.S. Department of Agriculture— Forest Service, Washington, D.C., variously paged.

——1994b, personnal communication, FAX, May 31, GMDigest Database, U.S. Department of Agriculture - Forest Service, Forest Health Protection, Morgantown, W. Va.

Vogel, T.M., Criddle, C.S., and McCarthy, P.L., 1987, Transformations of halogenated aliphatic compounds: *Environ. Sci. Technol.*, v. 21, no. 8, pp. 722-736.

Voldner, E.C., and Schroeder, W.H., 1989, Modelling of atmospheric transport and deposition of toxaphene into the Great Lakes ecosystem: *Atmos. Environ.*, v. 23, no. 9, pp. 1949-1961.

Walker, W.W., Cripe, C.R., Pritchard, P.H., and Bourquin, A., 1988, Biological and abiotic degradation of xenobiotic compounds in in vitro estuarine water and sediment/water systems: *Chemosphere*, v. 17, no. 12, pp. 2255-2270.

Waller, W.T., and Lee, G.F., 1979, Evaluation of observations of hazardous chemicals in Lake Ontario during the International Field Year for the Great Lakes: *Environ. Sci. Technol.*, v. 13, no. 1, pp. 79-85.

Wang, T.C., 1983, Toxic substance monitoring in the Indian River Lagoon, Florida: *Fla. Sci.*, v. 46, nos. 3-4, pp. 286-295.

——1991, Assimilation of malathion in the Indian River Estuary, Florida: *Bull. Environ. Contam. Toxicol.*, v. 47, no. 2, pp. 238-243.

Wang, T.C., Hoffman, M.E., David, J., and Parkinson, R., 1992, Chlorinated pesticide residue occurrence and distribution in mosquito control impoundments along the Florida Indian River Lagoon: *Bull. Environ. Contam. Toxicol.*, v. 49, no. 2, pp. 217-223.

Wang, T.C., Johnson, R.S., and Bricker, J.L., 1980, Residues of polychlorinated biphenyls and DDT in water and sediment of the Indian River lagoon, Florida—1977–1978: *Pest. Monit. J.*, v. 13, no. 4, pp. 141-144.

Wang, T.C., Krivan, J.P., Jr., and Johnson, R.S., 1979, Residues of polychlorinated biphenyls and DDT in water and sediment of the St. Lucie estuary, Florida, [USA], 1977: *Pest. Monit. J.*, v. 13, no. 2, pp. 69-71.

Wang, T.C., Lenahan, R.A., and Tucker, J.W., Jr., 1987b, Deposition and persistence of aerially-applied fenthion in a Florida estuary: *Bull. Environ. Contam. Toxicol.*, v. 38, pp. 226-231.

Wang, T.C., Lenahan, R.A., J. W. Tucker, Jr., and Kadlac, T., 1987a, Aerial spray of mosquito adulticides in a salt marsh environment: *Water Sci. Technol.*, v. 19, no 11, pp. 113-124.

Wangsness, D.J., 1983, Water and streambed-material data, Eagle Creek watershed, August 1980 and October and December 1982: U.S. Geological Survey Open-File Report 83-215, 41 p.

Wanner, O, Egli, T., Fleischmann, T., Lanz, R.P., and Schwarzenbach, R.P., 1989, Behavior of the insecticides disulfoton and thiometon in the river: A chemodynamic study: *Environ Sci. Technol.*, v. 23, np. 10, pp. 1232-1242

Ward, J.R., 1987, Surface-water quality in Pequea Creek basin, Pennsylvania, 1977-79: U.S. Geological Survey Water-Resources Investigations Report 85-4250, 66 p.

Warry, N.D., and Chan, C.H., 1981, Organic contaminants in the suspended sediments of the Niagara River: *J. Great Lakes Res.*, v. 7, no. 4, pp. 394-403.

Watson, V.J., Rice, P.M., and Monning, E.C., 1989, Environmental fate of picloram used for roadside weed control: *J. Environ. Qual.*, v. 18, no. 2, pp. 198-205.

Wauchope, R.D., 1978, The pesticide content of surface water draining from agricultural fields— a review: *J. Environ. Qual.*, v. 7, no. 4, pp. 459-472.

——1992, Environmental risk assessment of pesticides: Improving simulation model credibility: *Weed Technol.*, v. 6, pp. 753-759.

Wauchope, R.D., Buttler, T.M., Hornsby, A.G., Augustijn-Beckers, P.W.M., and Burt, J.P., 1992, The SCS/ARS/CES pesticide properties database for environmental decision-making: *Rev. Environ. Contam. Toxicol.*, v. 123, pp. 1-156.

Wauchope, R.D., Williams, R.G., and Marti, L.R., 1990, Runoff of sulfometuron-methyl and cyanazine from small plots: Effects of formulation and grass cover: *J. Environ. Qual.*, v. 19, no. 1, pp. 119-125.

Weaver, B.C.E., Gunnerson, C.G., Breidenbach, A.W., and Lichtenberg, J.J., 1965, Chlorinated hydrocarbon pesticides in major U.S. river basins: *Publ. Hea. Re.*, v. 80, no. 6, pp. 481-493.

Weber, J.B., 1970, Mechanisms of adsorption of s-triazines by clay colloids and factors affecting plant availability: *Agrochemia*, v. 10, no. 6, pp. 187-188.

Wehr, M.A., Mattson, J.A., Bofinger, R.W., and Sajdak, R.L., 1992, Ground-application trial of hexazinone on the Ottawa National Forest: U.S. Department Agriculture, Forest Service, North Central Forest Experiment Station Research Paper NC-308, 34 p.

West, S.D., Day, E.W., Jr., and Burger, R.O., 1979, Dissipation of the experimental aquatic herbicide fluridone from lakes and ponds: *J. Agric. Food Chem.*, v. 27, no. 5, pp. 1067-1072.

West, S.D., Langeland, K.A., and Laroche, F.B., 1990, Residues of fluridone and a potential photoproduct (N-methylformamide) in water and hydrosoil treated with the aquatic herbicide Sonar: *J. Agric. Food Chem.*, v. 38, no. 1, pp. 315-319.

Westerdahl, H.E. and Getsinger, K.D., 1988, Aquatic plants and susceptibility to herbicides, v. 2 *of* Aquatic plant identification and herbicide use guide: U.S. Army Corps of Engineers Waterways Experiment Station, Technical Report A-88-9, 104 p.

Whitmore, R.W., Kelly, J.E., and Reading, P.L., 1992, Executive summary, results and recommendations, v. 1 of National home and garden pesticide use survey: U.S. Environmental Protection Agency Report RTI/5100/17-01F68-WO-0032, 140 p.

Willis, G.H., and McDowell, L.L., 1982, Pesticides in agricultural runoff and their effects on downstream water quality: *Environ.Toxicol. Chem.*, v. 1, pp. 267-279.

Windholz, M., ed., 1976, *The Merck index* (9th ed.): Merck & Co., Rahway, N.J., variously paged.

Wnuk, M., Kelley, R., Breuer, G., and Johnson, L., 1987, *Pesticides in water supplies using surface water sources*: Iowa Department of Natural Resources, Des Moines, Iowa, 33 p.

Wojtalik, T.A., Hall, T.F., and Hill, L.O., 1971, Monitoring ecological conditions associated with wide-scale applications of DMA 2,4-D [dimethylamine salt of 2,4-D] to environments: *Pest. Monit. J.*, v. 4, no. 4, pp. 184-203.

Wolfe, N.L., Mingelgrin, U., and Miller, G.C., 1990, Abiotic transformations in water, sediments, and soil, *in* Cheng, H.H., ed., *Pesticides in the soil environment: Processes, impacts, and modeling*: Soil Science Society of America, Madison, Wis., pp. 103-168.

Wolman, M.G., 1971, The Nation's rivers: *Science*, v. 174, pp. 905-918.

Worthing, C.R., and Walker, S.B., eds., 1987, *The pesticide manual* (8th ed.): The British Crop Protection Council, Thornton Heath, United Kingdom, 1081 p.

Wotzka, P.J., 1994, personal communication, floppy disk, August, Minnesota Department of Agriculture, Agronomy Services Division, St. Paul, Minn.

Wotzka, P.J., Lee, J., Capel, P, Ma, L., 1994, Pesticide concentrations and fluxes in an urban watershed, *in* Pederson, G.L., ed., *Proceedings of the American Water Resources Association National Symposium on Water Quality*: American Water Resources Association technical publication series, no. TPS94-4, pp. 135-145.

Wu, T.L., 1981, Atrazine residues in estuarine water and the aerial atrazine into Rhode River, Maryland: *Water, Air, Soil Poll.*, v. 15, no. 2, pp. 173-184.

Wu, T.L., Correll, D.L., and Remenapp, H.E.H., 1983, Herbicide runoff from experimental watersheds: *J. Environ. Qual.*, v. 12, no. 3, pp. 330-336.

Wu, T.L., Lambert, L., Hastings, D., and Banning, D., 1980, Enrichment of the agricultural herbicide atrazine in the microsurface water of an estuary: *Bull. Environ. Contam. Toxicol.*, v. 24, no. 3, p. 411-14.

Yamane, C.M., and Lum, M.G., 1985, Quality of storm-water runoff, Mililani Town, Oahu, Hawaii, 1980-84: U.S. Geological Survey Water-Resources Investigations Report 85-4265, 63 p.

Yin, C., and Hassett, J.P., 1989, Fugacity and phase distribution of mirex in Oswego River and Lake Ontario waters: *Chemosphere*, v. 19, nos. 8-9, pp. 1289-1296.

Yoo, K.H., Touchton, J.T., and Walker, R.H., 1986, Sediment, nutrient and pesticide losses affected by tillage practices, in *Meeting the engineering challenges of agriculture—worldwide: Summer meeting, San Luis Obispo, California*: American Society of Agricultural Engineers, St. Joseph, Mich., pp. 1-15.

Yorke, T.H., Stamer, J., and Pederson, G.L., 1985, Effects of low-level dams on the distribution of sediment, trace metals, and organic substances in the lower Schuylkill River basin, Pennsylvania: U.S. Geological Survey Water-Supply Paper 2256-B, 53 p.

Yurewicz, M.C., Carey, W.P., and Garrett, J.W., 1988, Streamflow and water-quality data for three major tributaries to Reelfoot Lake, west Tennessee, October 1987-March 1988: U.S. Geological Survey Open-File Report 88-311, 20 p.

Zahnow, E.W., and Riggleman, J.D., 1980, Search for linuron residues in tributaries of the Chesapeake Bay: *J. Agric. Food Chem.*, v. 28, pp. 974-978.

Zitko, V., and McLeese, D.W., 1980, Evaluation of hazards of pesticides used in forest spraying to the aquatic environment: Fisheries and Environmental Sciences, Canada Department of Fisheries and Oceans, Biological Station (St.Andrews, N.B.), Canadian Technical Report of Fisheries and Aquatic Sciences series 985, 21 p.

Zoecon Corporation, 1990, Altosid XR: Zoecon Corporation, Dallas, TX, 4 p.

Zubkoff, P.L., 1992, The use of runoff and surface water transport and fate models in the pesticide registration process: *Weed Technol.*, v. 6, pp. 743-748.

Index

Printed and bound by CPI Group (UK) Ltd, Croydon, CR0 4YY

28/10/2024

01779981-0001